Advances in Steiner Trees

COMBINATORIAL OPTIMIZATION

VOLUME 6

Through monographs and contributed works the objective of the series is to publish state of the art expository research covering all topics in the field of combinatorial optimization. In addition, the series will include books which are suitable for graduate level courses in computer science, engineering, business, applied mathematics, and operations research.

Combinatoria (or discrete) optimization problems arise in various applications, including communications network design, VLSI design, machine vision, airline crew scheduling, corporate planning, computer-aided design and manufacturing, database query design, cellular telephone frequency assignment, constraint directed reasoning, and computational biology. The topics of the books will cover complexity analysis and algorithm design (parallel and serial), computational experiments and applications in science and engineering.

Advances in Steiner Trees

edited by

Ding-Zhu Du

Department of Computer Science,
University of Minnesota, U.S.A.

J.M. Smith

Department of Mechanical & Industrial Engineering,
University of Massachusetts, U.S.A.

and

J.H. Rubinstein

Department of Mathematics,
University of Melbourne, Australia

KLUWER ACADEMIC PUBLISHERS
DORDRECHT / BOSTON / LONDON

A C.I.P. Catalogue record for this book is available from the Library of Congress.

ISBN 978-1-4419-4824-3

Published by Kluwer Academic Publishers,
P.O. Box 17, 3300 AA Dordrecht, The Netherlands.

Sold and distributed in North, Central and South America
by Kluwer Academic Publishers,
101 Philip Drive, Norwell, MA 02061, U.S.A.

In all other countries, sold and distributed
by Kluwer Academic Publishers,
P.O. Box 322, 3300 AH Dordrecht, The Netherlands.

Printed on acid-free paper

Contents

Contents

Preface

The Volume on Advances in Steiner Trees is divided into two sections. The first section of the book includes papers on the general geometric Steiner tree problem in the plane and higher dimensions. The second section of the book includes papers on the Steiner problem on graphs.

The general geometric Steiner tree problem assumes that you have a given set of points in some d-dimensional space and you wish to connect the given points with the shortest network possible. The given set of points are

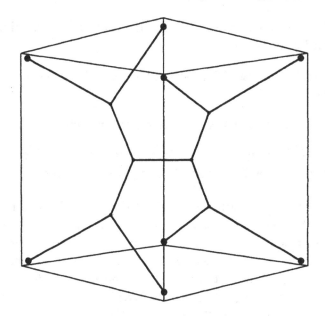

Figure 1: Euclidean Steiner Problem in E^3

usually referred to as *terminals* and the set of points that may be added to reduce the overall length of the network are referred to as *Steiner points*. What makes the problem difficult is that we do not know *a priori* the location and cardinality of the number of Steiner points. Thus, the problem on the Euclidean metric is not known to be in NP and has not been shown to be NP-Complete. It is thus a very difficult NP-Hard problem.

Figure 1 represents an instance in Euclidean E^3 where the given terminals are embedded in space as a unit cube and the Steiner points are interior to

the cube. The Steiner ratio $\rho = SMT/MST$ measures the overall quality
of the shortest network (i.e. SMT) relative to the minimum spanning tree
(i.e. MST) solution. This ratio is an important measure in the geometric
Steiner problem. For the unit cube illustrated here, $\rho = 0.885165$
The first section of the book includes the following six papers:

a) Albrecht, J. and D. Cieslik , "The Steiner Ratio of finite-dimensional
   Lp Spaces."

b) Booth, R., D.A. Thomas, and J.F. Weng, "Shortest Networks for One
   line and Two Points in Space."

c) Brazil, M, D. Thomas, and J. Weng, "Rectilinear Steiner Minimal
   Trees on Parallel Lines."

d) Jiang, T. and L. Wang, "Computing Shortest Networks with Fixed
   Topologies,"

e) Weng, Jia. "Steiner Trees, Coordinate Systems, and NP-Hardness."

f) Warme, D.M., P.Winter, and M. Zachariasen, "Exact Algorithms for
   Plane Steiner Tree Problems: A Computational Study,"

In the first chapter of Albrecht and Cieslik, they establish two new upper
bounds for the Steiner ratio in the d-dimensional space L_p. In particular,
the second upper bound has a corollary that for any fixed p, the limit of
the Steiner ratio in the d-dimensional space L_p as d goes to infinity is at
most $(1/2)^{1/p}$. This bound is tight for $p = 1$ and not tight for $d = 2$. It
leaves an interesting open problem to the reader.
In the second Chapter, Booth, Thomas, and Weng discuss the construction
of shortest networks for two points and a straight line in space. Algorithms
for the Steiner problem in E^3 are much more complex than in the plane,
and the authors present an interesting approach for solving this special
case of the 3-point problem and also how it relates the more difficult
4-point problem in E^3. They present a number of theoretical properties of
the solution to the problem and finally prove how the 3-point problem can
be solved by a certain quartic equation.
Chapter 3 by Brazil, Thomas, and Weng present a new dynamic
programming algorithm for solving the special case of the Rectilinear
Steiner problem in the plane in which all terminals lie on a bounded
number of horizontal lines. They improve on an old algorithm of Aho,

Garey and Hwang and contribute an algorithm that is much more practical.

Given a set of terminals with a tree topology, Jiang and Wang raise the issue of what is the computational complexity of constructing a shortest Steiner tree with this topology? In the Euclidean plane, it is well-known that such a Steiner tree can be constructed in time $O(n^2)$. However, Jiang and Wang in their chapter will demonstrate that in some spaces the problem is NP-hard. Actually, they provide a valuable survey on this problem. The reader will learn from this article what is the state-in-art on this problem.

The NP-hardness of the Euclidean Steiner tree problem was first proved by Garey, Graham, and Johnson in 1977. Rubinstein, Thomas, and Wormald in 1997 presented a simpler proof. This proof shows that the Euclidean Steiner tree problem for terminals on two parallel lines is still NP-hard, which solved an open problem proposed by Provan in 1988 whether the Euclidean Steiner tree problem for terminals on a convex polygon is polynomal-time solvable or not. (Provan showed that the rectilinear Steiner tree problem for terminals on a rectilinear convex polygon is polynomal-time solvable and the Euclidean Steiner tree problem for terminals on a convex polygon has a polynomial-time approximation scheme.) Weng in his chapter designs a proof based on the initial idea of Rubinstein, Thomas, and Wormald. This proof is clearer and more rigorous.

Warme, Winter and Zachariasen conclude the first part of the Volume with the state-of-the art article on exact algorithms for the Euclidean and Rectilinear Steiner problem in the plane. Their algorithms are the top performers in this field of algorithm design and the paper underscores their dominance with the extensive computational results provided in the Chapter.

In the second half of the book, we deal with SMTG problems. The SMTG problem on an undirected graph $G(Z, E, c)$ is defined as follows:

- *Given:* A connected undirected network $G(Z, E, c)$, with n vertices, m edges, a positive cost function c, and a subset $Q \subseteq Z$ of q vertices.

- *Find:* A minimum cost connected subnetwork T of G which spans all Q vertices.

In our case, a certain number of the terminals will be specified as the Q set and the others will act as the set $S = Z - Q$ or Steiner vertices which may or may not be part of the T solution. In contrast to the geometric Steiner problem, we do know the cardinality of the number of Steiner points. Nevertheless, the problem is still NP-Complete.

In the figures that follow, we have an initial graph from which we define the set of Q vertices then try to connect the Q vertices by using some of the given S vertices. In the first example, Figure 2 we have an instance of the SMTG problem with a euclidean metric while the second example Figure 3 illustrates a rectilinear metric.

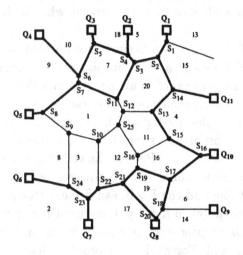

Figure 2: SMTG Network with Euclidean distances

The second section of the book includes the following six papers.

a) Bierman, P. and A. Zelikovsky, " On Approximation of the Power-p and Bottleneck Steiner Trees,"

b) Cheng, "Exact Steiner Trees in Graphs and Grid Graphs."

c) Colbourn, C. and G. Xue. "Grade of Service Steiner Trees in Series-Parallel Networks."

d) Duin, C. "Preprocessing the Steiner Problem in Graphs."

e) Provan, S. "A Fully-Polynomial Approximation Scheme for the Euclidean Steiner Augmentation Problem."

f) Wade, Austin and V. J. Rayward-Smith "Effective Local Search Techniques for the Steiner Tree Problem,"

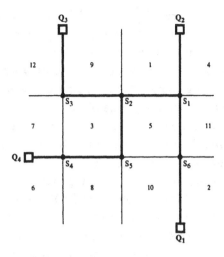

Figure 3: SMT network with Rectilinear Distances

g) Voss, S. "Modern Heuristic Search Methods for the Steiner Problem in Graphs,"

The second part of the Volume starts off with a Chapter by Berman and Zelikovsky. There are many variations of the Steiner tree problem with various application backgrounds. Berman and Zelikovsky in their chapter study a bottleneck Steiner tree problem and a power-p Steiner tree problem. The first problem is to minimize the maximum edge-weight and the second problem is to minimize the sum of the p-power of edge weights. In both problems, a Steiner point is sought to have degree at least three. This condition implies that the number of Steiner points must be upper-bounded by the number of terminals. They obtain two new upper bounds of approximation performance ratio for corresponding spanning trees.

The chapter of Cheng is a survey on graph-theoretic dynamic programming algorithms for computing exact optimal solution for the graphical Steiner tree problem. Several exponential time algorithms for general graphs and planar graphs, as well as two polynomial time algorithms for special cases in grid graphs are presented.

Colbourn and Xue present a new variant of the Steiner problem where more than one cost function of the edges of the network exists so that different grades(levels) of service are permitted. This type of generalization of the Steiner problem has important implications for communication

networks and is becoming a rapidly new and important topic. Colbourn
and Xue present a polynomially running time algorithm for the problem
which concerns series-parallel network topologies.

Cees Duin presents an important extensive body of work on preprocessing
in the Steiner problem in graphs. Preprocessing of the graph can represent
a significant impact on the running time of an algorithm and greatly
reduce the input size of the problem instance. This chapter already
compliments his extensive and original efforts to apply preprocessing
strategies to the Steiner problem in graphs.

J. Scott Provan present a fully polynomial approximation algorithm for the
Euclidean Steiner Augmentation Problem (ESAP). The ESAP has as input
a set of straight line segments and we seek to add an additional set of line
segments that will make the entire set 2-connected with the shortest total
Euclidean length. He presents a number of critical theoretical properties
necessary for the solution to the problem. Finally, he also present an
interesting collection of additional problems for which his algorithm would
be applicable.

Austin Wade and V.J. Rayward-Smith present a number of interesting
local search algorithm for the Steiner problem in Graphs. They provide
extensive computational results that demonstrate that a simulated
annealing algorithm has the best computational performance.

Finally, last but not least, Stefan Voss, provides a comprehensive survey
and overview of most all algorithmic approaches for the Steiner problem in
Graphs. Both constructive and improvement algorithms are examined. For
the Steiner problem in Graphs, a new reader to this area might wish to
start with Voss's Chapter, since his survey of the approaches is thorough
and very readable, and the bibliographic citations section is one of the
most comprehensive and up-to-date.

Ding-Zhu Du and J. MacGregor Smith, Editors
Minneapolis and Amherst

The Steiner Ratio of finite-dimensional $\mathcal{L}_p$-spaces

Jens Albrecht
Institute of Mathematics and C.S.
University of Greifswald, Germany

Dietmar Cieslik
Institute of Mathematics and C.S.
University of Greifswald, Germany
E-mail: cieslik@mail.uni-greifswald.de

Contents

1 Introduction

Starting with the famous book "What is Mathematics" by Courant and Robbins the following problem has been popularized under the name of Steiner: For a given finite set of points in a metric space find a network which connects all points of the set with minimal length. Such a network

1

D.-Z. Du et al. (eds.), Advances in Steiner Trees, 1-13.
© 2000 *Kluwer Academic Publishers.*

must be a tree, which is called a Steiner Minimal Tree (SMT). It may contain vertices other than the points which are to be connected. Such points are called Steiner points.[1] A classical survey of this problem in the Euclidean plane was given by Gilbert and Pollak [9]. An updated one can be found in [13].

If we don't allow Steiner points, that is if we connect certain pairs of given points only, then we refer to a Minimum Spanning Tree (MST). Starting with Boruvka in 1926 and Kruskal in 1956 Minimum Spanning Trees have a well-documented history [11] and effective constructions [1].

Obviously, these problems depend essentially on the way how the distances in the plane are determined. In the present paper we consider finite-dimensional spaces with p-norm, defined in the following way: Let A_d be the d-dimensional affine space. For the points $X = (x_1, ..., x_d)$ and $Y = (y_1, ..., y_d)$ of the space we define the distance by

$$\rho(X,Y) = \rho_p(X,Y) = \left(\sum_{i=1}^{d} |x_i - y_i|^p \right)^{1/p}$$

where $1 \leq p < \infty$ is a real number. If p runs to infinity we get the so-called Maximum distance

$$\rho_\infty(X,Y) = \max\{|x_i - y_i| : 0 \leq i \leq d\}$$

In each case we obtain a finite-dimensional Banach space written by $\mathcal{L}_p^d$.

A (finite) graph $G = (V, E)$ with the set V of vertices and the set E of edges is embedded in this space in the sense that

- V is a finite set of points in the space;

- Each edge $\underline{XY} \in E$ is a segment $\{tX + (1-t)Y : 0 \leq t \leq 1\}$, $X, Y \in V$; and

- The length of G is defined by

$$L(G) = L_p(G) = \sum_{\underline{XY} \in E} \rho_p(X,Y).$$

---

[1]The history of Steiner's Problem started with P.Fermat early in the 17th century and C.F.Gauß in 1836. At first perhaps with the famous book *What is Mathematics* by R.Courant and H.Robbins in 1941, this problem became popularized under the name of Steiner.

Now, **Steiner's Problem of Minimal Trees** is the following: Given a finite set N of points in the space $\mathcal{L}_p^d$. Find a connected graph $G = (V, E)$ embedded in the space such that $N \subseteq V$ and $L_p(G)$ is minimal as possible.

A solution of Steiner's Problem is called a Steiner Minimal Tree (SMT) for N in $\mathcal{L}_p^d$. The vertices in the set $V \setminus N$ are called Steiner points. We may assume that for any SMT $T = (V, E)$ for N the following holds: The degree of each Steiner point is at least 3 and $|V \setminus N| \leq |N| - 2$.

Whereas we have algorithms to construct SMT's in the plane $\mathcal{L}_p^2$, see [4], but which need exponential time, a method to find an SMT in the higher-dimensional spaces $\mathcal{L}_p^d$, $d2$, is still unknown. On the other hand, we said above that it is simple to find an MST.

The relative defect, which describes the length of an SMT divided by the length of an MST, is given in the **Steiner Ratio**:

$$m(d,p) = \inf \left\{ \frac{L(\text{SMT for } N)}{L(\text{MST for } N)} : N \subset \mathcal{L}_p^d \text{ finite set} \right\}.$$

$m(d,p)$ is a measure of how good an MST as an approximation of Steiner's Problem in the space $\mathcal{L}_p^d$ is. It is not hard to see that we have $1 \geq m(d,p) \geq 1/2$ for every dimension d, see [9]. In general, the exact value of the Steiner ratio is still unknown.

As an introductory example consider three points which form the nodes of an equilateral triangle of unit side length in the Euclidean plane. An MST for these points has length 2. An SMT uses one Steiner point and has length equals $3 \cdot \sqrt{1/3} = \sqrt{3}$. So we have an upper bound for the Steiner ratio of the Euclidean plane:

$$m(2,2) \leq \frac{\sqrt{3}}{2} = 0.86602....$$

Another example: In the d-dimensional affine space A_d, let $B(1)$ be the unit ball of ρ_1. $B(1)$ is the convex hull of

$$N = \{\pm(0, ..., 0, 1, 0, ..., 0) : \text{the i'th component is equal to } 1, i = 1, ..., d\}.$$

(1)

Then the set N contains $2d$ points. The distance of any two different points in N equals 2. Hence, an MST for N has the length $2(2d-1)$. Conversely, an

SMT for N with the Steiner point $(0, ..., 0)$ has the length $2d$. This implies

$$m(d,1) \le \frac{d}{2d-1}. \tag{2}$$

In view of (2) we see that the lower bound $1/2$ is the best possible for the Steiner ratio over the class of all spaces $\mathcal{L}_p^d$.[2]
Graham and Hwang [10] conjectured that equality holds, which is true in the planar case: $m(2,1) = \frac{2}{3} = 0.6666...$, see [12].[3]

Also the Euclidean Steiner ratio in the planar case is well-known, since Du and Hwang [7] showed

$$m(2,2) = \frac{\sqrt{3}}{2} = 0.86602... \tag{3}$$

Moreover, there are bounds for the Steiner ratio $m(2, 2k)$, see [2]. Particularly,

$$m(2,4) \le \frac{2}{3} \cdot \sqrt{\sqrt{2}} = 0.79280... \tag{4}$$

and

$$m(2,4) \ge \sqrt{\frac{3}{8}} \cdot \sqrt{\sqrt{2}} = 0.72823... \tag{5}$$

Our goal is to estimate the quantities $m(d, p)$ for the dimension d greater than 2 with help of investigations for configurations of points in a regular situation in the space $\mathcal{L}_p^d$.

2 The Steiner ratio of Euclidean spaces

$\mathcal{L}_2^d$ is the d-dimensional Euclidean space with Steiner ratio $m(d, 2)$.
First we consider the set N of $d + 1$ nodes of a simplex with edges of unit length. Then an MST for N has the length d. The sphere that circumscribes N has the radius $R(N) = \sqrt{d/(2d+2)}$. With the center of this sphere as Steiner point, we find a tree T interconnecting N with the length $(d + 1)R(N)$. Hence,

$$m(d,2) \le \sqrt{\frac{1}{2} + \frac{1}{2d}}. \tag{6}$$

---

[2] And also of all finite-dimensional Banach spaces.
[3] The methods given by Hwang in [12] do not seem to be applicable to proving the conjecture in the higher dimensional case.

We used a Steiner point of degree $d+1$. It is well-known that all Steiner points in an SMT in Euclidean space are of degree 3, compare [3] or [9]. Consequently, the tree T described above is not an SMT for N, if $d2$. Better estimates for the Euclidean Steiner ratio, based on investigations of finite sets, can be found by Chung, Gilbert [5] and Smith [20]:

dimension	bound by Chung, Gilbert	bound by Smith
$d = 2$	0.86602...	
$d = 3$	0.81305...	0.81119...
$d = 4$	0.78374...	0.76871...
$d = 5$	0.76456...	0.74574...
$d = 6$	0.75142...	0.73199...
$d = 7$	0.74126...	0.72247...
$d = 8$	0.73376...	0.71550...
$d = 9$	0.72743...	0.71112...
$d = 10$	0.72250...	
$d = 20$	0.69839...	
$d = 40$	0.68499...	
$d = 160$	0.67392...	
$d \to \infty$	0.66984...	

For each dimension $d > 2$, at present, the exact value of the quantity $m(d, 2)$ is not yet known. In particular, this is true for $d = 3$. Smith and McGregor Smith [21] conjectured $m(3, 2) = 0.78419\ldots$.

With help of the knowledge of good bounds for the Steiner ratio of Euclidean spaces and the embeddings of Euclidean in the spaces $\mathcal{L}_p^d$ there are estimates for $m(d, p)$.[4] For these facts see [3].

3 Using Equilateral Sets

We will use the idea of the existence of a regular simplex similar as in Euclidean spaces. If there is a regular simplex in $\mathcal{L}_p^d$ with $d+1$ vertices and with unit edge length then an MST has length d and an SMT has length at most $(d+1)R$, where R is the circumradius of the simplex. Hence,

$$m(d, p) \leq \left(1 + \frac{1}{d}\right) \cdot R.$$

---

[4]But such embeddings exist at most for spaces with $p = 2k$, where k is a positive integer, see [17] or [20].

In view of this fact, it is of interest to know something about equidistant points in finite-dimensional Banach spaces.

Of course, there is an equidistant set of $d+1$ points in the Euclidean space namely nodes of a regular simplex.[5] It is an open question whether there exist $d+1$ equidistant points in any d-dimensional Banach space, even if the unit ball is smooth and if $d = 4$. But it is true in each three-dimensional space, see [18].[6]

On the other hand, it is known that any set of equidistant points in a d-dimensional Banach space has at most the cardinality 2^d, and equality is attained only when the unit ball is affinely equivalent to the d-hypercube. Also, for sufficiently large dimension d in any d-dimensional affine space there exists a strictly convex unit ball B such that there is an equidistant set in the space $M_d(B)$ with at least $(1.02)^d$ points. For all these facts compare [15] and [8].

For our investigations we have the following facts: Let $1 < p < \infty$ and $d \geq 3$. Then there are in $\mathcal{L}_p^d$ at least $d+1$ equidistant points. This can be see with the following considerations: Consider d points, with exactly one coordinate equals 1, and all the other equal 0; that is for $i = 1, \ldots, d$ let

$$P_i = (x_{i,1}, \ldots, x_{i,d})$$

with

$$x_{i,j} = \begin{cases} 1 & : \quad \text{if } i = j \\ 0 & : \quad \text{otherwise} \end{cases}$$

It is

$$\rho(P_i, P_j) = 2^{1/p}$$

for all $1 \leq i < j \leq d$.

For the point $P = (x, \ldots, x)$ it holds

$$\rho(P, P_i) = \rho(P, P_j)$$

for all $1 \leq i, j \leq d$.

To create $\rho(P, P_i) = 2^{1/p}$ the value x has to fulfill the equation

$$((d-1)|x|^p + |1-x|^p)^{1/p} = 2^{1/p}.$$

---

[5]We used this fact above, to obtain the inequality (6).

[6]Moreover: In Euclidean spaces, regular simplices can be freely constructed. In general, however, a free construction of regular simplices is impossible as far as $d \geq 4$. Dekster [6] shows that if the Banach-Mazur distance between the unit ball B and the Euclidean ball $B(2)$ is small, a regular simplex can be freely constructed.

This we can realize by the fact that the function $f : [0,1] \to I\!R$ with

$$f(x) = ((d-1)x^p + (1-x)^p)^{1/p} - 2^{1/p}$$

has exactly one zero in $[0,1]$.

Theorem 3.1 *Let $1 < p < \infty$ and $d \geq 3$. Then*

$$m(d,p) \leq \frac{d+1}{2d} \cdot \left(\frac{d}{2}\right)^{1/p}.$$

Proof. Let $N = \{P_1, \ldots, P_d, P\}$ be a set with the points P_i and P constructed how above. An MST for N has the length

$$L_p(\text{MST for } N) = d \cdot 2^{1/p}. \tag{7}$$

Let $T = (N \cup \{S\}, E)$ be a tree with

$$S = (1/2, \ldots, 1/2)$$

and

$$E = \{\underline{SP_i} \mid i = 1 \ldots d\} \cup \{\underline{SP}\}.$$

Then

$$
\begin{aligned}
L_p(T) &= \sum_{i=1}^{d} \rho(S, P_i) + \rho(S, P) \\
&= d \cdot \frac{1}{2} \cdot d^{1/p} + |x - \frac{1}{2}| d^{1/p} \\
&\leq \frac{1}{2}(d+1) \cdot d^{1/p} \qquad \text{since } x \in [0,1]
\end{aligned}
$$

This gets the assertion. □

4 An Approach with Equilateral Sets and a Center

We will extend an idea which is given by Liu and Du [16] for the planar case.

Lemma 4.1 *Let $1 < p < \infty$. For $x > 2^{-1/p}$ we have*

$$h(x) = \frac{1}{2} - x^p + (x - 2^{-1/p})^p < 0.$$

Proof. It is

1.) $h(2^{-1/p}) = 0$.

2.) $h'(x) = p(-x^{p-1} + (x - 2^{-1/p})^{p-1}) < 0$ for $x > 2^{-1/p}$

Hence, for $x > 2^{-1/p}$ the function $h(x)$ is strictly monotonously decreasing and consequently we have $h(x) < 0$. □

Lemma 4.2 *Let $1 < p < \infty$. Then there are points $P_0^{(d)}, \ldots, P_d^{(d)}$ in the space $\mathcal{L}_p^d$ with*

$$\rho(P_i^{(d)}, P_j^{(d)}) = 1$$

for $0 \leq i < j \leq d$ and a point $Z^{(d)}$ with

$$\rho(P_i^{(d)}, Z^{(d)}) = \rho(P_j^{(d)}, Z^{(d)}) < 2^{-1/p}$$

for $0 \leq i, j \leq d$.

The proof used the idea to determine a "center" Z for $d+1$ equidistant points of the d-dimensional space and then move Z in the next dimension such that the point is in the desired position.

Proof. Induction over the dimension d
Let $d = 1$. Then the points $P_0^{(1)} = (0), P_1^{(1)} = (1)$ and $Z^{(1)} = (1/2)$ satisfy the assertion. Particularly, it is

$$\rho(P_i^{(1)}, Z^{(1)}) = \frac{1}{2} < 2^{-1/p},$$

for $i = 0, 1$.
Now assume that there are points $P_i^{(d)} = (x_{i,1}^{(d)}, \ldots, x_{i,d}^{(d)})$, $i = 0, \ldots, d$, and $Z^{(d)} = (z_1^{(d)}, \ldots, z_d^{(d)})$ in the d-dimensional space $\mathcal{L}_p^d$ with the properties

$$\rho(P_i^{(d)}, P_j^{(d)}) = 1 \quad \text{for } 0 \leq i < j \leq d \tag{8}$$

and

$$\rho(P_i^{(d)}, Z^{(d)}) = \rho(P_j^{(d)}, Z^{(d)}) \tag{9}$$

$$< 2^{-1/p} \quad \text{for } 0 \leq i, j \leq d. \tag{10}$$

Then we build up $d + 2$ equidistant points of the $(d + 1)$-dimensional space $\mathcal{L}_p^{d+1}$ in the following way:

$$P_i^{(d+1)} := (x_{i,1}^{(d)}, \ldots, x_{i,d}^{(d)}, 0) \text{ for } i = 0, \ldots, d$$

$$P_{d+1}^{(d+1)} := (z_1^{(d)}, \ldots, z_d^{(d)}, x)$$

with

$$x = (1 - \rho(P_0^{(d)}, Z^{(d)})^p)^{1/p}. \tag{11}$$

With help of (10) it follows

$$x > 2^{-1/p} \tag{12}$$

Now we determine the mutual distances between these points. It is

$$\rho(P_i^{(d+1)}, P_j^{(d+1)}) = \rho(P_i^{(d)}, P_j^{(d)})$$
$$= 1$$

for $0 \leq i < j \leq d$, using (8) and the construction of the points $P_0^{(d+1)}, \ldots, P_d^{(d+1)}$. Moreover

$$\rho(P_i^{(d+1)}, P_{d+1}^{(d+1)}) = \left(\sum_{j=1}^{d} |x_{i,j}^{(d)} - z_j^{(d)}|^p + x^p \right)^{1/p}$$
$$= (\rho(P_i^{(d)}, Z^{(d)})^p + 1 - \rho(P_0^{(d)}, Z^{(d)})^p)^{1/p} \quad \text{since (9)}$$
$$= 1 \quad \text{since (11)}$$

for $0 \leq i \leq d$.

Additionally, let $Z^{(d+1)} := (z_1^{(d)}, \ldots, z_d^{(d)}, z)$. Consequently we have for all $0 \leq i, j \leq d$ the equalities

$$\rho(P_i^{(d+1)}, Z^{(d+1)}) = \left(\sum_{k=1}^{d} |x_{i,k}^{(d)} - z_k^{(d)}|^p + z^p \right)^{1/p}$$
$$= (\rho(P_i^{(d)}, Z^{(d)})^p + z^p)^{1/p}$$
$$= (\rho(P_j^{(d)}, Z^{(d)})^p + z^p)^{1/p} \quad \text{since (9)}$$
$$= \left(\sum_{k=1}^{d} |x_{j,k}^{(d)} - z_k^{(d)}|^p + z^p \right)^{1/p}$$
$$= \rho(P_j^{(d+1)}, Z^{(d+1)}).$$

These facts are independent from the choice of z. Now we chose z such that the condition

$$\rho(P_i^{(d+1)}, Z^{(d+1)}) = \rho(P_{d+1}^{(d+1)}, Z^{(d+1)}) \tag{13}$$

for $0 \le i \le d$ is satisfied. Because z lies in $(0, x)$, the equation (13) is equivalent to

$$\sum_{j=1}^{d} \left| x_{i,j}^{(d)} - z_j^{(d)} \right|^p + z^p = (x - z)^p$$

$$\Leftrightarrow \rho(P_i^{(d)}, Z^{(d)})^p + z^p = (x - z)^p$$

$$\Leftrightarrow 1 - x^p + z^p = (x - z)^p \quad \text{since (9),(11).}$$

With other terms, z is a zero of the function $f(y) = 1 - x^p + y^p - (x - y)^p$. It is

$$f(x - 2^{-1/p}) = \frac{1}{2} - x^p + (x - 2^{-1/p})^p$$

$$< 0 \quad \text{since (12) and lemma 4.1}$$

$$\text{and} \quad f(x) = 1$$

Because the function f is continuous we have the desired zero z, and it holds

$$x - 2^{-1/p} < z < x \tag{14}$$

Moreover, for $0 \le i \le d+1$ we have

$$\rho(P_i^{(d+1)}, Z^{(d+1)}) = \rho(P_{d+1}^{(d+1)}, Z^{(d+1)}) \quad \text{since (13)}$$

$$= x - z$$

$$< 2^{-1/p} \quad \text{since (14)}$$

Hence, the mutual distance of the new constructed points $P_i^{(d+1)}$ equals 1 and the distance to $Z^{(d+1)}$ is bounded by $2^{-1/p}$.   □

Theorem 4.3 *Let $1 < p < \infty$. Then*

$$m(d, p) < \frac{d+1}{d} \cdot \left(\frac{1}{2}\right)^{1/p}.$$

Proof. Let N be the set with the $d+1$ points constructed above and let Z be the "center" of this construction. Then

$$L(\text{MST for } N) = d$$

and

$$L(\text{SMT for } N) \leq (d+1)2^{-1/p}.$$

These facts imply the assertion. $\qquad\qquad\qquad\qquad\qquad\qquad\square$

Also this bound is not sharp, since the estimation of the distance of the points to the center is to inefficient, at least for small dimensions. On the other hand, we only use one additional point, and it is to assume that more than one of such points decrease the length. For these assumption compare the investigations by Swanepoel [22] and [23].

5 Concluding Remarks

We find two upper bounds for the Steiner ratio $m(d,p)$ for $3 \leq d$ and $1 < p < \infty$. Both are not sharp.[7] Moreover, in the introduction we created similar bounds for the quantities $m(d,1)$ and $m(d,\infty)$.
On the other hand, in each case $m(d,p)$ is a monotonously decreasing function in the dimension d. Since the sequence $\{m(d,p)\}_{d=0,1,...}$ is also bounded there exists the limit

$$\underline{m}(p) = \lim_{d\to\infty} m(d,p). \qquad (15)$$

We know:

$$\underline{m}(1) = \underline{m}(\infty) = 1/2,$$

which follows from (2);

$$1/\sqrt{3} \leq \underline{m}(2) \leq 1/\sqrt{2},$$

which follows from [10] and theorem 4.3.
Moreover the theorem 4.3 give a common generalization of these facts in the inequality

$$\frac{1}{2} \leq \underline{m}(p) \leq \left(\frac{1}{2}\right)^{1/p} \qquad (16)$$

for any number p.

---

[7] If $d2^p$ then the second bound gives the better value.

References

[1] D. Cheriton and R.E. Tarjan, Finding Minimum Spanning Trees, *SIAM J. Comp.* 5(1976) pp. 724-742.

[2] D. Cieslik, The Steiner Ratio of $\mathcal{L}_p^2$, *3rd Twente Workshop on Graphs and Combinatorial Optimization*, eds. U.Faigle and C.Hoede, Universiteit Twente, 1993, Memorandum no 1132, pp. 31-34.

[3] D. Cieslik, *Steiner Minimal Trees*, to appear by Kluwer Academic Publishers, 1998.

[4] D. Cieslik and J. Linhart, Steiner Minimal Trees in L_p^2, *Discrete Mathematics*, (155)1996, pp. 39-48.

[5] F.R.K. Chung and E.N. Gilbert, Steiner Trees for the Regular Simplex, *Bull. Inst. Math. Ac. Sinica*, 4(1976), pp. 313-325.

[6] B.V. Dekster, Simplexes with Prescribed Edge Length in Minkowski and Banach Spaces, Preprint.

[7] D.-Z. Du and F.K.Hwang, A Proof of the Gilbert-Pollak Conjecture on the Steiner Ratio, *Algorithmica*, 7(1992), pp. 121-136.

[8] Z. Füredi, J. Lagarias, and F. Morgan, Singularities of minimal surfaces and networks and related extremal problems in Minkowski space, *DIMACS Series in Discrete Mathematics and Theoretical Computer Science*, (6)1991, pp. 95-106.

[9] E.N. Gilbert and H.O. Pollak, Steiner Minimal Trees, *SIAM J. Appl. Math.*, 16(1968), pp. 1-29.

[10] R.L. Graham and F.K. Hwang, A Remark on Steiner Minimal Trees, *Bull. of the Inst. of Math. Ac. Sinica*, (4)1976, pp. 177-182.

[11] R.L. Graham and P. Hell, On the History of the Minimum Spanning Tree Problem, *Ann. Hist. Comp.*, 7(1985), pp. 43-57.

[12] F.K. Hwang, On Steiner Minimal Trees with rectilinear distance; *SIAM J. Appl. Math.*, (30)1976, pp. 104-114.

[13] F.K. Hwang, D.S. Richards, and P. Winter, *The Steiner Tree Problem*, Ann. Discr. Math., (53)1992.

[14] J.B. Kruskal,On the shortest spanning subtree of a graph and the travelling salesman problem, *Proc. of the Am. Math. Soc.*, 7(1956), pp. 48-50.

[15] G. Lawlor and F. Morgan, Paired calibrations applied to soap films, immiscible fluids, and surfaces or networks minimizing other norms, *Pacific Journal of Mathematics*, (166)1994, pp. 55-82.

[16] Z.C. Liu and D.-Z. Du, On Steiner Minimal Trees with L_p Distance, *Algorithmica*, (7)1992, pp. 179-192.

[17] Y.I. Lyubich and L.N. Vaserstein, Isometric Embeddings between Classical Banach Spaces, Cubature Formulas, and Spherical Designs, *Geometriae Dedicata*, 47(1993), pp. 327-362.

[18] C.M. Petty, Equilateral sets in Minkowski space, *Proc. Amer. Math. Soc.*, (29)1971, pp. 369-374.

[19] J.J. Seidel, Isometric Embeddings and Geometric Designs, *Discrete Mathematics*, 136(1994), pp. 281-293.

[20] W.D. Smith, How To Find Steiner Minimal Trees in Euclidean d-Space, *Algorithmica*, 7(1992), pp. 137-178.

[21] W.D. Smith and J.M. Smith, On the Steiner Ratio in 3-Space, *J. of Combinatorial Theory*, A, 65(1995), pp. 301-332.

[22] K. Swanepoel, Extremal Problems in Minkowski Space related to Minimal Networks, *Proc. of the Am. Math. Soc.*, (124)1996, pp. 2513–2518.

[23] K. Swanepoel, *Combinatorial Geometry of Minkowski Spaces*, PhD, University of Pretoria, 1997.

Shortest Networks for One Line and Two Points in Space

R.S. Booth
Department of Mathematics and Statistics
The Flinders University of South Australia, GPO Box 2100, Adelaide
5001, Australia
E-mail: ray@maths.flinders.edu.au

D.A. Thomas
Department of Electrical and Electronic Engineering
The University of Melbourne, VIC 3052, Australia
E-mail: d.thomas@ee.unimelb.edu.au

J.F. Weng[1]
Department of Electrical and Electronic Engineering
The University of Melbourne, VIC 3052, Australia
E-mail: weng@ee.unimelb.edu.au

Contents

[1]Supported by a grant from the Australian Research Council.

D.-Z. Du et al. (eds.), Advances in Steiner Trees, 15-26.

References

Abstract

In this paper we study a new generalization of the Steiner tree problem: the shortest network interconnecting two points and a straight line in space. We prove that this shortest network, either with one access point on the line or the line being multi-accessible, can be computed exactly by solving a certain quartic equation. Moreover, we will show how Euclidean 3D-space can be partitioned into regions (point sets in space) which are defined by the topology of the Steiner minimal tree.

1   Introduction

Given a set of n points in the Euclidean plane, the Steiner tree problem asks for a minimum network T interconnecting the given points, possibly with some additional points to shorten the network [4]. The given points are referred to as *terminals*, the additional points are referred to as *Steiner points*, and T is called a *Steiner minimal tree*. The graph structure of T is called its *topology*. If the topology is known, then T can be found either by Melzak's geometric construction [7][6], or by the hexagonal coordinate method [13][5]. Both of these methods are time linear in n.

The Steiner tree problem in the plane has many generalizations. Firstly, instead of connecting points, it may be required to connect other objects. The objects can be any geometric configurations, for instance, finite or compact sets of points [3], smooth curves [14], or existing networks [12]. Secondly, there may exist some constraints on the terminals or edges, e.g. the terminals may lie on straight lines [2], or on smooth curves [9], or the edges may have maximum prescribed slope [1]. Thirdly, a connected set, if not a point, may have more than one *access point* [14]. In all of these generalizations, if the topology of T is known, then the construction of T can be investigated using either Melzak's method or the hexagonal coordinate method.

The construction of Steiner minimal trees becomes much more difficult when the Steiner tree problem is generalized to Euclidean 3D-space. The Steiner minimal tree for 3 points in space can be constructed by solving a quadratic equation since any 3 points must lie on a plane. On the other hand, Smith [10] has argued that the Steiner minimal tree for 4 points in space, the simplest non-planar Steiner minimal tree, cannot be found by any straight line and compass construction, nor by solving a finite sequence of equations of degree less than 8. In this paper we fill a gap by showing that the Steiner minimal tree for two points and one straight line, either with one access point on the line or the line being multi-accessible, can be computed exactly by solving a certain quartic equation. Moreover, we will show how how Euclidean 3D-space can be partitioned into regions (point sets in space) which are defined by the topology of the Steiner minimal tree.

To end this introduction we wish to point out the practical motivation for this research. In the mining industry, a ventilation and transportation network for mining ores is required to connect all underground orebodies to an existing vertical shaft. Our model is the simplest case of such mining networks. Some preliminary research on mining networks can be found in [1]. Another example is provided by an irrigation system consisting of a long feeder pipe and tributary pipes.

2   Possible Topologies and Notation

Let the given distinct points P, Q and the straight line L be such that neither P nor Q is on L. If only one access point, say R, is permitted on L, then the problem can be viewed as the Steiner tree problem with constraints. On the contrary, if multi-access points are permitted on L, then P and Q may separately join L like a conductive wire or a transport pipe. Hence T has four possible topologies in general which we should now define:

If P and Q join L separately at two access points on L, then T is referred to as T_0. If only one access point R is permitted and both P, Q directly join R, then T is referred to as T_1. If P and Q join a Steiner point S which in turn joins L, T is referred to as T_2. Finally, if P joins Q first and then one of P and Q joins L, T is referred to as T_3.

In the single-access point case the shortest possible network is either T_1, T_2 or T_3, while in the multi-access point case the shortest possible network is either T_0, T_2 or T_3, since $|T_0| < |T_1|$ if T_0 is permitted. It is well known that all angles at a Steiner point equal $120°$. Obviously, T_1 and T_3 are the degeneracies of T_2 where S collapses into R in T_1 and into P or Q in T_3.

The set of points Q in space for which T_i $(i = 0, 1, 2, 3)$ is the shortest network is referred to as the *optimal region* of T_i and is denoted by S_i.

By a transformation, we may assume that L is the Z-axis, $P = (p, 0, 0)$ and $Q = (r \cos \phi, r \sin \phi, q)$ with $0 < p \leq r$. Hence, Q lies on or outside the (half) cylinder

$$C_o : \quad x^2 + y^2 = p^2 \ (y \geq 0). \tag{1}$$

This condition implies that S cannot collapse into Q. By symmetry, we need only consider the case in which Q lies in the quarter-space, where $0 \leq \phi \leq 180°$, $0 \leq q$. We will employ the convention that '(vertically) above' means having a larger z-coordinate. Similarly, 'horizontal' means parallel to the OXY-plane.

3   Partition of Space in the Single-access Point Case

In general, the coordinates of a point A are denoted by (x_a, y_a, z_a). So the coordinates of R are $(0, 0, z_r)$. Because R is the optimal point on L in T_1, the two angles formed by PR, QR with OZ must be equal, by the variational argument [8]. That is, $p/z_r = r/(q - z_r)$. It follows that

$$z_r = \frac{pq}{p + r}, \quad q - z_r = \frac{rq}{p + r}. \tag{2}$$

On the other hand, since S collapses into R in T_1, $\angle PRQ \geq 120°$. Hence, $|PQ|^2 \geq |PR|^2 + |PR| \cdot |QR| + |QR|^2$. Using Equation (2) it is easy to derive from this inequality that $q \geq (p + r)\sqrt{1 + 2\cos\phi}$. This means that the boundary of S_1 and S_2 is C_{12} given by

$$C_{12} : \quad z = (p + \sqrt{x^2 + y^2})\sqrt{1 + \frac{2x}{\sqrt{x^2 + y^2}}} \ (y \geq 0). \tag{3}$$

Since S collapsing into P implies $\angle OPQ \geq 120°$, the boundary of S_2 and S_3 obviously is a $60°$ cone with OX as axis of rotation and with p as its vertex. Its equation is

$$C_{23} : \quad \frac{\sqrt{y^2 + z^2}}{x - p} = \tan 60° = \sqrt{3} \ (x \geq p, \ y \geq 0). \tag{4}$$

Clearly, since T_1 and T_3 are different degeneracies of T_2, C_{12} and C_{23} do not meet. We obtain the following theorem:

Theorem 3.1 *Let E^3 be the quarter-space such that $y \geq 0, z \geq 0$, and with the region bounded by the cylinder C_o removed. Let S_1 be the region in E^3 bounded by and above C_{12}. Let S_3 be the region in E^3 bounded by the cone C_{23} and containing the X-axis. Let $S_2 = E^3 - S_1 - S_3$. Then the optimal region of T_i $(i = 1, 2, 3)$ is S_i.*

Let l_i $(i = 0, 1, 2, 3)$ be the length of T_i. Trivially,

$$l_1 = |T_1| = |PR| + |QR| = \sqrt{p^2 + \frac{p^2q^2}{(p+r)^2}} + \sqrt{r^2 + \frac{r^2q^2}{(p+r)^2}},$$

and

$$l_3 = |T_3| = |OP| + |PQ| = p + \sqrt{(r\cos\phi - p)^2 + r^2\sin^2\phi + q^2}.$$

As for l_2, it is more complicated, and is derived in the next section.

4   Computing the Full Steiner Tree T_2

Let P, Q join L at R through the Steiner point S in T_2. For minimality, SR must be perpendicular to Z-axis, that is, SR must be a horizontal line. Let M be the midpoint of PQ and let $h = |PM|$. Then $h > 0$. Notice that $4h^2 = q^2 + w^2$. Suppose the other endpoint of the Simpson line [4] of T_2 is C. By Melzak's method, C lies on the circle Γ with centre M, radius $\sqrt{3}h$, in the plane orthogonal to PQ. Moreover, C satisfies the following conditions:

Condition 1 *RC intersects PQ at an interior point.*

Condition 2 *C is the point on Γ furthest from the Z-axis.*

Let Q_0 be the projection of Q on the OXY-plane. Unless $P = Q_0$, the coordinates of Q can be written as

$$Q = (p + w\cos\alpha, w\sin\alpha, q).$$

where $w = |PQ_0| \geq 0$, and $0 \leq \alpha \leq 180$.

We define three unit orthonormal vectors

$$
\begin{aligned}
\mathbf{u} &= (sin\alpha, -\cos\alpha, 0), \\
\mathbf{v} &= (q\cos\alpha, q\sin\alpha, -w)/(2h), \\
\mathbf{w} &= (w\cos\alpha, w\sin\alpha, q)/(2h).
\end{aligned}
$$

These unit vectors are defined so that $\mathbf{w}$ is in the direction of the segment PQ, and $\mathbf{u}$ is horizontal, oriented away from L and is also orthogonal to $\mathbf{w}$. Then $\mathbf{v}$ is the vector cross product $\mathbf{w} \times \mathbf{u}$, that is, orthogonal to both $\mathbf{w}$ and $\mathbf{u}$.

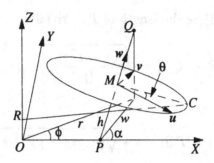

Figure 1: Diagrams of $\mathbf{u}, \mathbf{v}, \mathbf{w}$ and θ.

Let θ be the angle between $\overrightarrow{MC}$ and $\mathbf{u}$, measured from $\mathbf{u}$ to $\overrightarrow{MC}$ (Fig. 1). By the definition of $\mathbf{u}$ and Condition 2, $-90° \leq \theta \leq 90°$. Note that C lying on Γ implies that C satisfies the equation

$$(x_c, y_c, z_c) = (p, 0, 0) + h\mathbf{w} + h\sqrt{3}(\mathbf{u}\cos\theta + \mathbf{v}\sin\theta),$$

or

$$\begin{aligned}
x_c &= p + (w/2)\cos\alpha + h\sqrt{3}\sin\alpha\cos\theta + (\sqrt{3}/2)q\cos\alpha\sin\theta, \\
y_c &= (w/2)\sin\alpha - h\sqrt{3}\cos\alpha\cos\theta + (\sqrt{3}/2)q\sin\alpha\sin\theta, \\
z_c &= q/2 - (\sqrt{3}/2)w\sin\theta.
\end{aligned} \tag{5}$$

Since RC intersects PQ by Condition 1, there are scalars λ and μ $(0 \leq \lambda, \mu \leq 1)$ so that

$$(p, 0, 0) + \mu(w\cos\alpha, w\sin\alpha, q) = \lambda(x_c, y_c, 0) + (0, 0, z_c).$$

So the three equations

$$\begin{aligned}
\lambda x_c &= p + \mu w\cos\alpha, \\
\lambda y_c &= \mu w\sin\alpha, \\
z_c &= \mu q,
\end{aligned}$$

have a simultaneous solution for λ and μ. This implies

$$w z_c(x_c\sin\alpha - y_c\cos\alpha) = y_c pq.$$

By substitution of x_c, y_c and z_c, from (5), it follows that θ must satisfy the equation

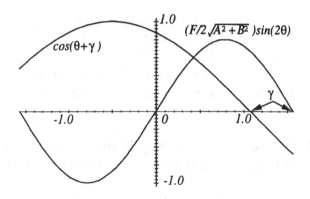

Figure 2: Two curves in the equation determining θ.

$$q \cos \theta (w + 2p \cos \alpha) - 4hp \sin \theta \sin \alpha = \sqrt{3} w^2 \sin \theta \cos \theta. \qquad (6)$$

This equation is equivalent to a quartic equation for θ.

From

$$l_2 = |T_2| = |RC| = \sqrt{x_c^2 + y_c^2} \qquad (7)$$

we obtain

$$
\begin{aligned}
l_2^2 &= (p + h\sqrt{3} \sin \alpha \cos \theta + (\sqrt{3}/2)q \sin \theta \cos \alpha + (w/2) \cos \alpha)^2 \\
&\quad + (-h\sqrt{3} \cos \alpha \cos \theta + (\sqrt{3}/2)q \sin \alpha \sin \theta + (w/2) \sin \alpha)^2 \\
&= 3h^2 \cos^2 \theta + (3/4)q^2 \sin^2 \theta + w^2/4 + p^2 + 2ph\sqrt{3} \sin \alpha \cos \theta \\
&\quad + qp\sqrt{3} \sin \theta \cos \alpha + pw \cos \alpha + (\sqrt{3}/2)wq \sin \theta. \qquad (8)
\end{aligned}
$$

Remark 4.1 Condition 2 implies $(d(l_2)^2/d\theta) = 0$. It is easy to show that for fixed w, q, α, as well as fixed p, $(d(l_2)^2/d\theta) = 0$ results in the same equation (6) as before.

Equation (6) can be rewritten as

$$A \cos \theta - B \sin \theta = F \sin \theta \cos \theta, \qquad (9)$$

where, obviously, $B = 4hp \sin \alpha \geq 0$ and $F = \sqrt{3} w^2 \geq 0$. Since

$$r^2 = p^2 + w^2 + 2wp \cos \alpha \geq p^2, \qquad (10)$$

we also have $A = q(w + 2p \cos \alpha) \geq 0$. Let $\cos \gamma = A/\sqrt{A^2 + B^2}$, $\sin \gamma = B/\sqrt{A^2 + B^2}$, $0 \leq \gamma \leq \pi/2$. Then Equation (9) is in the form of

$$\cos(\theta + \gamma) = \left(F/(2\sqrt{A^2 + B^2}) \right) \sin 2\theta.$$

The left side of the equation is a cosine curve of period 2π while the right side is a sine curve of period π. Hence, it is easy to see that the two curves have precisely one intersection in $[0, \pi/2]$ satisfying $-\pi/2 \leq \theta \leq \pi/2$ (Fig. 2). Thus C is determined by the unique solution θ of Equation (6) such that $0 \leq \theta \leq 90°$, and l_2 can be calculated from Equation (8). This proves the following theorem:

Theorem 4.1 *If Q lies in S_2, ie if the full Steiner tree T_2 exists, then T_2 can be constructed from a uniquely determined solution to quartic equation.*

In some special cases, Equation (9) has an explicit solution for θ. First, it is easily seen that if $A = 0$ then the solution is either $\theta = 0$, or

$$\theta = \arccos(\frac{-B}{F}) = \arccos(\frac{-4hp\sin\alpha}{\sqrt{3w^2}}).$$

Note that $A = 0$ means either $q = 0$ or $w + 2p\cos\alpha = 0$. The former case occurs if Q lies in the horizontal OXY-plane. The latter case occurs when $r = p$ because of Equality (10). That is, Q lies on the cylinder C_o.

Next, if $B = 0$, then the solution is either $\theta = 90$, or $\theta = -90$, or

$$\theta = \arcsin(\frac{A}{F}) = \arcsin(\frac{q(w + 2p\cos\alpha)}{\sqrt{3w^2}}).$$

Since $B = 0$ implies $\alpha = 0$, this is the special case that Q lies in the vertical OXZ-plane.

Finally, suppose $A = B$, ie Q lies on the surface defined by

$$q(w + 2p\cos\alpha) = 4hp\sin\alpha.$$

Then, if $x = \cos\theta - \sin\theta$, it is easy to derive from Equation (9) that x satisfies $Fx^2 - 2Ax + F = 0$. Therefore, Equation (9) can be decomposed into two consecutive quadratic equations. We leave the details of deriving the explicit expression of θ to the reader.

5   Partition of Space in the Multi-access Point Case

As we have stated, in the multi-access point case the shortest possible network is either T_0, T_2 or T_3 since always $|T_0| < |T_1|$. In T_0 the origin O is

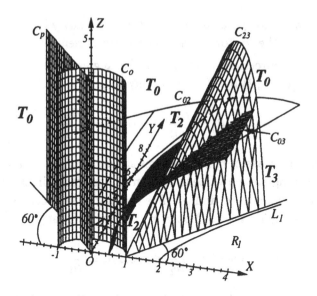

Figure 3: Partition of space (for $p = 1$).

the closest access point on OZ to P. Let D be the closest access point to Q. Then $T_0 = PO + QD$ and by minimality $QD \perp OZ$. Clearly,

$$l_0 = |T_0| = p + r.$$

It is easy to see from $l_0 = l_3$ that the surface separating the optimal regions of T_0 and T_3 is

$$C_{03} : z = \sqrt{2px - p^2} \ (x \geq p/2), \tag{11}$$

which is a parabolic cylinder with generators parallel to OY (Fig. 3). As we have shown in Section 3, the surface separating the regions S_2 and S_3 is the cone C_{23} determined by Equation (4). Since l_2 is determined by a quartic equation by Theorem 4.1, theoretically we can compute the exact surface C_{02} that separates the optimal region for T_0 from that for T_2. The surface C_{02} is determined by $l_0 = l_2$. Below we give a geometric description of this surface using the variational argument [8].

First let v_z be the vector describing the movement of Q in direction OZ. Let the angles between the vertical and the edges incident to Q in T_i $(i = 0, 2)$ be β_i respectively. Then obviously

$$0 = -\cos \beta_0 = \frac{\partial l_0}{\partial v_z} < -\cos \beta_2 = \frac{\partial l_2}{\partial v_z}.$$

That is, with the increase of q, T_0 remains constant while T_2 increases. Similarly, let v_r be the vector describing the movement of Q in the radial direction, ie in the direction of $\overrightarrow{DQ}$. Let the angles between DQ and the edges of Q in T_i $(i = 0, 2)$ be β_i^r respectively. Then

$$1 = -\cos\beta_0^r = \frac{\partial l_0}{\partial v_r} > -\cos\beta_2^r = \frac{\partial l_2}{\partial v_r}.$$

That is, with the increase of r, T_2 increases slower than T_0. Finally, let v_t be the vector describing the movement of Q in the tangent direction, ie in the direction perpendicular to v_z and v_r. Let the angles between this direction and the edges of Q in T_i $(i = 0, 2)$ be β_i^t respectively. Then it is also obvious

$$0 = -\cos\beta_0^t = \frac{\partial l_0}{\partial v_t} < -\cos\beta_2^t = \frac{\partial l_2}{\partial v_t}.$$

So we get a rough idea about the surface C_{02}: it is somewhat like an upwards cone and in the region above the surface C_{02}, T_0 is optimal.

Moreover, unlike C_{12}, C_{02} must meet the cone C_{23}. The reason is as follows. When Q lies in the OXZ-plane, $l_2 = x_c$ where x_c is the x-coordinate of the point C defined in Section 4. It is not hard to prove that $x_c = (r + p)/2 + \sqrt{3}q/2$ when Q lies on the OXZ-plane [15]. Hence the curve $l_0 = l_2$ in the OXZ-plane is determined by the equation $\sqrt{3}y - x - p = 0$. It follows that the surface C_{02}, the cone C_{23} and the parabolic cylinder meets at the same point $(2p, \sqrt{3}p, 0)$ in the OXZ-plane as shown in Fig. 3.

Let C_p be the vertical plane defined by

$$C_p : y = \sqrt{3}x \ (x \le 0). \tag{12}$$

Let R_1 be the region in the OXY-plane bounded by the X-axis and the hyperbolic curve

$$L_1 : y = \sqrt{3x^2 - 8px + 4p^2}$$

which is the projection of the intersection of the cone C_{23} and the parabolic cylinder C_{03}. Then summing up all arguments, we have the following conclusion:

Theorem 5.1 *Let E^3 be the quarter-space such that $y \ge 0$, $z \ge 0$, and with the region bounded by the cylinder C_o removed. Then,*

1. The optimal region S_0 for T_0 in E^3 consists of the angular region bounded by the plane C_p, the region outside the cone C_{23} and above C_{02}, and the region in the cone C_{23} but above the parabolic cylinder C_{03}; In particular, when Q lies above the parabolic cylinder C_{03} and its projection is in S_1, T_0 is optimal.

2. The optimal region S_2 for T_2 in E^3 is the region between the surface C_{02} and the cone C_{23};

3. The optimal region S_3 for T_3 in E^3 is space in the cone C_{23} and below the parabolic cylinder C_{03}.

References

[1] M.Brazil, D.A.Thomas and J.F.Weng, Gradient-constrained minimal Steiner trees, *Proceedings of the DIMACS Workshop on Network Design*, to appear.

[2] G.X.Chen, The shortest path between two points with a (linear) constraint (in Chinese), *Knowledge and Appl. Math.*, Vol. 4 (1980), pp. 1-8.

[3] E.J.Cockayne and Z.A.Melzak, Euclidean constructability in graph minimal problems, *Math. Mag.*, Vol. 42 (1969), pp. 206-208.

[4] F.K.Hwang, D.S.Richard and P.Winter, The Steiner tree problem, North-Holland, 1992.

[5] F.K.Hwang and J.F.Weng, Hexagonal coordinate systems and Steiner minimal trees, *Discrete Math.*, Vol. 62 (1986), pp. 49-57.

[6] F.K.Hwang, A linear time algorithm for full Steiner trees, *Oper. Res. Lett.*, Vol. 4 (1986), pp. 235-237.

[7] Z.A.Melzak, On the Steiner problem, *Canad. Math. Bull.* Vol. 4 (1961), pp. 143-148.

[8] J.H.Rubinstein and D.A.Thomas, A variational approach to the Steiner network problem, *Ann. Oper. Res.*, Vol. 33 (1991), pp. 481-499.

[9] J.H.Rubinstein, D.A.Thomas and N.C.Wormald, Steiner trees for terminals constrained to curves, *SIAM J. Discrete Math.*, Vol. 10 (1997), pp. 1-17.

[10] W.D.Smith, How to find Steiner minimal trees in Euclidean d-space, *Algorithmica*, Vol. 7 (1992), pp. 137-177.

[11] D.M.Y.Sommerville, *Analytical Conics*, G. Bell and Sons Ltd., London, 1933.

[12] D.Trietsch, Augmenting Euclidean networks — the Steiner case, *SIAM J. Appl. Math.*, Vol. 45 (1985), pp. 330-340.

[13] J.F.Weng, Generalized Steiner problem and hexagonal coordinate system (in Chinese), *Acta Math. Appl. Sinica*, Vol. 8 (1985), pp. 383-397.

[14] J.F.Weng, Shortest networks for smooth curves, *SIAM J. Optimization*, Vol. 7 (1997), pp. 1054-1068.

[15] J.F.Weng, Computing Steiner points, unpublished manuscript.

Rectilinear Steiner Minimal Trees on Parallel Lines[1]

Marcus Brazil
Department of Electrical and Electronic Engineering
The University of Melbourne, VIC 3052, Australia
E-mail: m.brazil@ee.unimelb.edu.au

Doreen A. Thomas
Department of Electrical and Electronic Engineering
The University of Melbourne, VIC 3052, Australia
E-mail: d.thomas@ee.unimelb.edu.au

Jia Feng Weng
Department of Electrical and Electronic Engineering
The University of Melbourne, VIC 3052, Australia
E-mail: weng@ee.unimelb.edu.au

Abstract

We consider the rectilinear Steiner tree problem for n points that lie on a bounded number of horizontal lines. We present a specific dynamic programming algorithm to construct a Steiner minimal tree for such a set of points, with time complexity $O(nk^3 10^k)$ and space complexity $O(nk5^k)$.

Contents

[1]Supported in part by a grant from the Australian Research Council.

D.-Z. Du et al. (eds.), Advances in Steiner Trees, 27-37.
© *2000 Kluwer Academic Publishers.*

1   Introduction

The rectilinear Steiner tree problem asks for a shortest network in the plane connecting n given points (referred to as *terminals*) in rectilinear metric. To accommodate the metric, we think of the edges of the network as being composed of horizontal and vertical line segments meeting only at their endpoints. The network may contain some additional vertices of degree 3 or 4 (called *Steiner points*) as well as additional points of degree 2 where horizontal and vertical line segments meet (called *corner points*) [6]. A shortest network must be a tree, called a *Steiner minimal tree* for the given points. The problem of finding such a tree has been proved to be NP-complete in general [4], but can be solved in polynomial time for some sets with special structures. In [1], Aho *et al.* showed that the rectilinear Steiner tree problem is linearly solvable if all terminals lie on two parallel lines, and briefly indicated how to extend this result to any fixed number of parallel lines, giving the following theorem.

Theorem 1.1 [Aho, Garey, Hwang] *The problem of constructing a rectilinear Steiner minimal tree for any set of terminals lying on a bounded number of horizontal lines can be solved in time linear in the number of terminals.*

The basic strategy in [1] for solving this theorem is to construct all possible forests of minimal Steiner trees for points on a vertical cutting line and the terminals to its left and gradually move this cutting line from left to right, building upon previous solutions. A slightly more general approach is as follows. The intersections of all vertical and horizontal lines through the terminals constitute the vertices (or *grid points*) of a grid graph, whose edges are all vertical and horizontal line segments between adjacent grid points. Hanan [5] has shown that there is always a rectilinear Steiner minimal tree lying on the grid graph of the terminals. One can now apply a divide-and-conquer algorithm by dividing the grid graph into at most n strips with a bounded number of grid points on the boundary and no internal grid points, constructing all Steiner forests on each of the strips, and recombining the strips in a suitable way so that the forests eventually form a single Steiner tree on the terminals. By only keeping track of the Steiner forests that are minimal with respect to the way that they connect up boundary points after each step, one can construct a Steiner minimal tree in linear time.

Clearly this technique can be extended to more general situations. An embedded planar graph is said to be *k-outerplanar* if the operation of successively removing all vertices on the infinite face and their adjacent edges terminates after k steps. Using the theorems of Baker [2], these ideas can be applied to solving the Steiner problem on k-outerplanar graphs, and, in particular, the rectilinear Steiner problem where the grid graph is k-outerplanar. Hence, we have the following corollary.

Corollary 1.2 *For a given constant k, the Steiner problem on a k-outerplanar graph can be solved in time linear in the number of terminals.*

Although the rectilinear Steiner tree problem for k horizontal lines can be solved in time linear in n the algorithm is exponential in k. Indeed, this appears inevitable, given that the problem is NP-hard. It can be shown that the algorithm of Aho *et al.* has time complexity $O(n16^k)$ and requires $O(n8^k)$ space. Furthermore, an implementation of the algorithm by Ganley and Cohoon [3], applied in the context of constructing thumbnail rectilinear Steiner trees, suggests that for $k \leq 7$ it is slower than a number of asymptotically inferior algorithms, while for larger k the space requirements make it impractical.

In this paper we take a similar approach to that of Aho *et al.*, but by carefully analyzing the patterns of connections are able to significantly reduce the degree of complexity with respect to k, which we believe will make our algorithm considerably more practical.

2   A Dynamic Programming Algorithm

2.1   Notation and Preliminaries

In this section we present a dynamic programming algorithm to implement Theorem 1.1. Let $A = \{a_1, a_2, ..., a_n\}$ be the given set of terminals. A line with maximal length in a rectilinear network is called *complete*. Since the length of a rectilinear network is a continuous function of the coordinates of terminals, by small perturbations of terminals we may assume that all terminals have different x-coordinates. A network that has a tree structure and has all vertices of degree no more that 3 is called a *Steiner tree*. A disjoint union of Steiner trees is called a Steiner forest. It has been shown, by sliding and flipping edges [6], that any Steiner tree T' can be transformed to a Steiner tree T that is no longer than T' and satisfies the following conditions:

1. *all Steiner points and corner points of T are grid points and each complete line of T has at least one terminal;*

2. *if all terminals in A have different x-coordinates, then any vertical complete line of T contains one and only one terminal, and consequently T contains no vertex of degree 4.*

Throughout the remainder of this paper we only consider Steiner trees satisfying these conditions.

Suppose there is a set of k horizontal lines $H_1, H_2, ..., H_k$ whose y-coordinates are $y_1, y_2, ..., y_k$ $(y_{i-1} > y_i)$, respectively such that all terminals in A lie on these lines and each H_i contains at least one terminal. Let $Y_H = \{y_1, y_2, ..., y_k\}$. For any point p, let $x(p)$ and $y(p)$ represent its x and y coordinates. We assume that at the preparation stage all terminals have been perturbed and sorted in such an order that $x(a_1) < x(a_2) < \cdots < x(a_n)$. Let $x_i = x(a_i)$ and $X_A = \{x_1, x_2, ..., x_n\}$. Hence, the grid point set to be used is $G = \{(x_i, y_j); x_i \in X_A, y_j \in Y_H\}$.

For convenience, a single terminal a_i in a Steiner tree T is regarded as the degenerate case of a vertical complete line if there are no other points of T lying on the vertical line through a_i. The following lemma comes directly from the facts stated above.

Lemma 2.1 *There is precisely one complete line on each vertical line through a terminal, and there is precisely one terminal on each vertical complete line.*

For $1 \le i \le n$, let h_i be the integer such that a_i lies on H_{h_i}. Then $a_i = (x_i, y_{h_i})$. Suppose T is a Steiner minimal tree on A. Define the *bar*, B_i, to be the vertical complete line in T containing a_i. For each i, the endpoints of B_i are $(x_i, y_{h'_i})$ and $(x_i, y_{h''_i})$, $1 \le h'_i \le h_i \le h''_i \le k$. For simplicity, we denote the interval $[(x_i, y_{h'_i})(x_i, y_{h''_i})]$ by $[h'_i, h''_i]_y$.

2.2   Subforests and their Patterns

Consider a Steiner tree T on A. Let F be the subforest of T lying between two given vertical cutting lines intersecting T. In general terms, by the *left pattern* (or *right pattern*) of F we mean a description of which points of T

intersect the cutting line on the left (or respectively right) of F and which of these cutting points belong to the same tree in F. If the leftmost of these two cutting lines passes through a_1, then F is said to be *relatively minimum* with respect to its right pattern if it is the shortest of all forests on the same terminals and rightmost cutting points as F, with the same right pattern as F, and such that every terminal is connected by a path to a rightmost cutting point. The key to our algorithm lies in efficiently recording the left and right patterns of given subforests, and building up forests that are minimum relative to these patterns.

For $1 \leq i \leq n$, let $A_i = \{a_1, a_2, \ldots, a_i\}$ and let R_i denote the smallest rectangle containing A whose sides are horizontal and vertical line segments. Let F_1 consist of the sole terminal a_1. Our algorithm will construct the relatively minimum Steiner forests F_i ($2 \leq i \leq n$) in R_i from F_{i-1} by adding one terminal a_i each time. Since the left pattern of F_i contains only a single cutting point, in constructing F it clearly suffices to find all F_i relatively minimum to the right pattern of F_i. The cutting lines we use are the vertical lines passing through the terminals. The advantage of this approach is that the forests $E_i = F_i - F_{i-1}$ have very simple structures.

First, to describe the right pattern $P^r(F_{i-1})$ of F_{i-1}, we use the following symbols to represent the grid points on the cutting line. If a grid point is not a cutting point, then its symbol is *blank* '⊔'. If a tree in F_{i-1} contains more than one cutting point, then we bracket the cutting points belonging to the tree by parentheses. If a grid point is the first or the last cutting point in the tree (counting from top to bottom), then it is symbolized by an *opening parenthesis* '(', or a *closing parenthesis* ')', respectively; otherwise, by an *asterisk* '*'. Finally, if a tree contains only one cutting point, then the grid point is expressed by a *shuttle*, '()' , an overlapped pair of opening and closing parentheses. Clearly, the five symbols ⊔, (,), * and () exhaust all possibilities of grid points. Thus, the right pattern of F_{i-1} can be represented by $P = P(H_1)P(H_2)\ldots P(H_k)$, a sequence of the 5 symbols. For instance, the forest F_{i-1} in Figure 1 has a right pattern

$$(\, () \,) \, \sqcup \, (\, * \,).$$

For this definition of the right pattern to be well defined, we need to show that a right pattern resulting from a particular forest uniquely partitions the cutting points into points belonging to different trees. This uniqueness is guaranteed by the lemma below, whose proof follow trivially from the planarity of the networks.

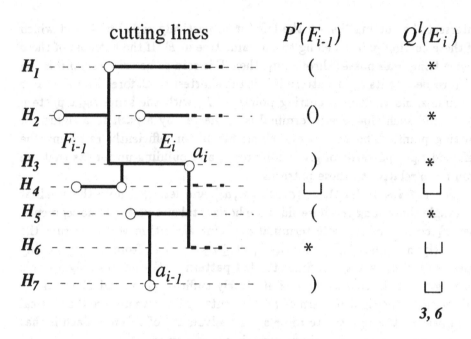

Figure 1: An example of some subforests of a Steiner tree and their patterns.

Lemma 2.2 *Suppose* (x_{i-1}, y_{j_1}) *and* (x_{i-1}, y_{j_2}) *are two cutting points belonging to the same tree* T' *in* F_{i-1}. *If another tree* T'' *in* F_{i-1} *has a cutting point in* $[j_1, j_2]_y$, *then all cutting points of* T'', *and in particular the opening and closing parentheses of* T'', *lie in* $[j_1, j_2]_y$.

Note that the lemma implies that during a scan of the right pattern of a given forest we can keep track of which connected component each cutting point is in by counting the number of opening and closing parentheses encountered.

By counting the number of possible right patterns P we have the following lemma.

Lemma 2.3 *There are at most* 5^k *possible right patterns of* F_{i-1}.

Next we study the structure of E_i. By Lemma 2.1 E_i is a union of horizontal line segments and the bar B_i. Together, the y-coordinates of the horizontal line segments and the endpoints of B_i uniquely specify the structure of E_i. It is convenient to think of the cutting points for the left pattern of E_i as being the leftmost points of the closure of E_i. We then represent the left pattern of E_i, denoted by $Q^l(E_i)$, by the triple (Q, h'_i, h''_i), where Q is a sequence of k symbols, corresponding to the k horizontal lines of the grid graph, each of which is either a ⊔ (denoting non-cutting points) or a ∗ (denoting cutting points), and h_i', h_i'' are integers such that $B_i = [h_i', h_i'']_y$. These definitions immediately give the following lemma.

Lemma 2.4 *The structure of E_i can be characterized by its left pattern $Q^l(E_i)$. The number of possible patterns $Q^l(E_i)$ is at most $k^2 2^k$.*

2.3 The Main Algorithm

The general strategy for constructing all suitable relatively minimum Steiner forests F_i is as follows. There are three arrays associated with all the relatively minimum Steiner forests F_{i-1} previously constructed, indexed by the set of right patterns. These are:

– *Exist*, which is set such that $Exist(P) = 1$ if $P = P^r(F_{i-1})$ for some relatively minimum Steiner forest F_{i-1}, and $Exist(P) = 0$ otherwise;

– *Topology*, which for each P such that $Exist(P) = 1$ stores the set of triples $Q^l(E_1), \ldots, Q^l(E_{i-1})$ used in determining the structure of the minimal forest F_{i-1} such that $P^r(F_{i-1}) = P$; and

– *Minlength*, where *Minlength*(P) stores the length of F_{i-1}.

To generate the F_is, we consider in turn every pattern P such that $Exist(P) = 1$. For each triple (Q, h_i', h_i''), representing a possible topology for E_i, the algorithm carries out a combinability check to ensure that F_{i-1} and E_i can be successfully combined into a forest F_i that can form part of a Steiner tree on A. If the check is successful, it then computes $P^r(F_i)$ from P. After carrying this out for all $P^r(F_{i-1})$ and all suitable triples (Q, h_i', h_i'') the relatively minimum forests in R_i can be determined for all right patterns, by combining the values in the array *Minlength* with the length of the forests in $R_i - R_{i-1}$. The arrays *Exist*, *Topology* and *Minlength* are updated, then i is increased by 1 and the procedure repeated until $i = n$ (see Figure 2).

A number of the above steps need to be considered in more detail, particularly the combinability check and the computation of the new right patterns. To successfully combine F_{i-1} and E_i into a forest F_i, P and (Q, h_i', h_i'') should satisfy the following lemma.

Lemma 2.5 *Let F be a rectilinear Steiner tree on A and let F_{i-1} be the subforest of F in R_{i-1}. For a given forest E_i in $R_i - R_{i-1}$ characterized by the triple (Q, h_i', h_i'') it is possible for $F_{i-1} \cup E_i$ to be the subforest of F in R_i if and only if:*

```
Input terminals and initialize all variables;
construct F₁;
for i = 2 to n do
begin
    Exist := Exist'; Exist' := 0;
    Minlength := Minlength'; Minlength' := ∞;
    for all P such that Exist(P) = 1 do
    begin
        obtain h'ᵢ₋₁, h''ᵢ₋₁ from Topology(P)
        for all (Q, h'ᵢ, h''ᵢ) such that h'ᵢ ≤ hᵢ ≤ h''ᵢ do
        begin
            if i = n and Q contains a * outside [h'ᵢ, h''ᵢ]_y then fail;
            check the combinability of P and (Q, h'ᵢ, h''ᵢ);
            m_e := the number of *s in Q;
            compute P', the new right pattern obtained by combining
            P and (Q, h'ᵢ, h''ᵢ);
            L := Minlength(P) + m_e(xᵢ − xᵢ₋₁) + (y_{h'ᵢ} − y_{h''ᵢ});
            if L < Minlength'(P') then
            begin
                Minlength'(P') := L;
                Exist'(P') := 1;
                Topology(P', i) := (Q, h'ᵢ, h''ᵢ)
            end
        end
    end
end;
for all P such that Exist'(P) = 1 do
    find  min{Minlength'(P)},  say  Minlength'(P');     Output:
Minlength'(P'), Topology(P').
(comment: The first output is the length of a minimal tree;
the second
is the construction of a minimal tree given by indicating the
positions of
horizontal and vertical edges in each Eᵢ in turn.)
```

Figure 2: The main algorithm

1. for each cutting point in Q, its corresponding grid point in P is also a cutting point; for each non-cutting point in Q, its corresponding grid point in P is also a non-cutting point unless it is in $[h'_{i-1}, h''_{i-1}]_y$;

2. for each cutting point in P but not in $[h'_{i-1}, h''_{i-1}]_y$, its corresponding grid point in Q is also a cutting point;

3. Q has no two cutting points in $[h'_i, h''_i]_y$ whose corresponding cutting points in P belong to the same tree in F_{i-1}; and

4. any tree in F_{i-1} has at least one cutting point in P whose corresponding grid point is a cutting point in Q.

Proof. It is clear that these conditions are sufficient. The necessity of the first two conditions is also obvious. The third condition guarantees the non-existence of cycles in F_i. If the last condition is not satisfied, then the tree in F_{i-1} becomes permanently isolated since it cannot be joined with other trees into a Steiner tree in any future steps.

The four conditions in the above lemma can be checked for a given (Q, h'_i, h''_i) by scanning P from top to bottom. We can use an array of integer variables to indicate whether each tree in F_{i-1} is linked to an edge of E_i outside $[h'_i, h''_i]_y$, or to an edge of E_i in $[h'_i, h''_i]_y$, or is not linked to E_i at all. It is then easy to show that the conditions in Lemma 2.5 can be checked in a single scan.

Note that it is only strictly necessary to check the last of the four conditions in Lemma 2.5, since if any of the first three conditions are not satisfied F will not be minimal. However, performing the extra checks does not increase computational complexity, and in practice decreases running time and space requirements.

Suppose this combinability check is passed for a given P and Q. Then F_i can be obtained by joining F_{i-1} and E_i, and we need to update P to construct the new right pattern P' of F_i. There are two steps to updating the pattern P. The first step is to compute the right pattern P' for all of F_i minus the bar B_i (as if the cutting line occurred slightly to the left of a_i). For all j outside the interval $[h'_{i-1}, h''_{i-1}]$ we set $P'(j) = P(j)$. For all j lying in $[h'_{i-1}, h''_{i-1}]$ we give P' the same pattern of cutting points and blanks as in Q. These cutting points are all $*$s, except the first cutting point is a (if $P(h'_{i-1}) = ($, the last cutting point is a) if $P(h''_{i-1}) = $), and if there is only one cutting point in the interval then it is a () if $P(h'_{i-1}) \neq *$ and $P(h''_{i-1}) \neq *$. It is easy to see that this gives the correct right pattern P' for $F_i - B_i$, which can be generated in a single scan of P. In the forest illustrated in Figure 1, for example, the new right pattern after this first step would be $(\,(\,)\,)$ ⊔ $(\,)$ ⊔ ⊔.

The second step is to incorporate the bar B_i. To do this, we scan P' to find the set of grid points G comprising all grid points on the interval $[h'_i, h''_i]$ and all cutting points of P' that belong to trees of $F_i - B_i$ which have a cutting point on P' in the interval $[h'_i, h''_i]$. If G contains only a single point then we leave P' unchanged. Otherwise we alter P' by changing the first element of P' corresponding to an element of G to a (, the last one to a), and all other corresponding elements to *s. Again, it is easy to see the new right pattern P' now correctly accounts for the presence of the bar B_i, and that this alteration can be achieved in a few scans of P'. After this second step the right pattern for the forest in Figure 1 would be $(\,(\,)\,*\,*\,*\,)$ ⊔.

The length of the new forest F_i is easily obtained by the formula:

$$Length(F_i) = Length(F_{i-1}) + m_e(x_i - x_{i-1}) + (y_{h'_i} - y_{h''_i}).$$

where m_e is the number of horizontal lines in E_i which can be computed in any scan of Q carried out before.

Obviously, different E_i yield different F_i. However, there may exist different F_{i-1} such that extending each by a given E_i results in forests F_i each of which has the same right pattern P'. Hence, to obtain a relatively minimum forest in R_i, we need to preserve a minimal length $Minlength(P)$ for each pattern P and compare it with each new obtained $Length(F_i)$. Therefore, joining all relatively minimum forests F_{i-1} in R_{i-1} and all possible forests E_i in $R_i - R_{i-1}$, at the final step, we obtain all relatively minimum Steiner trees in R_n. Comparing their lengths we can find the required Steiner minimum tree on A.

Clearly the main recursive part of the algorithm takes $O(k^3 10^k)$ time. Since, at any stage, there are at most 5^k relatively minimum Steiner forests under consideration and each forest has at most $(n-1)k$ edges, the largest space for the algorithm is $O(nk5^k)$. Summing up these arguments we have the following conclusion.

Theorem 2.6 *A rectilinear Steiner minimal tree for n terminals on k parallel lines can be found by a dynamic programming algorithm with time complexity $O(nk^3 10^k)$ and space complexity $O(nk5^k)$.*

References

[1] A.V. Aho, M.R. Garey and F.K. Hwang, Steiner Trees: Efficient special case algorithms, *Networks*, Vol. 7 (1977) pp. 37-58.

[2] B.S. Baker, Approximation algorithms for NP-complete problems on planar graphs, *J. of ACM*, Vol. 41 (1994) pp. 153-180.

[3] J.L. Ganley and J.P. Cohoon, Rectilinear Steiner trees on a checkerboard, *ACM Transactions on Design Automation of Electronic Systems*, Vol. 1 (1996) pp. 512-522.

[4] M.R. Garey and D.S. Johnson, The Steiner tree problem is NP-complete, *SIAM J. Appl. Math.*, Vol. 32 (1977) pp. 826-834.

[5] M. Hanan, On Steiner's problem with rectilinear distance, *SIAM J. Appl. Math.*, Vol. 14 (1966) pp. 255-265.

[6] F.K. Hwang, D.S. Richard and P. Winter, *The Steiner Tree Problem, (Annals of Discrete Math.)*, (Amsterdam, North-Holland, 1992).

Computing Shortest Networks with Fixed Topologies

Tao Jiang
Department of Computing and Software
McMaster University
Hamilton, Ont. L8S 4K1, Canada
E-mail: jiang@maccs.mcmaster.ca

Lusheng Wang
Department of Computer Science
City University of Hong Kong
83 Tat Chee Avenue, Kowloon, Hong Kong
E-mail: lwang@cs.cityu.edu.hk

Abstract

We discuss the problem of computing a shortest network interconnecting a set of points under a fixed tree topology, and survey the recent algorithmic and complexity results in the literature covering a wide range of metric spaces, including Euclidean, rectilinear, space of sequences with Hamming and edit distances, communication networks, etc. It is demonstrated that the problem is polynomial time solvable for some spaces and NP-hard for the others. When the problem is NP-hard, we attempt to give approximation algorithms with guaranteed relative errors.

1   Introduction

We know from the previous chapters that a *Steiner minimal tree (SMT)* is hard to compute because we have to deal with (i) a large number of possible

D.-Z. Du et al. (eds.), Advances in Steiner Trees, 39-62.
© 2000 Kluwer Academic Publishers.

tree topologies and (ii) a large number of possible Steiner points. Thus, it is natural to ask whether the problem becomes easier if a fixed tree topology is provided and we are only required to locate the Steiner points (in the space under consideration). This is the so called *fixed topology shortest network (FTSN)* problem.

More formally, let S be some metric space, R a set of *fixed* (also called *regular*) points in S, S a set of *moving* (also called *Steiner*) points in S, and T a (tree) topology interconnecting the set $V = R \cup S$. Our goal is to choose a location for each point in S so that the length of T is minimized, where the length of T is defined as the total length of the edges in T. For the convenience of presentation, we will sometimes also think of choosing locations for Steiner points as *labeling* them with elements of S. Note that here we allow two points to share a same location. Such sharings of locations may result in *networks* (after collapsing points with the same locations) with topologies different from T. In fact, the resulting networks may not be trees at all. These different topologies are usually referred to as *degeneracies* of T. Without loss of generality, we may assume that the regular points always form the external nodes of T and all internal nodes of T are Steiner points with degrees at least 3. It follows from the assumption that an instance of FTSN is sufficiently described by the (leaf labeled) topology T.

Not only is FTSN a natural and interesting variant of the SMT problem, it also has applications in many fields such as transportation networks [16], communication networks [37, 38], and computational molecular biology [34, 36, 27]. Some of the applications will be discussed in the sections to follow. Moreover, algorithms for FTSN are routinely used to help find SMT's, either exhaustively or heuristically [17, 5, 21].

In this chapter, we survey the efficient algorithms and complexity results for FTSN in various metric spaces including rectilinear, Euclidean plane, space of sequences with Hamming and edit distances, and communication networks with time-delay constraints. Our survey includes most of the results in the literature, especially the recent ones. It is shown that FTSN is indeed often easier than the SMT problem since the latter is NP-hard in all of the above mentioned spaces while the former is polynomial time solvable in several of them. We will also give a fairly generic *polynomial time approximation scheme (PTAS)* for the FTSN problem where the given topology has a bounded degree. [1] It is known that such a generic PTAS

---

[1] A PTAS is a family of algorithms $\{A_\epsilon | \epsilon > 0\}$ where algorithm A_ϵ achieves an approximation ratio of $1 + \epsilon$ (*i.e.* a relative error of ϵ) and runs in time polynomial in the size of

does not exist for the SMT problem [32, 33].

We also study a variant of FTSN in which we are given a topology with possibly some high-degree internal nodes and we are allowed to *split* a node into several nodes in the search for the shortest connection. In other words, we seek an optimal FTSN solution for any topology that is a *refinement* of the given the topology. The problem is called *Optimal Tree Refinement (OTR)* [40]. Observe that the OTR problem is actually a generalization of the SMT problem since the given topology can be a star with one internal node connecting to all regular points. We will consider OTR where the given topology has a bounded degree and present some complexity and approximation results.

The rest of the chapter is organized as follows. In the next section, we discuss the metric spaces in which the FTSN problem is polynomial time solvable. We then present the generic PTAS for FTSN with bounded degree topologies in Section 3. Sections 4 and 5 provide two applications of FTSN in the fields of communication networks and computational molecular biology (more precisely, sequence comparison). They also contain some interesting extensions of FTSN. Finally, the OTR problem is discussed in Section 6. We assume that the reader is familiar with standard terminology, notations and techniques in the analysis of algorithms, computational complexity and Steiner minimal trees. Consult [3, 17] for any missing information.

2 The Polynomial Time Solvable Cases

These cases mainly include three metric spaces: rectilinear space, space of sequences with Hamming distance, and Euclidean plane [2] We consider them in three separate subsections.

2.1 Sequences with Hamming Distance

For the ease of presentation, we start with the space of sequences with Hamming distance. Let n be any number and space S contain all binary sequences of length n. For any two binary sequences $x = x_1 \cdots x_n$ and

---

the input, for any *fixed* $\epsilon > 0$.

[2]Actually, the polynomial time algorithms for Euclidean plane only work for a special class of topologies. This will become clear later.

$y = y_1 \cdots y_n$, the Hamming distance between x and y is defined as

$$d(x, y) = \sum_{i=1}^{n} |x_i - y_i|.$$

The Hamming diatance FTSN problem has important applications in study of DNA and protein evolution [5, 24].

Let T be an instance of FTSN, where the external nodes of T represent regular points with given binary sequences of length n and the internal nodes of T correspond to Steiner points with unknown sequences. As observed in [6, 11, 27], to determine sequences for the Steiner points, it suffices to treat the n bit positions independently.

Consider an arbitrary bit position i. We root T at some arbitrary internal node. Then for each internal node v of T, we use the dynamic programming technique to calculate the shortest lengths (concerning only the i-th bit position) of the subtree rooted at v if the i-th bit of v is fixed as 0 or 1, starting from the *lowest* nodes. Once we find the shortest length of the whole tree at the root, a standard trace-back should reveal an optimal bit assignment at all internal nodes. The dynamic programming algorithm runs in linear time.

Theorem 1 *[11, 27] The Hamming distance FTSN problem can be solved in $O(n \cdot |T|)$ time.*

2.2   Rectilinear Space

Let n be a fixed number. The n-dimensional rectilinear space is also called the *n-dimensional Manhanttan space*. It includes all points in $\mathcal{R}^n$. For any two points $x, y \in \mathcal{R}^n$, where $x = (x_1, \ldots, x_n)$ and $y = (y_1, \ldots, y_n)$, the rectilinear distance between x and y is defined as

$$d(x, y) = \sum_{i=1}^{n} |x_i - y_i|.$$

The idea behind the polynomial time solution for rectilinear distance FTSN is similar to the one for Hamming distance FTSN. That is, to locate the Steiner points in $\mathcal{R}^n$, we can treat the n dimensions separately [4, 10, 14, 27]. Hence, it suffices to assume $n = 1$ and focus on the space $\mathcal{R}^1$. The only difference is that, instead of considering two bits 0 and 1, we now have to consider real intervals.

Again, we root T at some arbitrary internal node. Instead of using dynamic programming, we simply traverse the rooted T in post-order and calculate a closed interval for each internal node as the range of its location in an optimal solution, as follows.

Suppose that v is an internal node whose children are all external nodes. Let $x_1, \ldots, x_k$ be the children of v. Then we set the interval for v as follows:

$$I(v) = \begin{cases} [x_{(k+1)/2}] & \text{if } k \text{ is odd} \\ [x_{k/2}, x_{k/2+1}] & \text{if } k \text{ is even} \end{cases}$$

It is easy to see that we can always put v inside the interval $I(v)$ without compromising the optimality of a solution.

In general, suppose that v is an internal node with children $x_1, \ldots, x_k$, and that the intervals for nodes $x_1, \ldots, x_k$ have all been determined. Arrange the left end-points of the intervals $I(x_1), \ldots, I(x_k)$ into an ascending listing $l_1 \leq \cdots \leq l_k$. Similarly, arrange the right end-points of the intervals into an ascending listing $r_1 \leq \cdots \leq r_k$. Suppose that for some i, $r_1 < l_k, r_2 < l_{k-1}, \ldots, r_i < l_{k-i-1}$ and $r_{i+1} \geq l_{k-i-2}$. Note that such an i always exists uniquely. Then we define:

$$I(v) = [l_{k-i-2}, r_{i+1}].$$

Moreover, for each child x_j whose interval's left end-point equals l_h for some $k \geq j \geq k - i - 1$, we fix the location of x_j at l_h. Similarly, for each child x_j whose interval's right end-point equals r_h for some $1 \leq h \geq i$, we fix the location of x_j at r_h. For all the other children of v, we define their locations to be any points withint their respective intervals.

We repeat the above process until we have computed an interval for the root node. Finally, we fix the location of the root at any point with its interval. The following theorem states that the above algorithm indeed finds an optimal rectilinear FTSN.

Theorem 2 *[27] The above algorithm outputs a shortest tree in $O(n \cdot |T|)$ time.*

2.3 Euclidean Plane

Let (x_1, y_1) and (x_2, y_2) be two points in the plane $\mathcal{R}^2$. The Euclidean distance between them is defined as

$$d((x_1, y_1), (x_2, y_2)) = \sqrt{(x_1 - x_2)^2 + (y_1 - y_2)^2}.$$

The Euclidean planar SMT problem is perhaps the origin of all SMT problems. Recently, several polynomial time algorithms have been proposed for the Euclidean planar FTSN problem [15, 16, 44, 43]. Unfortunately, these algorithms only solve special cases of the problem, and it has been conjectured that Euclidean planar FTSN is in fact NP-hard [16]. In the following, we summarize the exact results in [15, 16, 44, 43] without giving the details of the algorithms.

Some standard definitions are necessary. A tree topology T is called a *Steiner topology* if every internal node of T has degree at most 3. A tree topology T is called a *full Steiner topology* if (i) it is a Steiner topology and (ii) all regular points are external nodes of T. Both Steiner and full Steiner topologies play crucial roles in the study of the Euclidean planar SMT problem. A *plane tree* (*i.e.* an embedding of some tree in the plane) interconnecting a set of regular and Steiner points is a *Steiner tree (ST)* if (i) every Steiner point has degree exactly three and (ii) no two adjacent edges form an angle less than 120°. Observe that every ST must have a Steiner topology. All the results discussed here assume that the input topology is a full Steiner topology although they can be trivially extended to Steiner topologies.

The first result is concerned with the case that there is actually an ST for the given regular points that has exactly the same topology as the input topology (note that it cannot even be a degeneracy of the input topology).

Theorem 3 *[15] Let T be a full Steiner topology. There is an $O(|T|)$ time algorithm that outputs an FTSN for T if the regular points in T have an ST with topology identical to T, and quits otherwise.*

Note that the above algorithm does *not* find a best solution among the trees with topology T in general. It works only if the input topology T satisfies the condition that it coincides with the topology of some ST for the input regular points. Otherwise it simply quits without outputting anything. The construction takes advantage of the special geometrical properties of ST's with full Steiner topologies, *e.g.* there are exactly $n - 2$ Steiner points and each Steiner point has degree exactly three with 120° subtending angles. The next result relaxes the condition slightly.

Theorem 4 *[16] Let T be a full Steiner topology. There is an $O(|T|^2)$ time algorithm that outputs an FTSN for T if the regular points in T have an ST whose topology is either T or a degeneracy of T, and quits otherwise.*

In other words, the algorithm can find an optimal solution for T if T or some degeneracy of T actually form the topology of an ST for the regular points. Again the construction works because of some special geometrical properties of ST's. The average time complexity of this algorithm has been improved to $O(|T| \log |T|)$ in [44]. It remains an interesting open problem if in general one can find an FTSN for a given full Steiner topology in polynomial time, although [16] conjectures that the problem is NP-hard. On the other hand, a polynomial time approximation scheme for the problem has been obtained very recently.

Theorem 5 *[43] Let T be a full Steiner topology. For any $\epsilon > 0$, there is an algorithm which can compute a network for T with length at most $1 + \epsilon$ times that of an optimal solution for T, in $O(|T| \sqrt{|T|} (\log(c(T)/\epsilon) + \log |T|))$ time, where $c(T)$ denotes the largest pairwise distance among the given regular points.*

The algorithm approaches the FTSN problem as a problem of minimizing a sum of Euclidean norms, and transforms it into a standard convex programming problem in conic form. Then it applies the interior-point method to find an approximate solution in polynomial time. The construction in fact works for any tree topology [45].

We note in passing that some other algorithms have been proposed for Euclidean planar FTSN in [23, 30] that either take more than polynomial time or do not work well for large inputs.

3　A Generic Polynomial Time Approximation Scheme

In this section, let us consider tree topologies with degrees bounded by some constant, *e.g.* 3. We are concerned with any metric space S that allows polynomial time computation of any constant size FTSN, *i.e.* there is a polynomial time algorithm which, when given a tree topology of constant size, can optimally locate the Steiner points in the space S. All of the specific metric spaces considered in this chapter have this property. We will present a *generic* PTAS that works for any such metric space S.

For the ease of presentation, we will from now on think of the FTSN problem as optimally labeling the Steiner points in topology T with elements of S. We first give a simple algorithm based on the *lifting* technique, achieving an approximation ratio of 2, and then extend the algorithm to a PTAS. The results were originally given in [19, 34].

3.1   A Ratio-2 Approximation Algorithm

For the purpose of computation, we arbitrarily root the given tree topology T at any internal node. The basic idea is to lift the regular points on the leaves of the given tree topology T to their ancestors. Define a *fully labeled* tree T' for T as a tree such that each of its internal node is also labeled with some point in S.

We need some notations. For a tree T, let $r(T)$ be the root of T, $L(T)$ the set of the leaves of T, and $c(T)$ the length of T (if T is fully labeled). For each node v in tree T, T_v denotes the subtree of T rooted at v. A leaf that is a descendant of node v is called a *descendant leaf* of v. Define $S(v)$ to be the set of labels of all descendant leaves of v. To simplify the presentation, we will assume without loss of generality that each internal node in T has exactly d children in our discussion.

Let $R = \{s_1, \ldots, s_m\}$ be the set of regular points in a given tree topology T. Let T^{min} denote an optimal solution for T. For each node v in T^{min}, the *closest descendant leaf* of v, denoted $l(v)$, is a descendant leaf of v such that the path from v to $l(v)$ is the shortest among all descendant leaves of v. For convenience, let $sl(v)$ denote the label of $l(v)$ (which is a point in space S). Define an *l-lifted* tree T^l as follows: for each internal node v in T, assign the point $sl(v)$ to v. The following lemma is known.

Lemma 6 *[34]* $c(T^l) \leq 2c(T^{min})$.

A tree is called a *lifted* tree if the label of each internal node equals the label of some child of the node. Obviously, T^l can be made as a lifted tree. From the above discussion, we can immediately conclude

Corollary 7 *There exists a lifted tree with length at most $2c(T^{min})$.*

Computing T^l is not easy since it is derived from the optimal tree T^{min}, which is unkown. However, in the following we describe a simple algorithm that constructs an optimal lifted tree T^*, *i.e.* one that has the smallest length among all the lifted trees. From the above corollary,

$$c(T^*) \leq c(T^l) \leq 2c(T^{min}).$$

The idea is again to use dynamic programming. For each $v \in T$, $i = 1, \ldots, m$ such that $s_i \in S(v)$, let $D[v, s_i]$ denote the length of an optimal lifted tree for T_v with v being assigned the point s_i. It is possible to compute

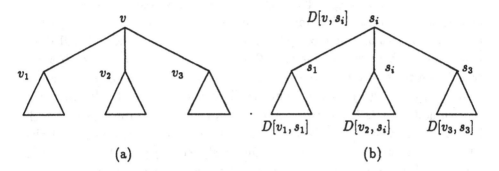

Figure 1: (a) The subtree T_v. (b) The lifted subtree, where $s_1 \in S(v_1)$, $s_i \in S(v_2)$, and $s_3 \in S(v_3)$.

$D[v, s_i]$ inductively. For each leaf v, we define $D[v, s_i] = 0$ if the label of v is s_i. Let v be an internal node, and $v_1, \ldots, v_d$ its children. Suppose that $s_i \in S(v_p)$, where $1 \le p \le d$. Clearly p is unique. Then $D[v, s_i]$ can be computed as follows: For each $q = 1, \ldots, d$ and $q \ne p$, find a j_q such that $s_{j_q} \in S(v_q)$ and $D[v_q, s_{j_q}] + d(s_i, s_{j_q})$ is minimized. Then

$$D[v, s_i] = D[v_p, s_i] + \sum_{\substack{1 \le q \le d \\ q \ne p}} D[v_q, s_{j_q}] + d(s_i, s_{j_q})$$

(See Figure 1.)

We compute the values $D[v, s_i]$ bottom up in T, and hence obtain:

Theorem 8 *[34] There is an algorithm that computes an optimal lifted tree in time $O(m^3 + m^2 f(S))$, where m is the number of leaves in T and $f(S)$ is the time required to compute the distance between any two points in the space S.*

Note that the above approximation algorithm in fact works for any input tree topology (*i.e.* its degree can be unbounded). The following extension, however, requires that the input topology has a bounded degree.

3.2   The PTAS

We can extend the above lifting algorithm to a PTAS by considering constant-size components of the tree T and augmenting the lifting technique with "local optimization".

Let $t > 0$ be an integer. For each $v \in T$, define a *depth-t component* $T_{v,t}$ as the subtree of T_v containing only the top $t + 1$ levels. Clearly, $T_{v,t}$ has at most d^t leaves and at most $(d^t - 1)/(d - 1)$ internal nodes. Recall that, if we label each node v in the tree T with the sequence $sl(v)$, we obtain a tree T^l which is at most twice as long as an optimal solution T^{min}. For a subtree $T_{v,t}, v \neq r(T)$, we can define a fully labeled subtree $T'_{v,t}$ by assigning to each node $u \in L(T_{v,t}) \cup \{v\}$ the points assigned to u in the tree T^l, which is $sl(u)$, and selecting the labels for the other nodes in $T_{v,t}$ such that the length of the subtree is minimized. Obviously, $c(T'_{v,t}) \leq c(T^l_{v,t})$, where $T^l_{v,t}$ is the depth-t subtree of T^l rooted at v. This hints us to partition the tree T into depth-t components, and construct an optimal fully labeled subtree $T'_{v,t}$ for each component $T_{v,t}$.

For a fixed t, we construct t different partitions P_0, P_1, ..., P_{t-1}, by slicing T horizontally as follows. For any fixed choice of q ($0 \leq q < t$), the top slice of partition P_q contains the top q levels of edges (and their associated nodes) in T. Each slice below the top one contains exactly t levels of edges in T, until at most t levels of edges remain. Note that two consecutive slices in a partition P_q share a common level of nodes of T. A *boundary* node for P_q is a node at a shared level in P_q.

For each partition P_q, a *depth-t component l-tree* T_q is a fully labeled tree for T obtained by assigning each boundary node u of P_q the label $sl(u)$ in the l-lifted tree T^l and optimally labeling the other nodes so that the length of the tree is minimized. The following lemma can be proven along the same lines of Lemma 6.

Lemma 9 *[34] There exists a depth-t component l-tree with length at most $(1 + \frac{3}{t})c(T^{min})$.*

Again, a depth-t component l-tree is hard to compute since we do not know an optimal solution. Define a *depth-t component tree* for P_q as any fully labeled tree whose boundary nodes have labels lifted from their decendent boundary nodes. Clearly, a depth-t component l-tree is a depth-t component tree. Below, we describe an algorithm to construct a *minimum-length* depth-t component tree $\hat{T}$, *i.e.* the length of $\hat{T}$ is the smallest among all the depth-t component trees for any P_q. The basic idea is to combine dynamic programming with local optimization.

Consider a $T_{v,t}$. Let u be a child of v and $L(T_{u,t-1}) = \{w_1, \ldots, w_k\}$ be the set of leaves in $T_{u,t-1}$. (Note that the component $T_{v,t}$ may not be full.) For each $i = 1, \ldots, k$, let $s_i \in S(w_i)$ be a label. Then for each s, $\hat{T}(v, t, u, s, s_1, \ldots, s_k)$ denotes the fully labeled subtree obtained from $T_{u,t-1} \cup \{(v, u)\}$ by labeling v with s and each node w_i with s_i, and selecting labels for the other nodes in $T_{u,t-1}$ so that the length of $\hat{T}(v, t, u, s, s_1, \ldots, s_m)$ is minimized.

The top subtree $T_{r(T),i}$, $0 \le i \le t - 1$, is treated similarly and the resulting fully labeled subtree is denoted $\hat{T}(r(T), i, u, s, s_1, \ldots, s_k)$.

Let v be a node of T, and $s \in S(v)$. Define $\hat{T}(v, s)$ as the minimum-length fully labeled subtree obtained from T_v such that the node v is labeled with s and $\hat{T}(v, s)$ forms a depth-t component tree for P_q. We use $D[v, s]$ to denote the length of $\hat{T}(v, s)$. Similar to the previous section, $D[v, s]$ can be computed recursively from bottom to top.

If v is a leaf, $D[v, s(v)] = 0$. Let v be an internal node and $u_1, \ldots, u_d$ the children of v. For each $i = 1, \ldots, d$, let $L(T_{u_i,t-1}) = \{w_{i,1}, \ldots, w_{i,k}\}$. Suppose that $s \in S(w_{p,r})$, where $1 \le p \le d$ and $1 \le r \le k$ are unique. Then $D[v, s]$ can be computed as follows: For each $i = 1, \ldots, d$ and each $j = 1, \ldots, k$, find an $s_{i,j}$ such that $s_{i,j} \in S(w_{i,j})$ if $i \ne p$ or $j \ne r$, and $s_{i,j} = s$ otherwise, and moreover

$$c(\hat{T}(v, t, u_i, s, s_{i,1}, \ldots, s_{i,k})) + \sum_{j=1}^{k} D[w_{i,j}, s_{i,j}]$$

is minimized. Then

$$D[v, s] \;=\; \sum_{i=1}^{d} c(\hat{T}(v, t, u_i, s, s_{i,1}, \ldots, s_{i,k})) \;+\; \sum_{i=1}^{d}\sum_{j=1}^{k} D[w_{i,j}, s_{i,j}].$$

For each $q = 0, \ldots, t - 1$, let $D[q]$ be the length of an optimal depth-t component tree for partition P_q. $D[q]$ can be computed from the top subtree $T_{r(T),q}$ and the values $D[v, s]$ of the nodes at level q of T in a way similar to above. Clearly, $\min\{D[q] | 0 \le q \le t - 1\} = c(\hat{T})$.

Theorem 10 *[34] For any t, there is a ratio-$1 + \frac{3}{t}$ approximation algorithm that runs in time $O(m^{d^{t-1}+2} M(d, t))$, where $M(d, t)$ is the time required to compute an optimal solution for a tree topology with degree d and depth t in the space S.*

4 Communication Networks With Time-Delay Constraints

We study an application of the FTSN problem in multicast communication, where a source wishes to send messages to a number of destinations and we have to find an optimal route for the messages. The results in this section originally appear in [37, 38].

Let $G = (V, E)$ be a graph representing some communication network, $c(e)$ the cost (*i.e.* length) on edge e in G, $delay(e)$ the time delay on edge e in G, $D \subseteq V$ the set of destinations, and T a rooted tree topology with $|D|$ leaves, each of which is labeled with a node in D. For simplicity, we assume that $delay(e)$'s are integers. The delay of a path in T is the sum of the delays of edges in the path. The problem is to assign a node $v \in V$ to every internal node of T such that the cost of the tree is minimized and the time delay parameter $Delay(T)$ is bounded by some pre-defined constant B. We will discuss three ways to define $Delay(T)$ [18].

In this section, we use $deg(i)$ to denote the degree of node i in T and i_{deg} the deg-th child of i.

4.1 Bounded Maximum Delay

In many real-time applications, there often exists a real-time constraint, *e.g.* the communication to any destination should be done within some specific time. Let s be the source (the node assigned to the root), v a destination, and $d(s, v)$ the time-delay of the path from s to v in T. The time-delay constraint for T is then defined as:

$$Delay(T) = \max_{v \in D} d(s, v) \leq B,$$

where B is the maximum delay specified by the users.

We will use a dynamic programming algorithm to solve the problem. Let $D[i, j, d]$ be the minimum cost of the subtree in T rooted at node i such that node $j \in V$ is assigned to i, and the longest delay for the subtree is d. Let $T(i, deg)$ be a subgraph of T that contains the subtree rooted at node i's deg-th child i_{deg}, and the edge (i, i_{deg}). (See Figure 2 (a).) Let $F[i, j, deg, d]$ be the minimum cost of the component $T(i, deg)$ such that node $j \in V$ is assigned to the node $i \in T$ and the longest delay in $T(i, deg)$ is d. Define

$$MF[i, j, deg, d] = \min_{d' \leq d} F[i, j, deg, d'].$$

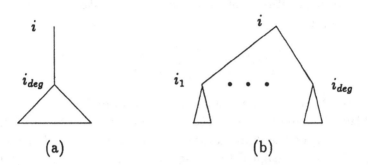

Figure 2: (a) An illustration of $T(i, deg)$.   (b) An illustration of $\hat{T}(i, deg)$.

We have

Lemma 11 *[37, 38]*

$$F[i, j, deg, d] = \min_{\substack{l \in V \text{ s.t.} \\ d_1 + delay(j,l) = d}} \{D[i_{deg}, l, d_1] + c(j, l)\}.$$

$$D[i, j, d] = \min_{deg=1}^{deg(i)} \{\{\sum_{k=1}^{deg(i)} MF[i, j, deg, d]\} - MF[i, j, k, d] + F[i, j, k, d]\}.$$

From the above recurrences, we can compute $D[i, j, d]$, $MF[i, j, deg, d]$ and $F[i, j, deg, d]$ inductively bottom up. Let

$$\text{MAX} = |V| \cdot \max_{u \in V \text{ and } v \in V} d(u, v)$$

be the maximum possible delay of any path in G and $deg(T)$ the degree of the tree T. The running time of the algorithm is $O(|T| \cdot |V| \cdot (deg(T)|V| + deg(T)^2) \cdot \text{MAX})$. The time is polynomial in terms of $|T|$, $|V|$ and $deg(T)$, but exponential in the input size of MAX, which is $\log \text{MAX}$. So, it is a pseudo-polynomial time algorithm.

4.2   Bounded Delay Variation

There are some network applications which require that the time difference between any two recipients seeing a multicast message should not exceed a certain bound δ. In this case, the time constraint is

$$\max_{u \in D \ \& \ v \in D} \{d(s, v) - d(s, u)\} \le \delta.$$

For the purpose of computation, we define the *shortest* and the *longest* delays of T to be

$$\min_{v \in D} d(s, v) \quad \text{and} \quad \max_{v \in D} d(s, v),$$

respectively. Again, we can use dynamic programming algorithm to solve the problem. However, we now need two indices to describe the time-delay. Let $D[i, j, d_1, d_2]$ be the minimum cost of the subtree rooted at node $i \in T$ such that $j \in V$ is assigned to i, and the shortest and the longest delays for the subtree are d_1 and d_2, respectively. Let $F[i, j, deg, d_1, d_2]$ be the minimum cost of the component $T(i, deg)$ such that $j \in V$ is assigned to $i \in T$, and the shortest and the longest delays are d_1 and d_2, respectively. Define

$$
\begin{aligned}
MF[i, j, deg, d_1, d_2] &= \min\{F[i, j, deg, d_1', d_2'], MF[i, j, deg, d_1 - 1, d_2 - 1], \\
&\quad MF[i, j, deg, d_1, d_2 - 1], MF[i, j, deg, d_1 - 1, d_2]\} \\
MMF[i, j, deg, d_1, d_2] &= \min\{MMF[i, j, deg, d_1 - 1, d_2], F[i, j, deg, d_1, d_2]\} \\
MFF[i, j, deg, d_1, d_2] &= \min\{MFF[i, j, deg, d_1, d_2 - 1], F[i, j, deg, d_1, d_2]\}
\end{aligned}
$$

Then we have

Lemma 12 *[37, 38]*

$$
F[i, j, deg, d_1, d_2] = \min_{\substack{l \in V \ \text{s.t.} \ d_1' + delay(j,l) = d_1 \\ \& \ d_2' + delay(j,l) = d_2}} D[i_{deg}, l, d_1', d_2'] + cost(j, l).
$$

$$
\begin{aligned}
D[i, j, d_1, d_2] = \min_{k=1}^{deg(i)} \min_{l=1}^{deg(i)} &\{\{ \sum_{deg=1}^{deg(i)} MF[i, j, deg, d_1, d_2]\} - MF[i, j, k, d_1, d_2] \\
&+ MMF[i, j, k, d_1, d_2]\} - MF[i, j, l, d_1, d_2] + MFF[i, j, l, d_1, d_2]\}.
\end{aligned}
$$

From Lemma 12 and the equations before it, we can compute the costs $D[i, j, d_1, d_2]$ bottom up and select the smallest $D[1, j, d_1, d_2]$ for each d_1 and d_2 among all j's. Thus the total running time of the algorithm is $O(|T| \cdot |V| \cdot \text{MAX}^2 \cdot (deg(T) \cdot |V| + deg(T)^3))$. This is again a pseudo-polynomial time algorithm.

4.3 Bounded Average Delay

The average time-delay constraint is defined as follows:

$$AveDelay(T) = \frac{1}{|D|} \sum_{u \in D} d(s, u) \le B. \tag{1}$$

For simplity, we consider the total delay constraint:

$$Total(T) = \sum_{u \in D} d(s, u) \le |D| \cdot B. \tag{2}$$

Let $D'[i, j, d]$ be the minimum cost (*i.e.* total delay) of any subtree rooted at i such that $j \in V$ is assigned to node i and the total delay of the subtree is d. The computation of $D'[i, j, d]$ seems to be difficult, especially, when the degree of the tree is large, since it requires the information of all i's children. In order to avoid trying all combinations, we define some auxiliary variables. Let $F'[i, j, d, deg]$ be the minimum cost of $\hat{T}[i, deg]$, the sub-component rooted at i which contains the deg subtrees rooted at the first deg children of i, such that the total delay is d. (See Figure 2 (b).) Let $G[i, j, d, deg]$ be the minimum cost of $T(i, deg)$, the sub-component rooted at i which contains the subtree rooted at the i_{deg}-th child of i such that the total delay is d. (See Figure 2 (a).) Obviously, $D'[i, j, d] = F'[i, j, d, deg(i)]$. The following recurrences are easiy to derive.

Lemma 13 *[37, 38]*

$$G[i, j, d, deg] = \min_{\substack{l \in V \ s.t. \\ d = d' + delay(j,l) \cdot count[i_{deg}]}} \{D'[i_{deg}, l, d'] + cost(j, l)\},$$

where $count[i_{deg}]$ is the number of leaves in the subtree rooted at i_{deg}.

$$F'[i, j, d, 1] = G[i, j, d, 1].$$

$$F'[i, j, d, deg] = \min_{d_1}\{F'[i, j, d_1, deg - 1] + G[i, j, d - d_1, deg]\}.$$

Again, Lemma 13 allows us to do the computation bottom up. The total time required is $O(|V| \cdot |T| \cdot \text{MAX} \cdot (deg(T) \cdot |V| + \text{MAX}))$.

All the above algorithms run in pseudo-polynomial time. In fact, we can show that

Theorem 14 *[37, 38] The FTSN problem under each of the three constraints, bounded maximum delay, bounded delay variation, and bounded average delay, is NP-hard.*

5   Space of Sequences with Edit Distance

In this section, we study the space containing all (finite) sequences over a fixed alphabet as its elements. The distance between two sequences is their *edit distance*. [3] The problem arises in computational molecular biology, in particular, the field of sequence comparison, under the names of *tree alignment* and *tree alignment with recombination*, respectively.

Multiple sequence alignment is the most critical cutting-edge tool for sequence analysis. It can help extracting, finding and representing biologically important commonalities from a set of sequences. These commonalities could represent some highly conserved subregions, common functions, or common structures. Multiple sequence alignment is also very useful in the inference of the evolutionary history of a family of sequences [2, 8, 29, 42]. One of the most important versions for multiple sequence alignment is *tree alignment*.

Tree alignment, as a form of FTSN, is defined as follows. Given a set X of m sequences and a semi-labeled (evolutionary) tree T with m leaves, where each leaf is associated with a given sequence, we want to reconstruct a sequence for each internal node to minimize the *cost* of T. Here, the cost of T is the sum of the edit distance of each pair of (given or reconstructed) sequences associated with an edge. The problem has been discussed in many papers, including [1, 8, 9, 26, 28, 29, 42].

Note that, a tree alignment computes a set of *reconstructed* sequences, one for each internal node of T. Once the internal sequences are computed, we can compute an optimal alignment [4] for each pair of sequences related by an edge of T using any (efficient) *pairwise alignment* algorithm, and easily induce a multiple alignment of the the given sequences in X from the pairwise alignments. This is how the problem got its name.

Sankoff gave an exact algorithm for tree alignment that runs in $O(n^m)$, where n is the length of the sequences and m is the number of leaf sequences [26, 29]. Since the problem is NP-hard [31, 33], it is unlikely to have a polynomial time algorithm. Some heuristic algorithms have also been considered in the past. Altschul and Lipman tried to cut down the compu-

---

[3] The edit distance between two sequences is the minimum number of edit operations (*i.e.* insertion, deletion and substitution) required to change one into the other.

[4] An alignment of two sequences is a correspondence between the letters in the sequences. It is usually shown as a 2-row matrix form by stacking one sequence on top of another with possibly some inserted spaces in both sequences. A multiple alignment is simply an extension of the notion to more than two sequences.

tation volume required by dynamic programming [1]. Sankoff, Cedergren and Lapalme gave an iterative improvement method to speed up the computation [28, 29]. Waterman and Perlwitz devised a heuristic method when the sequences are related by a binary tree [41]. Hein proposed a heuristic method based on the concept of a *sequence graph* [12, 13].

The first approximation algorithm with a guaranteed performance ratio was devised by Jiang, Lawler, and Wang [19, 34]. The algorithm is based on the lifting method described in Section 3, and achieves an approximation ratio of 2. The algorithm was also extended to a PTAS in [19, 34], along the lines sketched in Section 3. For any fixed t, the PTAS achieves an approximation ratio of $1 + \frac{3}{t}$ in time $O((m/d^t)^{d^{t-1}+2}M(2, t-1, n))$, where d is the degree of the given (rooted) tree T, n is the maximum length of the leaf sequences, and $M(d, t-1, n)$ is the time needed to optimally align a tree with $d^{t-1} + 1$ leaves, which is upper bounded by $O(n^{d^{t-1}+1})$ by the results in [26, 29]. Based on the analysis, to obtain an approximation ratio less than 2, exact solutions for depth-4 subtrees must be computed. For binary trees, this means optimally aligning 9 sequences. The time complexity, $O(n^9)$, would be impractical even for sequences of length 100 letters.

An improvement was given in [35], where the authors proposed a more efficient ratio-2 algorithm and a new PTAS for the case where the given tree is a regular d-ary tree (*i.e.* each internal node has exactly d children), which is much much faster than the one in [34]. For a fixed t, the performance ratio of the new PTAS is $1 + \frac{2}{t} - \frac{2}{t2^t}$ and the running time is $O(\min\{2^t, m\}mhM(d, t-1, n))$, where h is the depth of the tree T. Since it is possible to optimally align up to 5 sequences ($t = 3$) of length 200 letters in practice using ideas from [7, 20], we expect that solutions with costs at most 1.583 times the optimum can be obtained in practice for sequences of length 200 letters.

For tree alignment, the given tree is typically a binary tree. Recently, Wang, Jiang and Gusfield designed a PTAS for binary trees [36]. The new approximation scheme adopts a more clever partition strategy and has a better time efficiency. For any fixed r, where $r = 2^{t-1} + 1 - q$ and $0 \le q \le 2^{t-2} - 1$, the new PTAS runs in time $O(mhn^r)$ and achieves an approximation ratio of $\frac{2^{t-1}}{2^{t-2}(t+1)-q}$. Here the parameter r represents the "size" of local optimization. In particular, when $r = 2^{t-1} + 1$, its approximation ratio is simply $\frac{2}{t+1}$.

In the following, we briefly sketch the ratio-2 algorithm originally from [35] to illustrate some of the new ideas.

For simplicity, we assume that T is a binary tree, *i.e.* each node of T has at most a *left* and a *right* children. A *uniformly lifted tree* for T is a lifted tree where the labels of the nodes at each level are either all lifted from their left children or from their right children. Clearly, in a binary tree of depth h, there are only 2^h possible uniformly lifted trees. Moreover, since there is exactly one leaf sequence s that is lifted all the way to the root in any lifted tree, and the path from the leaf s to the root is unique, a uniformly lifted tree is completely determined by the leaf sequence lifted to the root.

The following lemma strengthens Lemma 6.

Lemma 15 *[35] There exists a uniformly lifted tree with a cost at most twice the optimum.*

Now, let us focus on the computation of an optimal uniformly lifted tree. We will first give an algorithm that works for full binary trees (*i.e.* binary trees in which evey internal node has exactly two children and all leaves are at the same level), and then generalize it to work for an arbitrary binary tree.

Suppose that T is a full binary tree. For each $v \in V(T) \cup L(T)$ and each label s in $S(r)$, $C[v, s]$ denotes the cost of the uniformly lifted tree $l(T_v^s)$. We can compute $C[v, s]$ recursively. For each leaf v, we define $C[v, s_i] = 0$ if the label of v is s_i. Let v be an internal node, and v_1, v_2 its children. Suppose $s_i \in S(v_p)$ and $s_i' \in S(v_q)$, where $1 \le p \le 2$, $q \in \{1, 2\} - \{p\}$, and s_i and s_i' are in the same position of T_{v_i}'s ($i = 1, 2$). Then $C[v, s_i]$ can be computed as follows:

$$C[v, s_i] = C[v_p, s_i] + C[v_q, s_i'] + d(s_i, s_i'), \tag{3}$$

where $d(s_i, s_i')$ is the edit distance between s_i and s_i'. Since the sizes of both $V(T) \cup L(T)$ and $S(r)$ are bounded by $O(m)$, we can compute all the $C[r, s_i]$'s in $O(m^2)$ time if the pairwise edit distances have been pre-computed. A better bound can be obtained by a more careful analysis. A pair of leaf sequences (s, s') is a *legal pair* if s and s' are assigned at the ends of an edge in a uniformly lifted tree. It is easy to see that a leaf sequence s can form at most h legal pairs, where h is the depth of the tree. Thus, there are at most mh legal pairs in total. Therefore, the running time of our new algorithm is $O(mh + mhn^2) = O(mhn^2)$, where n is the maximum length of the given sequences.

The above algorithm works only for full binary trees. For an arbitrary binary tree T, equation (3) may not make sense since s_i' may not be unique.

However, using a data structure called *extended trees* in [35], the algorithm can be modified to work for an arbitrary binary tree with the same time complexity.

Theorem 16 *[35] There is a ratio-2 algorithm for tree alignment on binary trees that runs in $O(mhn^2)$ time.*

We note in passing that Ravi and Kececioglu has recently designed an approximation algorithm with performance ratio $\frac{d+1}{d-1}$ for *regular d-ary* trees. Some extensions of tree alignment taking into account the *recombination* of sequences are studied in [22].

6 Optimal Tree Refinement

Since the SMT problem is computationally very challenging, in some applications such as phylogenetic inference people often use heuristic methods to compute a rough estimate of the topology of some SMT and then proceed to refine the topology until a satisfactory answer is obtained. It has been found the rough estimate is often a *contraction* of the correct topology because the input data may not provide enough clear evidence for the heuristic methods to determine every node of the correct topology and many heuristic methods tend to work conservatively, *i.e.* they only output nodes that they are very sure about. Hence, it would be interesting to consider the following extension of the FTSN problem.

A tree T_1 is a *contraction* of another tree T_2 if T_1 can be obtained from T_2 by contracting some of the edges. If T_1 is a contraction of T_2 then T_2 is a *refinement* of T_1. The *Optimal Tree Refinement (OTR)* problem is that, given a topology T whose external nodes represent regular points and internal nodes represent Steiner points, find an FTSN for some refinement T' of T so that the length of the FTSN is minimized over all possible refinements of T. Observe that if topology T' is a refinement of topology T, then the FTSN length of T' is always less than or equal to that of T.

The OTR problem is first introduced in [40] for the space of sequences with Hamming distance. It is shown that the problem is NP-hard when the degree of the input topology T is unbounded [40]. When the input topology T is required to have bounded degrees, the complexity of OTR is open. Several polynomial time approximation algorithms have been constructed for OTR with bounded-degree topologies in [40], based on the techniques developed in [19, 34] for tree alignment. Below we give a summary of these approximation results. The results in fact hold for any metric space satisfying the condition described in Section 3.

Suppose that the degree of topology T is bounded by some constant $d > 0$. The lifting technique described in Section 3 can be easily extended to also consider the possible ways of splitting an internal node. A fully labeled tree (*i.e.* a solution for FTSN) T' is a *lifted labeled refinement* of T if (i) T' refines T and (ii) each internal node of T' is labeled with the label of one of its children. The following lemma is a trivial variant of Lemma 6 in Section 3.

Lemma 17 *[40] There exists a lifted labeled refinement of T of length at most twice the length of an optimal OTR solution for T.*

Since there is only a constant number of ways to split (or refine) an internal node of T, a straightforward modification of the lifting algorithm in Section 3, based on dynamic programming, can find an optimal lifted labeled refinement. The basic idea is that, to decide an optimal refinement at an internal node v while doing lifting, we consider every possible refinement of v and for each such refinement, we compute the optimal cost of lifting a particular label to v under the refinement. This results in a polynomial time approximation algorithm with ratio 2.

Theorem 18 *[40] There is an approximation algorithm for OTR with ratio 2 running in time $O(|T|^3 + |T|^2 \cdot f(S))$ on input topologies with degrees bounded by a constant d, where $f(S)$ denotes the time required for computing the distance between any two points in the space S.*

The approximation can be improved to a PTAS following the same lines as in Section 3.

Theorem 19 *[40] The OTR problem restricted to bounded degree topologies has a PTAS.*

Acknowledgement. The authors would like to thank Guoliang Xue for many helpful discussions. Financial supports by NSERC Operating Grant OGP0046613, Canadian Genome Analysis and Technology (CGAT) Research Grant GO-12278, Hong Kong RGC 9040297, 9040352, and CityU Strategic 7000693 are greatly appreciated.

References

[1] S. Altschul and D. Lipman. Trees, stars, and multiple sequence alignment, *SIAM Journal on Applied Math.*, 49 (1989), pp. 197-209.

[2] S. C. Chan, A. K. C. Wong and D. K. T. Chiu. A survey of multiple sequence comparison methods, *Bulletin of Mathematical Biology*, 54, 4 (1992), pp. 563-598.

[3] T.H. Cormen, C.E. Leiserson and R. Rivest. *Introduction to Algorithms*, The MIT Press, Cambridge, MA, 1990.

[4] J.S. Farris. Methods for computing wagner trees, *Systematic Zoology* 19, 1970, pp. 83-92.

[5] J. Felsenstein. *PHYLIP version 3.5c (Phylogeny Inference Package)*, Department of Genetics, University of Washington, Seattle, WA, 1993.

[6] W.M. Fitch. Towards defining the course of evolution: minimum change for a specific tree topology, *Systematic Zoology* 20, 1971, pp. 406-416.

[7] S. Gupta, J. Kececioglu, and A. Schaffer. Making the shortest-paths approach to sum-of-pairs multiple sequence alignment more space efficient in practice, *Proceedings of the 6th Symposium on Combinatorial Pattern Matching, Springer LNCS 937*, 1995, pp. 128-143.

[8] D. Gusfield. *Algorithms on Strings, Trees, and Sequences: Computer Science and Computational Biology*, Cambridge University Press, 1997.

[9] D. Gusfield. Efficient methods for multiple sequence alignment with guaranteed error bounds, *Bulletin of Mathematical Biology*, 55 (1993), pp. 141-154.

[10] M. Hanan. On Steiner problem with rectilinear distance, *SIAM Journal on Applied Mathematics* 14, 1966, pp. 255-265.

[11] J.A. Hartigan. Minimum mutation fits to a given tree, *Biometrics* 29, 1973, 53-65.

[12] J. Hein. A tree reconstruction method that is economical in the number of pairwise comparisons used, *Mol. Biol. Evol.*, 6, 6 (1989), pp. 669-684.

[13] J. Hein. A new method that simultaneously aligns and reconstructs ancestral sequences for any number of homologous sequences, when the phylogeny is given, *Mol. Biol. Evol.*, 6 (1989), pp. 649-668.

[14] F.K. Hwang, On Steiner minimal trees with rectilinear distance, *SIAM J. Appl. Math.* 30, 1976, pp. 104-114.

[15] F.K. Hwang. A linear time algorithm for full Steiner trees, *Oper. Res. Lett.* 4, 1986, pp. 235-237.

[16] F.K. Hwang and J.F. Weng. The shortest network under a given topology, *Journal of Algorithms* 13, 1992, pp. 468-488.

[17] F.K. Hwang and D.S. Richards. Steiner tree problems, *Networks* 22, 1992, pp. 55-89.

[18] X. Jia and L. Wang, Group multicast routing using multiple minimum Steiner trees, *Journal of Computer Communications*, pp. 750-758, 1997.

[19] T. Jiang, E. L. Lawler and L. Wang, Aligning sequences via an evolutionary tree: complexity and approximation, *Proc. 26th ACM Symp. on Theory of Computing*, pp. 760-769, 1994.

[20] J.Lipman, S.F. Altschul, and J.D. Kececioglu. A tool for multiple sequence alignment, *Proc. Nat. Acid Sci. U.S.A.*, 86, pp.4412-4415, 1989.

[21] F. Liu and T. Jiang. *Tree Alignment and Reconstruction (TAAR) V1.0*, Department of Computer Science, McMaster University, Hamilton, Ontario, Canada, 1998. The software is available via WWW at http://www.dcss.mcmaster.ca/~fliu/taar_download.html

[22] B. Ma, L. Wang and M. Li. Fixed topology alignment with recombination, *Proc. 9th Annual Combinatorial Pattern Matching Conf.*, 1998.

[23] W. Miehle. Link length minimization in networks, *Oper. Res.* 6, 1958, pp. 232-243.

[24] G.W. Moore, J. Barnabas and M. Goodman. A method for constructing maximum parsimony ancestral amino acid sequences on a given network, *Journal of Theoretical Biology* 38, 1973, pp. 459-485.

[25] R. Ravi and J. Kececioglu. Approximation algorithms for multiple sequence alignment under a fixed evolutionary tree, *Proc. 5th Annual Symposium on Combinatorial Pattern Matching*, 1995, pp. 330-339.

[26] D. Sankoff. Minimal mutation trees of sequences, *SIAM Journal of Applied Mathematics*, 28 (1975), pp. 35-42.

[27] D. Sankoff and P. Rousseau. Locating the vertices of a Steiner tree in an arbitrary metric space, *Mathematical Programming* 9, 1975, pp. 240-246.

[28] D. Sankoff, R. J. Cedergren and G. Lapalme. Frequency of insertion-deletion, transversion, and transition in the evolution of 5S ribosomal RNA, *J. Mol. Evol.* 7 (1976), pp. 133-149.

[29] D. Sankoff and R. Cedergren. Simultaneous comparisons of three or more sequences related by a tree, in D. Sankoff and J. Kruskal, editors, *Time warps, string edits, and macromolecules: the theory and practice of sequence comparison*, pp. 253-264, Addison Wesley, 1983.

[30] W.D. Smith. How to find Steiner minimal trees in Euclidean d-space, *Algorithmica* 7, 1992, pp. 137-177.

[31] E. Sweedyk and T. Warnow, The tree alignment problem is NP-complete, *Manuscript*, 1994.

[32] L. Trevisan. When Hamming meets Euclid: the approximability of geometric TSP and MST, *Proc. 29th ACM STOC*, 1997, pp. 21-29

[33] L. Wang and T. Jiang. On the complexity of multiple sequence alignment, *Journal of Computational Biology* 1, 1994, pp. 337-348.

[34] L. Wang, T. Jiang and E.L. Lawler. Approximation algorithms for tree alignment with a given phylogeny, *Algorithmica* 16, 1996, pp. 302-315.

[35] L. Wang and D. Gusfield. Improved approximation algorithms for tree alignment, *Journal of Algorithms* 25, 1997, pp. 255-173.

[36] L. Wang, T. Jiang, and Dan Gusfield. A more efficient approximation scheme for tree alignment, *Proc. 1st Annual International Conference on Computational Molecular Biology*, 1997, pp. 310-319.

[37] L. Wang and X. Jia, Fixed topology Steiner trees and spanning forests with application in network communication, *Proc. 3rd Annual Computing and Combinatorics Conf.*, 1997, pp. 373-382.

[38] L. Wang and X. Jia, Fixed topology Steiner trees and spanning forests, *Theoretical Computer Science*, to appear.

[39] H. T. Wareham, A simplified proof of the NP-hardness and MAX SNP-hardness of multiple sequence tree alignment, *Journal of Computational Biology* 2, pp. 509-514, 1995.

[40] M. Bonet, M. Steel, T. Warnow, and S. Yooseph. Better methods for solving parsimony and compatibility, *Proc. 2nd Annual International Conference on Computational Molecular Biology*, 1998, pp. 40-49.

[41] M.S. Waterman and M.D. Perlwitz. Line geometries for sequence comparisons, *Bull. Math. Biol.*, 46 (1984), pp. 567-577.

[42] M.S. Waterman. *Introduction to Computational Biology: Maps, sequences, and genomes*, Chapman and Hall, 1995.

[43] G. Xue and Y. Ye. An efficient algorithm for minimizing a sum of Euclidean norms with applications, *SIAM J. Optim.* 7, 1997, pp. 1017-1036.

[44] G. Xue and D.Z. Du. An $O(n \log n)$ average time algorithm for computing the shortest network under a given topology, to appear in *Algorithmica*, 1997.

[45] G. Xue, private communication, 1998.

Steiner Trees, Coordinate Systems and NP-Hardness

J. F. Weng[1]
Department of Mathematics and Statistics
and Department of Electrical and Electronic Engineering
  (*E-mail:* weng@ms.unimelb.edu.au),
The University of Melbourne, Parkville, VIC 3052, Australia

Contents

1   Introduction

Given a set A of points $a_1, a_2, \ldots,$ in the Euclidean plane, the Steiner tree problem asks for a minimum network $T(A)$ (or T if A is not necessarily mentioned) interconnecting A with some additional points to shorten the network [6]. The given points are referred to as *terminals* and the additional points are referred to as *Steiner points*. Trivially, T is a tree, called the

---

[1]Supported by a grant from the Australian Research Council.

D.-Z. Du et al. (eds.), Advances in Steiner Trees, 63-80.
© 2000 *Kluwer Academic Publishers.*

Euclidean Steiner minimal tree (ESMT) for A. It is well known that Steiner minimal trees satisfy an *angle condition*: all angles at the vertices of Steiner minimal trees are not less than 120° [6]. A tree satisfying this angle condition is called a *Steiner tree*. Therefore, a Steiner minimal tree must be a Steiner tree.

By the topology of a network we mean its graph structure. The topology of a Steiner tree is called a *Steiner topology*. By minimality, for any Steiner topology there exists at most one Steiner tree with a given topology [6]. Although the Steiner tree with a given topology is easily constructed [7],[5], in general, the Steiner tree problem is NP-hard [4].

By the direction of an edge we mean the direction angle from the horizontal to the edge. A Steiner tree is called *full* if every terminal is of degree one. Then the angle condition implies a trivial lemma:

Lemma 1.1 *All angles in a full Steiner tree are exactly equal to* 120°. *Therefore, the edges in such a tree have only three directions.*

Since any Steiner tree is an edge disjoint union of its full components [6], constructing full Steiner trees is of significant importance in the Steiner tree problem. Clearly, any algebraic approach of the Steiner tree problem involves a certain coordinate system. By Lemma 1.1, a coordinate system with three axes 120° apart, called a *hexagonal coordinate system*, is more natural than the common Cartesian coordinate system for constructing full Steiner trees. In Section 2 we study the use of hexagonal coordinates in the Steiner tree problem. Motivated also by Lemma 1.1, in Section 3 we define a generalization of Steiner trees, called *APE-Steiner trees*. Along the same line, the APE-Steiner tree problem can be solved using a general three coordinate system. In the last section, using hexagonal coordinate systems we give a direct proof of the NP-hardness of the Euclidean Steiner tree problem.

2   Hexagonal Coordinate Method

Let O be the origin. The hexagonal coordinate system has three axis OU, OV, OW such that OU is the 0° line, OV is the 120° line, and OW is the 240° line. There are three possible definitions of coordinates in hexagonal coordinate systems and we use the following one [9]. For any point p, let u, v and w be the intersections of OU, OV and OW with the lines through p and parallel to OW, OU and OV, respectively. (Note the order!)

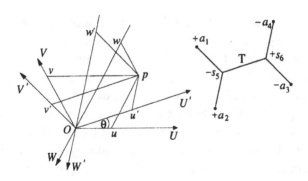

Figure 1: Hexagonal coordinates.

Then, the distances from O to u, v and w are defined to be the coordinates of p. For simplicity, we will also use u, v and w — the same letters for the intersections — to denote these coordinates of p (Figure 1).

Clearly, u, v and w satisfy the following equation, called the *closure of coordinates*:

$$u + v + w = 0. \tag{1}$$

Let $\rho = 1/\sqrt{3}$. It is easy to derive the relations between the hexagonal coordinates and the Cartesian coordinates:

$$u = x - \rho y, \quad v = 2\rho y, \quad w = -x - \rho y, \tag{2}$$

$$x = u + \frac{v}{2}, \quad y = \frac{v}{2\rho}. \tag{3}$$

Suppose T is a full Steiner tree. If the edges of T are not just parallel to the axes, then we can rotate the axes OU, OV, OW anticlockwise at a certain angle θ so that the new axes OU', OV', OW' are parallel to the edges of T. Let $l = \cos\theta, k = \rho\sin\theta$, then by the Pythagorean Theorem,

$$l^2 + 3k^2 = 1. \tag{4}$$

It is not hard to show [9] that the transformations between the old and the new coordinates are:

$$\begin{bmatrix} u' \\ v' \\ w' \end{bmatrix} = \begin{bmatrix} l & k & -k \\ -k & l & k \\ k & -k & l \end{bmatrix} \begin{bmatrix} u \\ v \\ w \end{bmatrix}, \tag{5}$$

where u', v', w' are new coordinates of p.

Now look at an example. Figure 1 shows a full Steiner tree T spanning 4 terminals a_1, a_2, a_3, a_4, and two Steiner points s_5, s_6. Suppose, after the rotation, the new axes OU', OV' and OW' are parallel to $s_5 s_6, a_1 a_5$ and $a_2 a_5$, respectively. Since $a_1 s_5 / OV'$ and $a_2 s_5 / OW'$, we have $w_5' = w_1'$ and $u_5' = u_2'$ by the definition of coordinates. Then, by Equation (1) the third coordinate of s_5 is $v_5' = -u_2' - w_1'$. This shows that the coordinates of a Steiner point can be directly computed from the coordinates of its two adjacent points and from the closure of coordinates. As in the Melzak geometric construction [7], the next step is to compute the coordinates of s_6 from the coordinates of s_5 and a_3: $u_6' = -v_5' - w_3' = u_2' + (w_1' - w_3')$. Finally, $u_6' = u_4'$ gives an equation

$$(u_2' - u_4') + (w_1' - w_3') = 0.$$

Because this equation describes the topology of the full Steiner tree T, we call it the *characteristic equation* of the tree. Again, since $a_1 s_5 / OV'$, $|a_1 a_5| = v_1' - v_5'$. The lengths of other edges of T can be obtained in a similar way. Adding them and using the closure of coordinates of the Steiner points, we obtain the length of T

$$|T| = (v_1' - v_3') + (w_2' - w_4').$$

It is not hard to prove that, in the general case, the characteristic equation can be formulated as follows. Suppose T is a full Steiner tree spanning n terminals a_i ($i = 1, 2, \ldots, n$) and $n - 2$ Steiner points s_i ($i = n + 1, n + 2, \ldots, 2n - 2$) with coordinates u_i', v_i', w_i' ($i = 1, 2, \ldots, 2n - 2$), respectively. The indices i of terminals a_i can be classified in two ways:

1. T is a tree and the vertices of T can be partitioned into two disjoint sets (marked by positive or negative signs in Figure 1) so that the endpoints of any edge belong to different sets. Thus, the indices of terminals that belong to one set compose an index set A_1, and the indices of all other terminals compose another index set A_2.

2. The edge incident to a terminal is parallel either to OU', or to OV', or to OW'. Let U, V, W be three disjoint index sets so that the index of a terminal is in U, or V, or W, if its incident edge is parallel to OU', or OV', or OW', respectively.

Now we can prove that the coordinates of terminals satisfy the following *characteristic equation*, which completely describes the topology of T:

$$\sum_{i \in W} \epsilon_i u_i' + \sum_{i \in U} \epsilon_i v_i' + \sum_{i \in V} \epsilon_i w_i' = 0, \qquad (6)$$

where $\epsilon_i = +1$ if $i \in A_1$, and $\epsilon_i = -1$ if $i \in A_2$.

Remark 2.1 Alternatively we can define $\epsilon_i = -1$ if $i \in A_1$ and $\epsilon_i = +1$ if $i \in A_2$. Moreover, the characteristic equation can also be written as

$$\sum_{i \in V} \epsilon_i u_i' + \sum_{i \in W} \epsilon_i v_i' + \sum_{i \in U} \epsilon_i w_i' = 0.$$

These variant definitions do not change the solution of the problem.

Substituting u', v', w' from (5), the characteristic equation becomes

$$0 = \sum_{i \in W}(l\epsilon_i u_i + k\epsilon_i v_i - k\epsilon_i w_i) + \sum_{i \in U}(-k\epsilon_i u_i + l\epsilon_i v_i + k\epsilon_i w_i)$$

$$+ \sum_{i \in V}(k\epsilon_i u_i - k\epsilon_i v_i + l\epsilon_i w_i)$$

$$= l\left(\sum_{i \in W} \epsilon_i u_i + \sum_{i \in U} \epsilon_i v_i + \sum_{i \in V} \epsilon_i w_i\right)$$

$$+ k\left(\sum_{i \in W}(\epsilon_i v_i - \epsilon_i w_i) + \sum_{i \in U}(\epsilon_i w_i - \epsilon_i u_i) + \sum_{i \in V}(\epsilon_i u_i - \epsilon_i v_i)\right).$$

Then, by the transformations (2), the characteristic equation is of the form

$$F_l(x_i, \rho y_i) \cdot l + F_k(x_i, \rho y_i) \cdot k = 0, \tag{7}$$

where F_l and F_k are linear functions of x_i and ρy_i, with integer coefficients. Combining this equation with Equation (4), we obtain

$$l = \frac{F_k(x_i, \rho y_i)}{\sqrt{3F_l^2(x_i, \rho y_i) + F_k^2(x_i, \rho y_i)}}, \quad k = \frac{-F_l(x_i, \rho y_i)}{\sqrt{3F_l^2(x_i, \rho y_i) + F_k^2(x_i, \rho y_i)}}. \tag{8}$$

Similarly to the characteristic equation, it is also easy to prove that the length of T is

$$|T| = \left|\sum_{i \in U} \epsilon_i u_i' + \sum_{i \in V} \epsilon_i v_i' + \sum_{i \in W} \epsilon_i w_i'\right|. \tag{9}$$

Again, by (5) and (2) this equation becomes

$$|T| = \left| l\left(\sum_{i \in U} \epsilon_i u_i + \sum_{i \in V} \epsilon_i v_i + \sum_{i \in W} \epsilon_i w_i\right) \right.$$

$$+ k\left(\sum_{i \in U}(\epsilon_i v_i - \epsilon_i w_i) + \sum_{i \in V}(\epsilon_i w_i - \epsilon_i u_i) + \sum_{i \in W}(\epsilon_i u_i - \epsilon_i v_i)\right)\right|$$

$$= |lG_l(x_i, \rho y_i) + kG_k(x_i, \rho y_i)|,$$

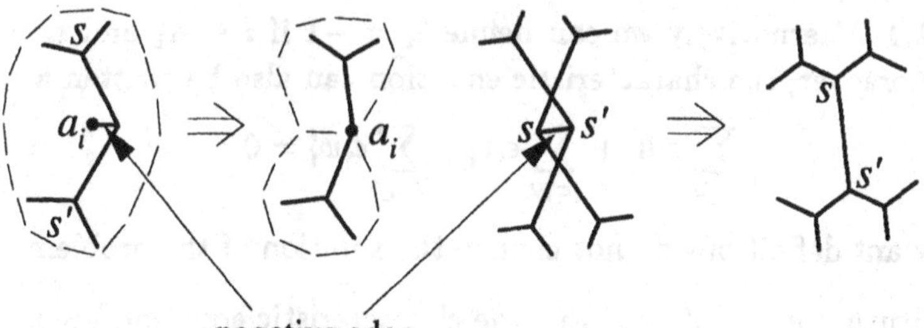

negative edge

Figure 2: Negative edges.

where G_l and G_k are also linear functions of x_i and ρy_i, with integer coefficients. Thus, substituting l and k from (8), finally we obtain the following formula of the tree length:

$$|T| = \left| \frac{G_l(x_i, \rho y_i) F_k(x_i, \rho y_i) - G_k(x_i, \rho y_i) F_l(x_i, \rho y_i)}{\sqrt{3F_l^2(x_i, \rho y_i) + F_k^2(x_i, \rho y_i)}} \right|. \tag{10}$$

As to the coordinates of Steiner points, they can be determined by the coordinates of terminals as shown in the above example.

We must point out that as in Melzak's construction [7] or in any similar methods, the tree obtained by the hexagonal coordinate method might not be the real shortest tree because of the existence of *negative edges*. Suppose T is a Steiner tree with a Steiner point s such that the three edges of s lie in an angle no more than 180°. Then the intermediate edge e of s becomes a negative edge (Figure 2). When $|T|$ is calculated by Formula (10), the edge e contributes a negative value $-|e|$ to $|T|$. In this case Formula (10) does not give the right solution of the problem, however, the occurrence of negative edges indicates how the tree could be shortened. There are two cases. If a negative edge is incident to a terminal a_i, then decomposing T at a_i could produce a shorter tree (Figure 2(1)). If a negative edge joins two Steiner points s and s', then exchanging the connections of s, s' with their adjacent vertices could produce a shorter tree (Figure 2(2)).

3   APE-Steiner Trees and General Three Coordinate Systems

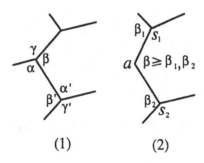

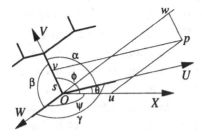

Figure 3: APE-Steiner trees.

Figure 4: General three coodinate systems.

Consider a network T with a full Steiner topology. Suppose s and s' are two adjacent Steiner points of T. Let α, β, γ and α', β', γ' be the angles, listed in counterclockwise order around s and s' respectively, such that γ and γ' are the angles not involving the edge ss' as shown in Figure 3(1). If $\alpha = \alpha', \beta = \beta'$ and $\gamma = \gamma'$ for every pair of adjacent Steiner points s and s' in T, then the edges in any zigzag path of T (*ie* any path in which one alternates between turning left and right at each vertex) are alternately parallel. Such a tree is called a full *Steiner tree with alternately parallel edges*, or a full *APE-Steiner tree* for short [11]. This definition can be extended to a network T with a non-full Steiner topology. Note that a degree 3 terminal can be regarded as a Steiner point and a degree 2 terminal can be regarded as a degeneracy of a Steiner point. Thus, T is an APE-Steiner tree if it is an edge disjoint union of full APE-Steiner trees, and if for any degree 2 terminal a the (smaller) angle β of a is no less than the corresponding alternate angles β_1, β_2 of its adjacent Steiner points s_1, s_2, respectively (Figure 3(2)).

By the definition, a full APE-Steiner tree satisfies the following lemma, which is parallel to Lemma 1.1:

	ϕ	ψ	θ	α	β	Examples
Type 1			?	✓	✓	classical Steiner trees
Type 2	✓	✓	?			gradient-constrained networks
Type 3			✓	?	✓	

Table 1: Three types of full APE-Steiner trees.

Lemma 3.1 *The edges in a full APE-Steiner tree have only three directions.*

Let θ, ϕ and ψ be the three edge directions in a full APE-Steiner tree (Figure 4). By the definition of full APE-Steiner trees, $\alpha + \beta + \gamma = 360°$ and $\phi + \psi + \beta = 360°$ (Fig. (4)). For a given topology and a set of terminals, if any two of the six variables $\alpha, \beta, \gamma, \theta, \phi$ and ψ are known, then a unique APE-steiner tree can be constructed if it exists. Therefore, there are three non-equivalent types of full APE-Steiner trees as shown in Table 1, where the ticks indicate the two initially specified angles. For example, the gradient-constrained Steiner trees [1] are APE-Steiner trees in which two edge directions are known. Hence, the gradient-constrained Steiner trees are APE-Steiner trees of Type 2.

To solve the APE-Steiner tree problem we define a general three coordinate system as follows. Let O be the origin, OU, OV and OW be the axes which are in the directions θ, ϕ and ψ, respectively. As in the hexagonal coordinate system, the coordinates u, v, w of a point p in the general three coordinate system are defined to be the distances from O to the intersections of OU, OV, OW, with the lines through p that are parallel to OW, OU, OV, respectively. Let

$$c_u = \frac{1}{\sin \beta}, \quad c_v = \frac{-1}{\sin(\alpha + \beta)}, \quad c_w = \frac{1}{\sin \alpha}. \tag{11}$$

Then, we can prove that in the general three coordinate system

i. corresponding to Equation (1), the closure of coordinates is

$$c_u u + c_v v + c_w w = 0; \tag{12}$$

ii. corresponding to Equation (6), the characteristic equation is

$$c_u \sum_{i \in W} \epsilon_i u_i + c_v \sum_{i \in U} \epsilon_i v_i + c_w \sum_{i \in V} \epsilon_i w_i = 0; \tag{13}$$

iii. corresponding to Equation (9), the length of T is

$$|T| = \left| \sum_{i \in U} \epsilon_i u_i + \sum_{i \in V} \epsilon_i v_i + \sum_{i \in W} \epsilon_i w_i \right|. \tag{14}$$

The hexagonal coordinate system is a special three coordinate system. In general, since α, β and γ are not equal, we cannot apply the rotation technique described in Section 2 to APE-Steiner trees. Instead, we have a direct approach as follows. Clearly, to determine any type of full APE-Steiner trees we need only to find one more angle, which has been indicated by a question mark in Table 1. It is not hard to establish the following transformation formulae between u, v, w coordinates and the Cartesian coordinates, which correspond to Equations (2) and (3):

$$u = \frac{x \sin(\alpha + \beta + \theta) - y \cos(\alpha + \beta + \theta)}{\sin(\alpha + \beta)},$$

$$v = \frac{y \cos \theta - x \sin \theta}{\sin \alpha},$$

$$w = \frac{-x \sin(\alpha + \theta) + y \cos(\alpha + \theta)}{\sin \beta}, \tag{15}$$

and

$$x = \frac{v \sin \alpha \cos(\alpha + \theta) - w \sin \beta \cos \theta}{\sin \alpha},$$

$$y = \frac{v \sin \alpha \sin(\alpha + \theta) - w \sin \beta \sin \theta}{\sin \alpha}. \tag{16}$$

First we study Type 1. After substituting $c_u, c_v, c_w, u_i, v_i, w_i$ from (11) and (15), and after simplification, the characteristic equation (13) becomes

$$\left(\sum_{i \in W} (\epsilon_i x_i \sin(\alpha + \beta + \theta) - \epsilon_i y_i \cos(\alpha + \beta + \theta)) \right) \sin \alpha$$

$$+ \left(\sum_{i \in U} (\epsilon_i x_i \sin \theta - \epsilon_i y_i \cos \theta) \right) \sin \beta$$

$$- \left(\sum_{i \in V} (\epsilon_i x_i \sin(\alpha + \theta) - \epsilon_i y_i \cos(\alpha + \theta)) \right) \sin(\alpha + \beta) = 0. \tag{17}$$

Solving this equation, we obtain the solution θ for Type 1:

$$\theta = \arctan \left(\frac{-B(\alpha, \beta, x_i, y_i)}{A(\alpha, \beta, x_i, y_i)} \right),$$

where

$$A(\alpha,\beta,x_i,y_i) = \sum_{i\in W}\epsilon_i x_i \sin\alpha\cos(\alpha+\beta)+\sum_{i\in U}\epsilon_i x_i \sin\beta-\sum_{i\in V}\epsilon_i x_i \cos\alpha\sin(\alpha+\beta)$$

$$+(\sum_{i\in W}\epsilon_i y_i - \sum_{i\in V}\epsilon_i y_i)\sin\alpha\sin(\alpha+\beta),$$

$$B(\alpha,\beta,x_i,y_i) = (\sum_{i\in W}\epsilon_i x_i - \sum_{i\in V}\epsilon_i x_i)\sin\alpha\sin(\alpha+\beta)$$

$$-\sum_{i\in W}\epsilon_i y_i \sin\alpha\cos(\alpha+\beta) - \sum_{i\in U}\epsilon_i y_i \sin\beta + \sum_{i\in V}\epsilon_i y_i \cos\alpha\sin(\alpha+\beta).$$

Now we turn to Type 2. Since ϕ and ψ are known in Type 2, OX can be so selected that $\psi = -\phi$, ie the horizontal line bisects the angle β. Let

$$\omega = \frac{\beta}{2} = \pi - \frac{\phi+\psi}{2},$$

then $\alpha = \pi-\omega-\theta$, $\alpha+\beta+\theta = \pi+\omega$, $\alpha+\theta = \pi-\omega$ and $\alpha+\beta = \pi-\omega-\theta$. Substituting them in Equation (17), we have the characteristic equation of Type 2 as follows:

$$\left(\sum_{i\in W}(\epsilon_i x_i\sin\omega - \epsilon_i y_i\cos\omega)\right)\sin(\omega+\theta)+\left(\sum_{i\in U}(\epsilon_i y_i\cos\theta - \epsilon_i x_i\sin\theta)\right)\sin 2\omega$$

$$-\left(\sum_{i\in V}(\epsilon_i x_i\sin\omega + \epsilon_i y_i\cos\omega)\right)\sin(\omega-\theta) = 0. \quad (18)$$

Solving the equation we obtain the solution θ for Type 2:

$$\theta = \arctan\left(\frac{-B(\omega,x_i,y_i)}{A(\omega,x_i,y_i)}\tan\omega\right),$$

where

$$A(\omega,x_i,y_i) = \left(\sum_{i\in W}\epsilon_i x_i - 2\sum_{i\in U}\epsilon_i x_i + \sum_{i\in V}\epsilon_i x_i\right)\sin\omega-\left(\sum_{i\in W}\epsilon_i y_i - \sum_{i\in V}\epsilon_i y_i\right)\cos\omega,$$

$$B(\omega,x_i,y_i) = \left(\sum_{i\in W}\epsilon_i x_i - \sum_{i\in V}\epsilon_i x_i\right)\sin\omega-\left(\sum_{i\in W}\epsilon_i y_i - 2\sum_{i\in U}\epsilon_i y_i + \sum_{i\in V}\epsilon_i y_i\right)\cos\omega.$$

Finally, because θ is known in Type 3, by rotating OX we can make $OX = OU$, ie $\theta = 0$. Hence, the characteristic equation (17) is simplified as follows:

$$\left(\sum_{i\in W}(\epsilon_i x_i \sin(\alpha+\beta) - \epsilon_i y_i \cos(\alpha+\beta))\right)\sin\alpha - \left(\sum_{i\in U}\epsilon_i y_i\right)\sin\beta$$

$$-\left(\sum_{i\in V}(\epsilon_i x_i \sin\alpha - \epsilon_i y_i \cos\alpha)\right)\sin(\alpha+\beta) = 0. \quad (19)$$

The solution of Type 3 given by this equation is

$$\alpha = \left(\arctan\frac{A(\beta,x_i,y_i)}{2B(\beta,x_i,y_i)} - \arcsin\frac{A(\beta,x_i,y_i)+2C(\beta,x_i,y_i)}{\sqrt{A^2(\beta,x_i,y_i)+4B^2(\beta,x_i,y_i)}}\right)\Big/2,$$

where

$$A(\beta,x_i,y_i) = \left(\sum_{i\in W}\epsilon_i x_i - \sum_{i\in V}\epsilon_i x_i\right)\cos\beta + \left(\sum_{i\in W}\epsilon_i y_i - \sum_{i\in V}\epsilon_i y_i\right)\sin\beta,$$

$$B(\beta,x_i,y_i) = \left(\left(\sum_{i\in W}\epsilon_i x_i - \sum_{i\in V}\epsilon_i x_i\right)\sin\beta - \left(\sum_{i\in W}\epsilon_i y_i - \sum_{i\in V}\epsilon_i y_i\right)\cos\beta\right)\Big/2,$$

$$C(\beta,x_i,y_i) = \left(\sum_{i\in V}\epsilon_i y_i - \sum_{i\in U}\epsilon_i y_i\right)\sin\beta.$$

In all three types of APE-Steiner trees, once θ (or α) is found, then we obtain the coordinates u_i, v_i, w_i of all terminals from Equations (15), and obtain $|T|$ from Equation (14). The coordinates of the Steiner points can be derived from the coordinates of terminals as explained in the example in Section 2.

4 Hexagonal Coordinates and the NP-hardness of the Steiner Tree Problem

Given three directions any two of which meet an angle of 120°, a network in the Euclidean plane with a Steiner topology that interconnects a terminal set A and has all line segments parallel to the three directions is called a *hexagonal Steiner tree* and denoted by $H(A)$ (or simply by H). If the length

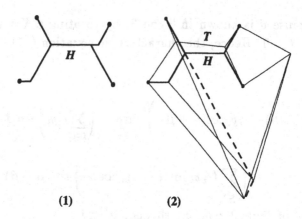

(1)   (2)

Figure 5: Hexagonal Steiner trees.

of H (the sum of lengths of all line segments in H) is minimal, then H is called a *hexagonal Euclidean Steiner minimal tree* (H-ESMT). An edge in H is called a *straight* edge if it contains only one straight line segment, otherwise a *non-straight* edge. In particular, an edge consisting of two line segments is referred to as a *bow leg*. If all terminals of H are of degree one, then H is full. As in the rectilinear Steiner tree problem, it is not hard to prove the following theorem [10]:

Theorem 4.1 *A full hexagonal Steiner minimal tree H can be transformed into an equally long tree so that all edges in a full component of the new tree are straight except at most one edge being a bow leg and incident to a terminal.*

A hexagonal Steiner tree satisfying the condition in this theorem is called *canonical*. Below we consider only canonical hexagonal Steiner trees. By this theorem we can use Melzak's construction to expand a full hexagonal Steiner tree into a *Simpson line* which consists of at most two straight line segments. Figure 5(1) depicts a full hexagonal Steiner tree H. In Figure 5(2) the bold lines show the canonical form of H and its Simpson line while the thin lines are the corresponding (classical) Steiner tree T and the Simpson line of T [6].

Hexagonal Steiner trees have some useful properties. One property is the inequality $|T(A)| \geq (\sqrt{3}/2)|H(A)|$, which is easy to prove [10] and has been employed in the proof of the Steiner ratio conjecture [3]. Here we discuss another property and its application to the proof of the NP-hardness of the Steiner tree problem.

A point is referred to as an *integer point* (with respect to a coordinate system) if all coordinates of the point are integers. Because of the closure of coordinates, a point with two integer coordinates in a hexagonal coordinate system is an integer point. The following theorem is also easily seen.

Theorem 4.2 *Suppose p and q are two integer points in a hexagonal coordinate system. Then, the third vertex of the equilateral triangle whose base is pq is an integer point. Therefore, if all terminals in a full hexagonal Steiner tree H are integer points, then all Steiner points in H are integer points. Moreover, H can be expanded into a Simpson line whose endpoints are integer points, and the length of H is an integer.*

Now we discuss the computational complexity of the Steiner tree problem. The Euclidean Steiner minimal tree (ESMT) problem can be described as a decision problem as follows [6]:

GIVEN: A set A of terminals in the Euclidean plane and an integer l.
DECIDE: Is there a Steiner tree T spanning A such that $|T| \leq l$?

There are two difficulties in the ESMT problem. One involves numerical computations of irrationals (on finite-precision machines) and another involves the combinatorial property of the problem, *ie* the superexponential number of Steiner topologies. To separate the two aspects and to show that the computational hardness of the Steiner tree problem substantially comes from the latter, people are used to discussing the discretized version of the original problem. To discuss the computational complexity we may assume that all given terminals are integer points. Then discretizing the problem has two steps: first discretizing the network by moving all Steiner points to the closest integer points, and then discretizing the metric by using the integer distance $|ab|_d = \lceil |ab| \rceil$ where a, b are two integer points. Garey, Graham and Johnson [4] first proved that the discretized Euclidean Steiner minimal tree (DESMT) problem is NP-complete. To show the NP-hardness of the ESMT problem we need only find a special class of Steiner minimal trees so that the computation of such trees is NP-hard. In 1997, Rubinstein, Thomas and Wormald [8] found such a class where the terminals are constrained to lie on two parallel lines. The Steiner tree problem for such sets is called the PALIMEST (parallel line minimum Euclidean Steiner tree) problem. Rubinstein *et al* proved that the discretized PALIMEST problem is NP-complete since it can be polynomially transformed into a known NP-complete problem, the SUBSET SUM problem. Given a set $S = \{d_1, d_2, \ldots, d_n\}$ of integers and an integer s, the SUBSET SUM problem asks if there is a subset J

such that $\sum_{i \in J} d_i = s$. By sufficiently large scale-up, the NP-hardness of the discretized PALIMEST problem implies the NP-hardness of the original (non-discretized) version. Consequently, the following theorem holds as argued above:

Theorem 4.3 *The ESMT problem is NP-hard.*

Remark 4.1 Recently Rubinstein *et al*'s proof has been improved and a bound of the scale-up has been found in [2].

Note that the computational complexity of a problem is irrelevant to the choice of coordinate systems. As we explained in Section 2, hexagonal coordinate systems are more natural and helpful in the study of the ESMT problem. Hence, we define integer points to be the points having integer hexagonal coordinates. By this definition of integer points, we give a new proof of the NP-hardness of the Steiner tree problem without discretization.

First we show the H-ESMT problem is NP-complete using the configuration designed by Rubinstein *et al* in the PALIMEST problem. A hexagonal Euclidean Steiner minimal tree with terminals lying on two parallel lines is referred to as an H-PALIMEST. The H-PALIMEST problem is defined as follows:

INSTANCE: A set A_H of integer points in the Euclidean plane lying on two parallel lines, and an integer l_H.

QESTION: Is there a hexagonal Steiner tree H spanning A_H such that $|H| \leq l_H$?

Let d_i $(i = 1, 2, \ldots, n)$ and $s \leq \sum_{i=1}^{n} d_i = D$ be a given instance of the SUBSET SUM problem. Construct four vertical lines l_1, l_1', l_2', l_2 ordered from left to right in the hexagonal coordinate system $OUVW$ so that the distances between them satisfy

$$d(l_1', l_2') >> d(l_1, l_1') = d(l_2', l_2) >> D, \tag{20}$$

where $d(l_1', l_2')$ and $d(l_1, l_1')$ are integers. Let v_i, v_i' $(i = 1, 2, \ldots, n)$ be integer points on l_1 as shown in Figure 6(1) such that $|v_i v_i'| = 2\sqrt{3} d_i$. Let H_0 be the full hexagonal Steiner tree spanning v_i, v_i' $(i = 1, 2, \ldots, n)$, a point v on l_1 and some points u_0, u_i, $(i = 1, 2, \ldots, n)$ on l_2, so that its topology is as shown in Figure 6(1), and that it has some edges being horizontal. It is easy to see that all Steiner points and u_0, u_i, $(i = 1, 2, \ldots, n), v$ are integer points because v_i, v_i' $(i = 1, 2, \ldots, n)$ are integer points and because $d(l_1', l_2')$ and $d(l_1, l_1')$ are integers. Let vp be the Simpson line of H_0 as

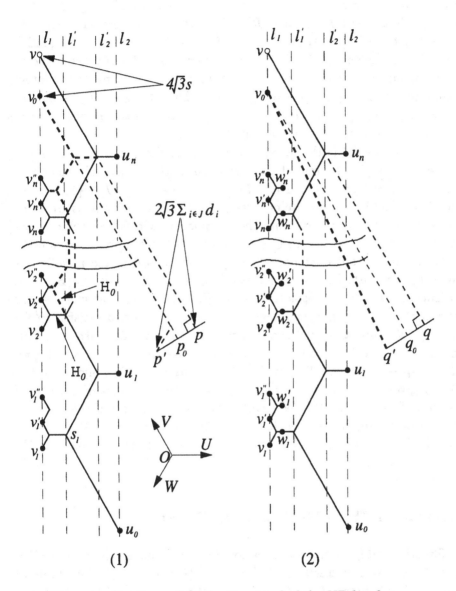

Figure 6: Configurations for the proof of the NP-hardness.

shown in Figure 6(1), then $L_H = |vp| = |H_0|$ is an integer by Theorem 4.2. Now let v_i'' be the integer points on l_1 above v_i' and satisfying $|v_i'v_i''| = |v_iv_i'| = 2\sqrt{3}d_i$ $(i = 1, 2, \ldots, n)$. Define a hexagonal tree $H_1 = \bigcup_i v_i'v_i''$ where each $v_i'v_i''$ is a bow leg. Clearly, $H = H_0 \cup H_1$ is a hexagonal Steiner tree whose length is $|H| = |H_0| + |H_1| = L_H + 4D$. Because of Inequalities (20), H is a Steiner minimal tree for its terminals. Finally, let v_0 be the integer point on l_1 below v such that $|v_0v| = 4\sqrt{3}s$. Let p_0 be the integer point such that $v_0p_0 \parallel vp$ and $p_0p \perp vp$. Then $|v_0p_0| = L_H - 6s$. Let $A_H = \{v_0, u_0, v_i, v_i', v_i'', u_i \ (i = 1, 2, \ldots, n)\}$ which is a set of integer points and can be constructed in polynomial time.

Now suppose H' is the hexagonal Steiner minimal tree on A_H. Again, by Inequalities (20) H' must be decomposed as $H' = H_0' \cup H_1'$ where H_1' is a union of bow legs spanning $v_iv_i', i \in J$ for some subset $J \subset \{1, 2, \ldots, n\}$, and H_0' is a full hexagonal Steiner tree with the same topology as H_0 except that v is replaced by v_0, and that the terminals v_i and v_i' are replaced by v_i' and v_i'' respectively for each $i \in J$. (In Figure 6(1) we suppose the subset J contains 2 and n but does not contain 1, and H_0 is depicted in solid lines while H_0' is in dashed lines.) It follows that $|H'| = |H_0'| + |H_1'| = |H_0'| + 4D$. By Theorem 4.2 the Simpson line v_0p' of H_0' (consisting of at most two line segments) has an integer endpoint p' that lies on p_0p and satisfies $|p'p| = 2\sqrt{3}\sum_{i \in J}d_i$. Note that v_0p' is a straight line and coincides with v_0p_0 if and only if there is a J such that $\sum_{i \in J}d_i = s$. Otherwise v_0p' is a bow leg and $|H'| > L_H - 6s + 4D$. Clearly, H' can be constructed in polynomial time for any given J. So, let $l_H = L_H - 6s + 4D$, then the answer to the given instance of the SUBSET SUM problem is YES if and only if the answer to the corresponding instance of the H-PALIMEST problem is YES. It follows that the H-PALIMEST problem, and hence the H-ESMT problem, is NP-hard. The H-ESMT problem is obviously in the class NP. The following theorem is proved:

Theorem 4.4 *The H-ESMT problem is NP-complete.*

Slightly modifying the configuration described above, we can prove Theorem 4.3 without discretization. Let w_i, w_i' $(i = 1, 2, \ldots, n)$ be the points that are $d_i + 1$ distant from l_1, and lie on the right part of the bisectors of $v_iv_i', v_i'v_i''$ respectively (Figure 6(2)). Clearly, they are integer points. Now these points are added into the terminal set. Let T_0 be the full Steiner tree spanning $\{v, u_0, v_i, v_i', w_i, u_i \ (i = 1, 2, \ldots, n)\}$, T_1 be the union of the trees spanning all triangles $v_i'v_i''w_i' \ (i = 1, 2, \ldots, n)\}$. Then all edges in T_0 and T_1

have integer lengths. Let L be the length of T_0, which is an integer and equal to the length of the Simpson line vq as shown in Figure 6(2). Similarly, let v_0q_0 be the line parallel to vq and perpendicular to qq_0, then $|v_0q_0| = L - 6s$. The length of T_1 is $\sum_{i=1}^{n}(4d_i + 1) = 4D + n$.

Let $A = A_H \cup \{w_i, w'_i \ (i = 1, 2, \ldots, n)\}$. Suppose T' is the Euclidean Steiner minimal tree on A. For the same reason, T' is decomposed into a union as $T' = T'_0 \cup T_1$, where T'_0 is a full Steiner minimal tree with the same topology as T_0 except that v is replaced by v_0, and that the terminals w_i are replaced by w'_i for each i in some subset $J \subset \{1, 2, \ldots, n\}$. Again, the Simpson line v_0q' of T'_0 (a straight line) has an integer endpoint q' so that $q'q \perp vq$. Similarly, v_0q' coincides with v_0q_0 if and only if there is a subset J such that $\sum_{i \in J} d_i = s$. Otherwise, $|T| = |T'_0| + |T_1| = |v_0q'| + |T_1| > L - 6s + 4D + n$. Hence, the NP-hardness of the ESMT problem is deduced from the NP-completeness of the SUBSET SUM problem by setting $l = L - 6s + 4D + n$. This completes the proof of Theorem 4.3.

Remark 4.2 Since we do not know if the ESMT problem is in the class NP, it is still open whether the ESMT problem is NP-complete or not.

References

[1] M. Brazil, D.A. Thomas and J.F. Weng, Gradient-constrained minimal Steiner trees, *DIMACS Series in Discrete Math. and Theoretical Comput. Science*, Vol. 40, pp. 23-38.

[2] M. Brazil, D.A. Thomas and J.F. Weng, On the complexity of the Steiner problem, preprint.

[3] D.Z. Du and F.K. Hwang, A proof of the Gilbert-Pollak conjecture on the Steiner ratio, *Algorithmica*, Vol. 7 (1992), pp. 121-135.

[4] M.R. Garey, R.L. Graham and D.S. Johnson, The complexity of computing Steiner minimal trees, *SIAM J. Appl. Math.*, Vol. 32 (1977), pp. 835-859.

[5] F.K. Hwang and J.F. Weng, Hexagonal coordinate systems and Steiner minimal trees, *Discrete Math.*, Vol. 62 (1986), pp. 49-57.

[6] F.K. Hwang, D.S. Richard and P. Winter, *The Steiner tree problem*, North-Holland Publishing Co., Amsterdam, 1992.

[7] Z.A.Melzak, On the problem of Steiner, *Canad. Math. Bull.*, Vol. 4 (1961), pp. 143-148.

[8] J.H.Rubinstein, D.A.Thomas and N.C.Wormald, Steiner trees for terminals constrained to curves, *SIAM J. Discrete Math.*, Vol. 10 (1997), pp. 1-17.

[9] J.F. Weng, Generalized Steiner problem and hexagonal coordinate system (Chinese, English summary), *Acta. Math. Appl. Sinica*, Vol. 8 (1985), pp. 383-397.

[10] J.F. Weng, Steiner problem in hexagonal metric, unpublished manuscript.

[11] J.F.Weng, A new model of generalized Steiner trees and 3-coordinate systems, *DIMACS Series in Discrete Math. and Theoretical Comput. Science*, Vol. 40, pp. 415-424.

Exact Algorithms for Plane Steiner Tree Problems: A Computational Study

D.M. Warme
System Simulation Solutions, Inc.
Alexandria, VA 22314, USA
E-mail: warme@s3i.com

P. Winter
Department of Computer Science
University of Copenhagen, Denmark
E-mail: pawel@diku.dk

M. Zachariasen
Department of Computer Science
University of Copenhagen, Denmark
E-mail: martinz@diku.dk

Abstract

We present a computational study of exact algorithms for the Euclidean and rectilinear Steiner tree problems in the plane. These algorithms — which are based on the generation and concatenation of full Steiner trees — are much more efficient than other approaches and allow exact solutions of problem instances with more than 2000 terminals. The full Steiner tree generation algorithms for the two problem variants share many algorithmic ideas and the concatenation part is identical (integer programming formulation solved by branch-and-cut). Performance statistics for randomly generated instances, public library instances and "difficult" instances with special structure are presented. Also, results on the comparative performance on the two problem variants are given.

D.-Z. Du et al. (eds.), Advances in Steiner Trees, 81-116.

Contents

1 Introduction

The Euclidean and rectilinear Steiner tree problems in the plane are by far
the most studied geometric Steiner tree problem variants. The classical Euclidean problem has roots more than two centuries back while the rectilinear
was first considered by Hanan [13] in the 1960's. The interest in the latter came mainly from applications in, e.g., VLSI design. In the 1970's both
problems were shown to be NP-hard [11, 12] and this virtually shattered the
hope of finding efficient (polynomial time) algorithms for these problems.

Informally, we ask for a shortest interconnection — a Steiner minimum
tree (SMT) — of a set Z of n terminals (points in the plane) with respect
to a given distance function. Let $u = (u_x, u_y)$ and $v = (v_x, v_y)$ be a pair
of points in the Cartesian plane $\Re^2$. The distance in the L_p-metric, $1 \leq$

$p \leq \infty$, between u and v (or simply the L_p-distance) is $\|uv\|_p = (|u_x - v_x|^p + |u_y - v_y|^p)^{1/p}$. In this work we consider the Steiner tree problem with the Euclidean L_2-distance and the rectilinear (or Manhattan) L_1-distance. Hanan [13] proved that for the rectilinear version, it is always possible to find an SMT with edges belonging to the *Hanan grid*. The Hanan grid is obtained by drawing horizontal and vertical lines through all terminals. As a consequence, Steiner points appear at intersections of these lines only.

Euclidean SMTs (ESMTs) and rectilinear SMTs (RSMTs) are unions of *full Steiner trees* (FSTs) whose terminals are incident with one FST-edge each. An Euclidean FST (EFST) and a rectilinear FST (RFST) spanning k terminals, $2 \leq k \leq n$, has $k - 2$ Steiner points (except when $k = 4$; RFSTs can then have one Steiner point incident with four edges). Steiner points in EFSTs have three incident edges meeting at 120°. Steiner points in RSMTs are incident with 3 edges (except for the already mentioned case).

A minimum spanning tree (MST) for the terminals in Z is a shortest network spanning Z without introducing Steiner points. Euclidean MSTs (EMSTs) and rectilinear MSTs (RMSTs) for Z can be constructed in $O(n \log n)$ time [18]. The length of an EMST (resp. RMST) exceeds the length of an ESMT (resp. RSMT) by at most a factor of $2/\sqrt{3}$ (resp. 3/2) [15].

First exact algorithms for the Euclidean and rectilinear Steiner tree problems (see Hwang et al. [15] for references) are based on a straightforward common framework. Subsets of terminals are considered one by one. For each subset, all its FSTs are determined one by one, and the shortest is retained. Several tests can be applied to these shortest FSTs in order to identify and prune away those that cannot be in any SMT. Surviving FSTs are then concatenated in all possible ways to obtain trees spanning all terminals. The shortest of them is an SMT.

The bottleneck of this approach is the generation of FSTs. Winter [25] suggested a departure from the above general framework. He observed that substantial improvements are available if EFSTs are generated across various subsets of terminals. Retained EFSTs are not necessarily minimal. However, pruning tests are so powerful that only very few EFSTs survive. Similar strategy was recently applied to the generation of RFSTs by Zachariasen [27]. Spectacular speed-ups have been achieved in both cases. As a consequence, the concatenation of FSTs became a bottleneck of FST-based algorithms for both the Euclidean and the rectilinear Steiner tree problems.

Recent results of Warme [24] improved the concatenation dramatically. He noticed that the concatenation of FSTs can be formulated as a problem of finding a minimum spanning tree in a hypergraph with terminals as vertices and subsets spanned by FSTs as (hyper)edges. He solved this problem using branch-and-cut. Instances of the Euclidean and rectilinear Steiner tree problem with as many as 2000 terminals can today be solved in a reasonable amount of time. In Figure 1 we illustrate the dramatic progress in the performance of exact algorithms.

The purpose of this work is to present computational results using the EFST generator of [25, 26] and the RFST generator of [27] with the FST concatenator of [24]. We present the main algorithmic ideas, but the reader is referred to the above mentioned papers for further details. The work is organized as follows: The algorithms for generating EFSTs and RFSTs are described in Section 2. A survey of FST concatenation algorithms is presented in Section 3. Computational results are given in Section 4 and concluding remarks in Section 5.

2   Generating Full Steiner Trees

The algorithms given by Winter [25] and Zachariasen [27] for generating EFSTs and RFSTs, respectively, share many algorithmic ideas. The simpler RFST algorithm is given first. The EFST algorithm is then described using the same framework.

Before we present the general framework, we note that 2-terminal FSTs in any SMT can be restricted to edges of an arbitrarily chosen MST [15]. Thus we only describe the generation of FSTs with three or more terminals.

2.1   Generation of RFSTs

We assume that terminals are in general position, i.e., no pair of terminals has the same x- or y-coordinates. This assumption simplifies the description by avoiding some straightforward but tedious special cases.

Hwang [14] proved that there always exists an RSMT in which every RFST has a very restricted shape shown in Figure 2. Any such RFST spanning k terminals has a *root terminal* z_0 and a *tip terminal* z_t. The root and the tip are connected by a *backbone*. The backbone consists a *long leg* (incident with the root) and a perpendicular *short leg* (incident with the tip). The remaining terminals are attached to the long leg by alternating straight line segments called *branches*. At most one branch can be attached to the short leg (away from the root) as shown in Figure 2b. Steiner points are all on the backbone at points where branches are attached.

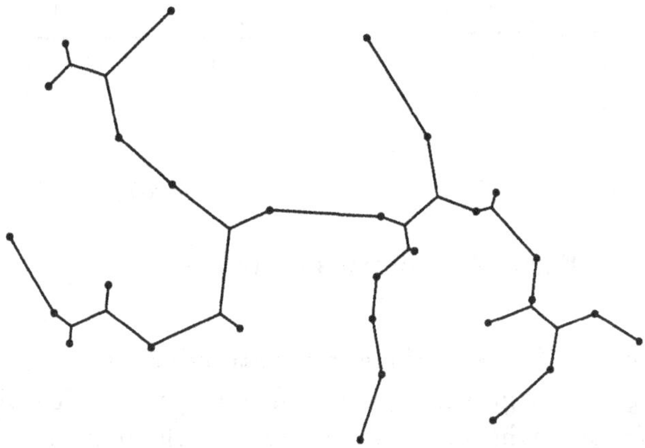

29 cities in the United States and Canada

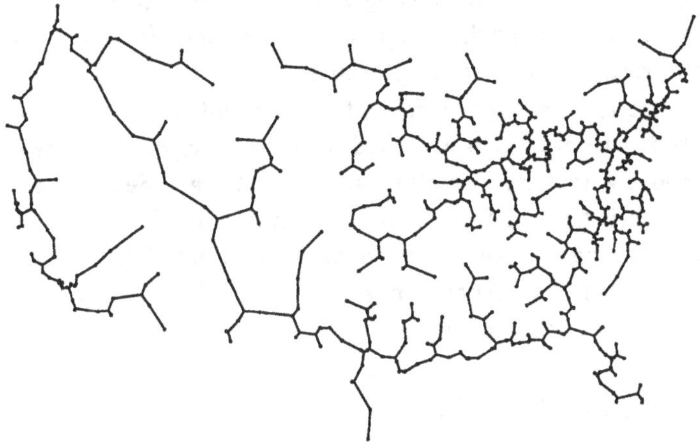

532 cities in the United States

Figure 1: Euclidean Steiner minimum tree examples. The top instance was published in Scientific American in 1989 and was at that time "close to the limit of computing capabilities" [2]. The bottom instance, att532 from the TSPLIB collection, was solved in a few hours on a workstation using the algorithm presented in this paper.

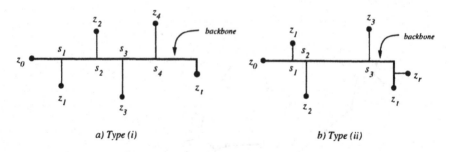

Figure 2: Two types of RFSTs.

The generation of RFSTs with three or more terminals is as follows. For a given root z_0, long legs are grown in each of the four possible directions. For a given direction (to the right, say), only terminals with greater x-coordinate are considered.

Suppose that the long leg from z_0 has been extended to the x-coordinate x_k of the terminal z_k, and the branches alternate below and above the long leg. Several tests (described in Subsection 2.3 and in [27]) can be applied to check whether such an alternating sequence can appear in at least one RSMT. If not, the z_k-branch is replaced by another (alternating) branch farther to the right or backtracking occurs.

Assume that the long leg from z_0 to z_k cannot be pruned away. Each terminal z_t to the right of z_k is attached to the long leg as a tip (by extending the long leg and attaching the alternating short leg). Several tests can be applied to check if this RFST can occur in any RSMT. Furthermore, each terminal z_r to the right of the tip with its y-coordinate between y_0 and y_t is attached to the short leg. Again, most of such RFSTs can be pruned away by one of several efficient pruning tests.

2.2 Generation of EFSTs

Consider an EFST for k terminals, $3 \le k \le n$, as shown in Figure 3. A path between 2 arbitrary terminals (one considered as a root and the other as a tip) goes through one or more Steiner points and can be considered as a backbone; the portion of the backbone from the root to the last Steiner point is the long leg. At each Steiner point, the long leg turns 60° either to the left or to the right. Hence, contrary to the rectilinear case, the long leg does not have a fixed direction. Furthermore, in the rectilinear case, the branches are line segments appearing in the alternating fashion. This is not so in the Euclidean case. Branches can involve more than one terminal and consecutive branches do not need to alternate. In conclusion, the generation of EFSTs is more complicated than the generation of RFSTs. On the other hand, pruning tests for the Euclidean case are more efficient so that the number of surviving EFSTs is (on average) smaller than the number of surviving RFSTs (for the same number of terminals).

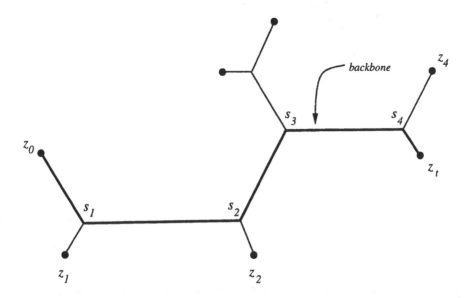

Figure 3: EFST.

In order to explain how EFSTs with three or more terminals are generated, we restrict our attention to 1-terminal branches. Assume that z_0 is selected as the root, and k terminals z_1, z_2, ..., z_k have been attached to the long leg (in that order) making 60° left turns at each Steiner point (Figure 4).

The first complication (when compared with the rectilinear case) arises due to the fact that the locations of Steiner points are not known (Figure 4). The first Steiner point s_1 on the long leg must be located somewhere on the *Steiner arc* $\widehat{z_0 z_1}$ of z_0 and z_1. It is determined as follows. Consider the equilateral triangle with the line segment $z_0 z_1$ as one of its sides, and with its third corner to the right of $z_0 z_1$ (when looking from z_0 toward z_1). This third corner is referred to as the *equilateral point* and is denoted by e_1. Consider the circle $C(z_0, z_1, e_1)$ circumscribing this equilateral triangle. The arc from z_0 to z_1 (clockwise) is the Steiner arc $\widehat{z_0 z_1}$.

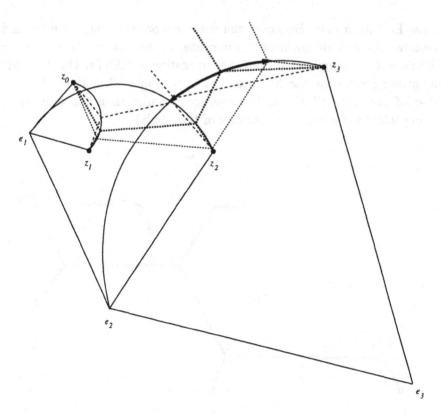

Figure 4: Long leg construction. Different types of dotted lines are examples of partial FSTs for different locations of the Steiner point on $\widehat{e_2 z_3}$.

Possible locations of the Steiner point s_2 adjacent to s_1 and z_2 (with the long leg making a 60° left turn at s_2) are on the Steiner arc $\widehat{e_1 z_2}$. The corresponding equilateral point is denoted by e_2. The Steiner arc $\widehat{e_1 z_2}$ can be reduced; the edge connecting s_1 and s_2 must overlap with the line segment from e_1 to s_2. Hence, only projections of feasible locations of s_1 seen from e_1 are feasible locations of s_2 on $\widehat{e_1 z_2}$.

Steiner arcs for Steiner points $s_3, s_4, ..., s_k$ are determined in analogous manner. Consider the feasible subarc of the Steiner arc $\widehat{e_{k-1} z_k}$ ($k = 3$ in Figure 5). Let e_k denote the associated equilateral point. Consider the region R_k bounded by half-lines rooted at e_k through the extreme points of the feasible subarc of the Steiner arc $\widehat{e_{k-1} z_k}$ (Figure 5). Any terminal $z_t \in R_k \setminus C(e_{k-1}, z_k, e_k)$ yields an EFST. More precisely, the intersection of the line segment $z_t e_k$ with $\widehat{e_{k-1} z_k}$ is the location of s_k. Given the location of

s_k, the location of s_{k-1} is given as the intersection of the line-segment $s_k e_{k-1}$ with $e_{k-2}\widehat{z}_{k-1}$. Locations of $s_{k-2}, s_{k-3}, ..., s_1$ can be determined successively in the same manner.

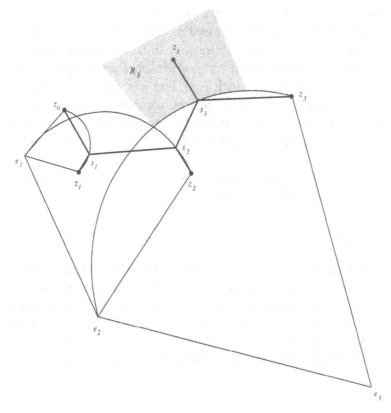

Figure 5: EFST construction.

It can be shown that the length of the EFST is equal to the length of the line-segment $z_t e_k$. Hence, the locations of Steiner points could in principle be determined only for EFSTs belonging to the ESMT.

Once all EFSTs for a given long leg have been identified, the long leg is extended by adding one more branch. If no extension is possible, the current long leg is redirected by requiring 60° right turn at s_k. This calls for the recomputation of e_k (now to the left of the line segment from e_{k-1} to z_k). Other equilateral points $e_1, e_2, ..., e_{k-1}$ are not affected. Once all extensions of this long leg have been considered, z_k is replaced by another terminal. If all terminals have been considered, backtracking occurs.

The efficiency of the above approach stems from the fact that when long legs are expanded as well as when tips are attached, very powerful pruning tests can be applied. Most long legs never involve more than 2-3 Steiner points. Steiner arcs for these long legs are usually very narrow so that at most one or two tips can be attached (if any).

A serious complication when generating EFSTs is that it is not sufficient to grow long legs with 1-terminal branches. In fact, branches with arbitrary many terminals can be attached. These branches are long legs. We omit a detailed description of how such long legs are generated. Suffice it to say that whenever a long leg is not pruned away, it is saved. Long legs are extended either by attaching a terminal (as described above) or by attaching a saved disjoint long leg. Only very few EFSTs of the latter kind survive the pruning tests.

Another rather technical issue which we do not cover here is how to avoid multiple generation of the same EFST.

2.3   Pruning

In this subsection we give a brief description of some tests that make it possible to prune away the majority of FSTs. We focus on pruning tests that are common for both the Euclidean and the rectilinear case. The description is kept on a general level. The reader is referred to Winter and Zachariasen [26] and to Zachariasen [27] for details. Furthermore, some additional tests, not described below, can be found in these two papers.

2.3.1   Lune Property

A *lune* L_{uv} of a line segment uv is the intersection of two circles both with radius $\|uv\|_p$ and centred at u and v, respectively (Figure 6). A necessary condition for the line segment uv to be in any SMT is that its lune contains no terminals.

Edges in rectilinear backbones are known since the locations of Steiner points are fixed. As backbones are expanded, it is straightforward to check if the lunes of the end-points of added line segments are empty. The same applies when tips (possibly with side branches) are added.

The situation is more complicated in the Euclidean case. Locations of Steiner points are not fixed. Fortunately, they are restricted to Steiner arcs. Consider the most recent Steiner arc of the long leg. If the corresponding Steiner point s_k is assumed to be one of the extreme points of this Steiner arc, then the location of s_{k-1} on the long leg can be determined. If one of

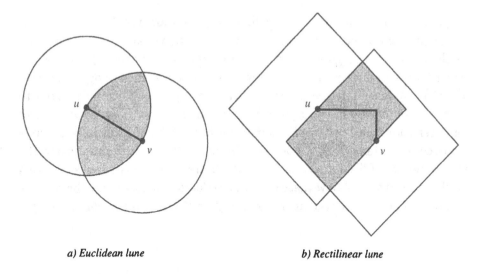

a) Euclidean lune   *b) Rectilinear lune*

Figure 6: Euclidean and rectilinear lunes.

the lunes $L_{s_k s_{k-1}}$ or $L_{s_k z_k}$ contains a terminal z, the Steiner arc $e_{\widehat{k-1}z_k}$ can be narrowed until none of them contains z.

Several other tests involving more complex empty regions are available, in particular for the rectilinear case. These tests will not be covered here; the reader is referred to Zachariasen [27] for details.

2.3.2 Bottleneck Steiner Distances

Construct an MST for the set of terminals Z. Let $b_{z_i z_j}$ denote the length of the longest edge on the unique path from a terminal z_i to a terminal z_j. We refer to $b_{z_i z_j}$ as the *bottleneck Steiner distance*. Bottleneck Steiner distances for all pairs of terminals can be determined in $O(n^2)$ time. Consider an Euclidean or rectilinear SMT for Z. It is straightforward to show that no edge on the path between a pair of terminals z_i and z_j can be longer than $b_{z_i z_j}$.

When growing long legs in the rectilinear case, longest edges on paths from the terminal of the added branch to other terminals can be determined. If any of these edges has length greater than the corresponding bottleneck Steiner distance, the long leg can be pruned away. Similar arguments apply when a tip (possibly with a branch attached to the short leg) is added to the long leg.

The use of bottleneck Steiner distances when generating EFSTs is more complicated since the locations of Steiner points are not fixed. However, as already mentioned, they are restricted to Steiner arcs. It is therefore possible to determine good lower and upper bounds for the lengths of edges incident with Steiner points. If a lower bound for such an edge is greater than the bottleneck Steiner distance between a pair of terminals forced to use this edge, then the long leg can be pruned away. If an upper bound is greater than the bottleneck Steiner distance, then it is usually possible to narrow the Steiner arc (upper bounds are typically computed using the extreme points of Steiner arcs). It is outside the scope of this work to give a detailed description of how these bounds are determined and how the arcs are narrowed. The interested reader is referred to Winter and Zachariasen [26].

2.3.3   Upper Bounds

Several good heuristics for both the rectilinear and the Euclidean Steiner tree problems are available (see Hwang et al. [15]). They can be used to exclude some long legs and FSTs. Once again, the situation is simpler for the rectilinear case than for the Euclidean case.

Consider a rectilinear long leg with s_k as the most recent Steiner point. Use one of the rectilinear heuristics to determine a tree spanning s_k and the terminals attached to the long leg. If this tree is shorter than the tree given by the long leg, then the long leg cannot occur in the RSMT. Similarly, when a tip is added to the long leg (possibly with a branch attached to the short leg), and the resulting RFST can be shown to be non-optimal for its terminals (by generating a shorter tree using a heuristic), then the RFST can be pruned away. Similarly, if the MST for the terminals (using bottleneck Steiner distances between terminals) is shorter than the RFST, then the RFST can be pruned away. This test can be very efficient since bottleneck Steiner distances between pairs of terminals are often much smaller than L_1-distances.

Similar tests are available for the Euclidean case when a tip is added to a long leg. However, when a long leg is expanded, the location of the most recent Steiner point s_k is not fixed, and the determination of upper bounds becomes more complicated. Once again, the reader is referred to Winter and Zachariasen [26] for a description of how upper bounds are determined and how they are used to prune away long legs or to narrow the Steiner arcs.

3 Concatenating Full Steiner Trees

In contrast to the FST generation problem, the FST concatenation problem is purely combinatorial and essentially metric-independent. Given a set of FSTs $\mathcal{F} = \{F_1, F_2, ..., F_m\}$, known to contain a subset whose union is an SMT for Z, the problem is to identify a subset $\mathcal{F}^* \subseteq \mathcal{F}$ such that the FSTs in $\mathcal{F}^*$ interconnect Z and have minimum total length.

This concatenation problem has been solved using simple backtrack search (Section 3.1) and by applying dynamic programming (Section 3.2). However, backtrack search has a very steep running time growth and can only be applied for moderately sized problems. Dynamic programming has better worst-case running time properties, but is in practice even slower (and consumes much more memory).

Warme [24] presented an integer programming formulation for the concatenation problem which was solved by branch-and-cut (Section 3.3). Warme proved that the concatenation problem is equivalent to finding a minimum spanning tree (MST) in the hypergraph $H = (Z, \mathcal{F})$ and showed that the general hypergraph MST problem is NP-hard. This algorithm has increased the solvable range by more than an order of magnitude (see computational results in Section 4).

3.1 Backtrack Search

Backtrack search is a simple, yet reasonably effective "dumb" algorithm. Starting with a partial solution consisting of a single FST $F_i \in \mathcal{F}$, we seek the shortest-length tree interconnecting Z and containing F_i. This is done by recursively adding FSTs to the solution until it interconnects Z or it can be concluded that it cannot be optimal, e.g., if a cycle is created. In this case the search backtracks and some other FSTs are added. Obviously, it is only necessary to try FSTs spanning a particular terminal as the initial FST.

The cut-off tests applied during the search determine the practical behaviour of the algorithm, which otherwise runs in time $\Theta(2^m)$. Winter [25] used only relatively simple cut-off tests and observed that the concatenation phase began to dominate the generation phase already at $n \approx 15$. Cockayne and Hewgill [6, 7] improved the concatenation phase considerably by applying *problem decomposition*, *FST compatibility* and *FST pruning*. Similar ideas were also applied to the rectilinear problem by Salowe and Warme [20] but with substantially less success. More recently, Winter and Zachariasen

[26] improved FST compatibility and pruning substantially and this allowed the exact solution of randomly generated Euclidean instances with up to 140 terminals (on average one hour on a workstation).

Problem decomposition reduces the initial concatenation problem to several smaller concatenation problems. Define $G(\mathcal{F})$ to be an undirected graph with Z as its vertices; two nodes (terminals) z_i and z_j of $G(\mathcal{F})$ are adjacent if and only if there exists an FST in $\mathcal{F}$ spanning z_i and z_j. If $G(\mathcal{F})$ has an articulation point, it can be split into two subproblems at that node [6] — each biconnected component corresponds to a subproblem for which concatenation may be done separately. More sophisticated decomposition methods, based on vertex separators, have been proposed [6, 20, 23]. Problem decomposition has been the single most important algorithmic improvement for concatenation based on backtrack search, but the effect of decomposition diminishes for larger problem instances, in particular for the rectilinear problem. Also, for problem instances with special structure, such as regular lattices for the Euclidean problem, decomposition has virtually no effect.

The notion of *FST compatibility* was introduced by Cockayne and Hewgill [7] (a slightly different definition is given here). Two FSTs $F_i, F_j \in \mathcal{F}$ are *incompatible* if they cannot appear simultaneously in any SMT (e.g., if they span two or more common terminals), and otherwise they are *compatible*. In a preprocessing phase, a series of compatibility tests are applied to every pair of FSTs; the result is stored in a compatibility matrix which is used during backtrack search, as described later.

Let $F_i, F_j \in \mathcal{F}$ be two FSTs sharing at least one terminal. In the current implementation we used the following series of compatibility tests:

- F_i and F_j share at most one common terminal z_c.

- The angle between the two edges incident with z_c is at least 120° (resp. 90°) for the Euclidean (resp. rectilinear) problem.

- $F_i \cup F_j$ passes longest edge test (described in Section 2.3.2).

- $F_i \cup F_j$ passes MST-BSD test (described in Section 2.3.3).

These tests can be performed with a fairly low computational overhead and they dominate many of the tests proposed in the literature, some of which are apparently metric-dependent [7, 20, 26].

The notion of FST compatibility can be used to remove FSTs from candidacy. Sophisticated variants of *FST pruning* can remove more than half of the initial list of FSTs [9, 26] at moderate computational effort —

and have approximately doubled the solvable range. One effective pruning technique is the following: Consider an FST $F_i \in \mathcal{F}$ and the set $\mathcal{F}_i$ of FSTs compatible to this FST. If the graph $G(\{F_i\} \cup \mathcal{F}_i)$ is disconnected, then F_i cannot be augmented to a tree spanning Z and may be eliminated from further consideration.

Furthermore, FST compatibility can be used during backtrack search. At any point during the search only FSTs which are compatible to all FSTs in the current solution need to be considered. This reduces the search tree drastically without any significant computational overhead (recall that the compatibility matrix is computed in a preprocessing phase). Salowe and Warme [20] gave a strategy for scheduling the FSTs to add during the search in a "most promising" fashion; these ideas were elaborated by Warme [23]. However, the effect of using various scheduling techniques has been much less prominent than the effect of problem decomposition and FST compatibility.

3.2 Dynamic Programming

One of the earliest exact algorithms for the Steiner tree problem in graphs is the dynamic programming algorithm of Dreyfus and Wagner [8]. By reducing the rectilinear problem to the Steiner tree in graphs this algorithm solves the rectilinear problem in time $O(n^2 3^n)$. Apart from the divide-and-conquer algorithm of Smith [21] which runs in time $n^{O(\sqrt{n})}$ (but with a huge constant factor), dynamic programming algorithms provide the best worst-case bounds for the rectilinear problem (the running time bounds given in this section do not generalize to the Euclidean problem as explained below).

Ganley and Cohoon [10] gave the first FST based dynamic programming algorithm. It uses the well-known fact that an SMT is either an FST or can split into two SMTs joined at a terminal; note that at least one of these smaller SMTs can again be assumed to be an FST. Subsets of Z are enumerated in order of increasing cardinality — SMTs for each subset are computed and stored in a lookup table. Since rectilinear FSTs can be computed in linear time[1] finding the shortest joined SMT dominates the running time which is $O(n 3^n)$. By proving an $O(n 1.62^n)$ bound on the number of FSTs with Hwang topology, Ganley and Cohoon reduced the running time to $O(n^2 2.62^n)$. More recently, Fößmeier and Kaufmann [9] improved the former bound to $O(n 1.38^n)$ by using additional necessary conditions on rectilinear FSTs, thereby improving the latter bound to $O(n^2 2.38^n)$.

---

[1] Assuming that the terminals have been sorted in x- and y-direction in a preprocessing phase.

Although dynamic programming provides the best worst-case bounds for the concatenation problem the practical behaviour seems to be inferior to backtrack search [9]. In addition, huge memory requirements make the approach impractical for instance sizes larger than $n \approx 40$.

These running time bounds do not generalize to the Euclidean problem, since no upper bound except from $O(2^n)$ is known for the total number of Euclidean FSTs. Furthermore, no polynomial time algorithm is known for computing a shortest FST for a set of terminals — this may be compared to the linear time algorithm for the rectilinear problem.

3.3 Integer Programming

In this section an integer programming formulation is given for the concatenation problem. Fundamental valid inequalities and polytope properties are presented and important details on the branch-and-cut algorithm are highlighted. For further details, we refer to the paper by Warme [24].

Consider a *hypergraph* $H = (Z, \mathcal{F})$ with the set of terminals Z as its vertices and the set of FSTs $\mathcal{F}$ as its hyperedges, that is, each FST $F_i \in \mathcal{F}$ is considered to be a subset of Z which is denoted by $Z(F_i)$. A *chain* in H from $z_0 \in Z$ to $z_k \in Z$ is a sequence $z_0, F_1, z_1, F_2, z_2, \ldots, F_k, z_k$ such that all vertices and hyperedges are distinct and $z_{i-1}, z_i \in Z(F_i)$ for $i = 1, 2, \ldots, k$. A spanning tree in H is is a subset of hyperedges $\mathcal{F}' \subseteq \mathcal{F}$ such that there is a *unique* chain between every pair of vertices $z_i, z_j \in Z$. The uniqueness implies that there can be no pair of hyperedges $F_i, F_j \in \mathcal{F}'$ that share two or more vertices, i.e., we have $|Z(F_i) \cap Z(F_j)| \leq 1$ for all $F_i, F_j \in \mathcal{F}'$. The problem of finding a minimum spanning tree (MST) in H where each hyperedge $F_i \in \mathcal{F}$ has weight equal to its length $|F_i|$ is equivalent to solving the concatenation problem.

The hypergraph MST problem is NP-hard, even when all hyperedges span at most a constant number $K \geq 3$ of vertices. In fact, Tomescu and Zimand [22] have shown that the existence of a spanning tree in a K-uniform hypergraph (in which all hyperedges span exactly K vertices) is an NP-complete problem for $K \geq 3$. For $K = 2$ the problem reduces to the MST problem in ordinary graphs, which can be solved in polynomial time.

Now we give the integer programming (IP) formulation. Let c be a vector in $\Re^m$ whose components are $c_i = |F_i|$. Denote by x an m-dimensional *binary* vector. The IP formulation is then

$$\min cx \qquad\qquad (1)$$

$$\text{s.t.} \quad \sum_{F_i \in \mathcal{F}} (|Z(F_i)| - 1)x_i \ = \ n - 1 \tag{2}$$

$$x_i + x_j \ \leq \ 1, \quad F_i \text{ incompatible with } F_j \tag{3}$$

$$\sum_{F_i \in \mathcal{F}} \max(0, |Z(F_i) \cap S| - 1)x_i \ \leq \ |S| - 1, \quad \forall S \subseteq Z, \ 2 \leq |S| < n \tag{4}$$

The objective (1) is to minimize the total length of the chosen FSTs subject to the following constraints: Equation (2) enforces the correct number and cardinality of hyperedges to construct a spanning tree. Constraints (3) add compatibility information to the formulation (Section 3.1). Finally, constraints (4) eliminate cycles by extending the standard notion of subtour elimination constraints; these constraints also ensure, in conjunction with equation (2), that the chosen FSTs interconnect Z.

Warme [24] proved several fundamental properties of the corresponding polytope. Let $\mathcal{K}_n$ be the complete hypergraph with n vertices. This graph has $2^n - n - 1$ hyperedges (all hyperedges span two or more vertices). Let $ST_n \subset \{0,1\}^{2^n-n-1}$ denote the set of incidence vectors of spanning trees of $\mathcal{K}_n$. That is, each spanning tree of $\mathcal{K}_n$ defines a binary vector in which an element is 1 if and only if the corresponding hyperedge is chosen.

Then $conv(ST_n)$, the convex hull of the incidence vectors of all spanning trees, is the *spanning tree in hypergraph polytope*, STHGP(n). Warme proved that $conv(ST_n)$ has dimension $2^n - n - 2$, that all spanning trees are extreme points and that the cycle elimination constraints (4) are facet defining for $n \geq 3$. In particular the last result explains the strength of this formulation (as evidenced by the computational results in Section 4).

The integer program is solved via branch-and-cut. Lower bounds are provided by linear programming (LP) relaxation, i.e., by relaxing integrality of every component x_i of x to $0 \leq x_i \leq 1$. Obviously the number of compatibility constraints (3) is bounded by $O(m^2)$, but the number of cycle elimination constraints (4) is exponential in n. The latter constraints are dynamically added by separation methods. This separation problem can be solved in polynomial time (in n and m) by finding minimum cuts in certain graphs [24]. However, heuristic separation methods are also used whenever applicable in order to speed up convergence to LP-optimum.

Enumeration is done using *best choice* branching. Let x_i be a possible non-integral branch variable. Assume that the two LP-subproblems corresponding to $x_i = 0$ and $x_i = 1$ yield objective values z_i^0 and z_i^1, respectively. The x_i that maximizes the value of $\min(z_i^0, z_i^1)$ is chosen as the new branch variable. Note that solving the $x_i = 1$ subproblem can be skipped if z_i^0

is already too low to be maximum. Furthermore, the process can terminate immediately if both z_i^0 and z_i^1 meet or exceed the best known integer solution.

Nodes in the branching tree are chosen using a *best node first* selection strategy, such that the outstanding node having the lowest objective value is processed next. This may (theoretically) produce a large number of outstanding nodes, but in practice the tightness of the formulation produces only a fairly small number of nodes.

4   Computational Experience

4.1   Experimental Conditions

The computational study was made on an HP9000 workstation[2]. The rectilinear and Euclidean FST generators were programmed in C++ using the class library LEDA version 3.4.1 [17]; the random_source class in LEDA was used for generating pseudo-random numbers. The FST concatenator was programmed in C using CPLEX version 5.0 to solve all the linear programming (LP) relaxations.

The test bed mainly consists of three sets of problem instances. The first set is a collection of 60 randomly generated instances from the *OR-Library* [1], 15 instances for each problem size 100, 250, 500 and 1000. All instances in the *OR-Library* with 100 and fewer terminals have previously been solved as Euclidean instances [26] and all instances with 1000 and fewer terminals have previously been solved as rectilinear instances [24].

The second set is a selection of 26 public library instances taken from *TSPLIB* [19] (ranging from 198 to 7397 terminals). These are the same instances as studied by Zachariasen [27]; none of these have previously been solved as rectilinear or Euclidean problems. *TSPLIB* is a collection of instances for the Traveling Salesman Problem (TSP), mainly plane real-world Euclidean problem instances. All instances are given as (the coordinates of) points in the plane.

Thirdly, the average behaviour of the exact algorithm is studied on a large set of randomly generated instances. Fifty instances were generated for each size 100, 200, 300, 400 and 500. Terminals were drawn with uniform distribution from the unit square.

---

[2]Machine: HP 9000 Model C160. Processor: 160 MHz PA-RISC 8000. Main memory: 256 MB. Performance: 10.4 SPECint95 and 16.3 SPECfp95. Operating system: HP-UX 10.20. Compiler: GNU C++ 2.7.2.1 (optimization flag -O3).

Computational results on FST generation, FST concatenation and SMT properties for these three sets are presented in Sections 4.2, 4.3 and 4.4, respectively. In addition, we present results for two small sets of instances (one for the rectilinear and one for the Euclidean problem) for which the full Steiner tree approach performs relatively badly — at least when compared to the average case. These results are given in Section 4.4.

In order to ease the comparison between the rectilinear and the Euclidean problem, statistics for both problems are presented in the same table. Also, we use the same table layout for each of the three main instance classes. However, while results for *OR-Library* and *TSPLIB* instances are presented for each instance, only averages (and standard deviations) are given for the randomly generated instances. Thus it is possible both to study algorithmic behaviour and solution properties for specific instances and to identify general tendencies for increasing problem sizes.

4.2 Generation

As observed in previous studies (e.g. [20, 26]) the number of FSTs surviving the pruning tests described in Section 2 is almost *linear*. For randomly generated instances (Table 1 and 3; Figure 7), approximately $4.0n$ rectilinear and $2.3n$ Euclidean FSTs survive. For problems with more structure (in particular instances with many co-linear and equidistant terminals) *fewer* rectilinear FSTs survive (Table 2). These instances are on the other hand very difficult to solve as Euclidean problems; for many of these instances the Euclidean FST generation did not finish within the allotted CPU time of approximately one week.

Generated rectilinear FSTs on average span less than four terminals while largest FSTs span 14 terminals (Table 3). Euclidean FSTs on average only span three terminals and at most 9 terminals. Figure 10 illustrates this more clearly: For a given FST size, approximately twice as many rectilinear FSTs are generated — the ratio increases for larger FSTs.

We also present statistics on incompatibility based on the tests described in Section 3.1. These tests are essentially metric-independent. Recall that two FSTs are incompatible if they cannot appear simultaneously in any SMT and that tests for incompatibility are only applied to FST pairs which share at least one terminal. The percentage of FST pairs sharing at least one terminal which are incompatible will be used as "measure of incompatibility". Interestingly, this measure is extremely independent of the metric (rectilinear/Euclidean), terminal set distribution and instance size. More than

		Rectilinear				Euclidean		
n	Count	Size	Incomp	CPU	Count	Size	Incomp	CPU
100 (1)	397	3.60 (11)	66.95	1.9	239	3.05 (7)	69.37	208.9
100 (2)	508	3.84 (17)	66.48	4.6	239	2.97 (8)	65.86	264.7
100 (3)	331	3.42 (12)	68.98	1.3	219	2.97 (8)	66.34	214.7
100 (4)	349	3.48 (10)	71.42	1.4	231	2.96 (6)	70.67	193.2
100 (5)	337	3.41 (9)	67.15	1.4	211	2.85 (5)	65.51	177.5
100 (6)	495	3.91 (12)	67.87	4.9	225	3.05 (8)	65.80	233.7
100 (7)	490	3.89 (11)	69.22	4.1	245	2.99 (7)	64.46	333.1
100 (8)	361	3.51 (9)	69.32	1.5	221	2.96 (8)	68.14	177.2
100 (9)	396	3.69 (11)	68.24	2.6	237	2.95 (7)	67.38	206.1
100 (10)	365	3.46 (9)	65.21	1.7	225	2.93 (6)	63.49	188.3
100 (11)	336	3.28 (8)	67.14	1.2	235	3.01 (6)	65.89	153.8
100 (12)	357	3.37 (9)	65.32	1.4	210	2.82 (6)	66.74	114.7
100 (13)	401	3.87 (13)	72.13	3.0	221	2.95 (8)	68.30	135.3
100 (14)	312	3.15 (9)	63.44	0.8	224	2.92 (8)	66.38	179.9
100 (15)	356	3.39 (10)	67.90	1.4	233	3.09 (7)	68.05	240.8
250 (1)	936	3.65 (21)	69.00	6.5	550	2.87 (7)	65.00	1303.4
250 (2)	889	3.34 (10)	65.35	3.9	527	2.84 (8)	64.74	935.3
250 (3)	870	3.37 (12)	66.53	3.8	575	3.01 (9)	71.43	1279.2
250 (4)	911	3.48 (11)	68.34	4.4	537	2.84 (7)	63.29	1129.4
250 (5)	914	3.45 (11)	66.16	4.3	537	2.88 (8)	67.53	1031.2
250 (6)	909	3.40 (10)	66.76	4.1	501	2.80 (7)	65.22	975.3
250 (7)	914	3.47 (11)	68.21	4.0	533	2.87 (7)	65.36	1054.2
250 (8)	1102	3.65 (17)	66.81	8.2	631	3.09 (7)	67.46	2046.8
250 (9)	922	3.43 (12)	66.01	4.2	537	2.85 (8)	64.00	1228.6
250 (10)	1139	3.72 (14)	67.60	8.8	573	2.99 (9)	66.47	1518.6
250 (11)	980	3.67 (17)	65.74	7.0	590	3.00 (8)	66.31	1280.4
250 (12)	951	3.62 (18)	68.77	5.7	557	2.89 (7)	66.23	1085.7
250 (13)	1009	3.49 (9)	66.43	5.1	627	3.06 (8)	67.83	1565.5
250 (14)	992	3.51 (12)	67.82	5.3	575	2.98 (7)	68.28	1152.3
250 (15)	988	3.53 (12)	65.47	6.0	574	2.99 (8)	68.11	1301.6
500 (1)	1927	3.57 (15)	67.10	14.0	1271	3.16 (10)	71.51	5461.8
500 (2)	2249	3.86 (15)	69.79	20.7	1184	2.98 (7)	68.15	6620.6
500 (3)	2159	3.77 (15)	69.51	15.8	1248	3.14 (10)	70.22	6572.3
500 (4)	1891	3.62 (15)	68.04	11.1	1159	3.01 (8)	66.39	5768.3
500 (5)	1875	3.46 (15)	68.03	9.5	1080	2.89 (7)	67.12	4112.7
500 (6)	2070	3.68 (14)	67.40	14.3	1191	3.06 (9)	69.22	6378.0
500 (7)	1953	3.51 (10)	67.61	10.1	1118	2.91 (7)	65.42	4738.5
500 (8)	2027	3.59 (14)	67.95	12.0	1092	2.92 (8)	65.78	5027.0
500 (9)	2011	3.63 (13)	66.91	13.2	1191	3.01 (8)	69.30	5445.7
500 (10)	1953	3.48 (11)	67.45	10.3	1208	3.00 (8)	67.20	5326.4
500 (11)	2068	3.55 (12)	67.21	12.2	1150	2.96 (8)	67.52	6330.9
500 (12)	1912	3.59 (13)	68.77	12.2	1086	2.88 (8)	65.69	5255.1
500 (13)	1879	3.48 (17)	66.71	9.6	1077	2.86 (6)	65.10	4085.0
500 (14)	2126	3.68 (17)	67.71	14.3	1305	3.14 (10)	69.40	7511.8
500 (15)	1962	3.61 (13)	67.24	12.4	1176	2.99 (9)	68.48	5277.4
1000 (1)	4176	3.70 (12)	68.70	31.1	2197	2.90 (9)	65.26	20933.1
1000 (2)	4012	3.61 (17)	67.75	25.5	2280	2.96 (10)	66.73	23147.0
1000 (3)	4023	3.58 (16)	67.11	26.7	2185	2.93 (9)	65.98	22887.5
1000 (4)	4092	3.62 (18)	67.96	26.8	2476	3.07 (10)	69.53	24932.6
1000 (5)	4025	3.69 (16)	67.73	32.5	2214	2.94 (10)	67.11	23218.7
1000 (6)	4336	3.81 (21)	69.91	31.8	2313	2.98 (10)	67.80	24286.8
1000 (7)	3998	3.64 (14)	68.50	29.3	2236	2.91 (8)	66.48	20443.2
1000 (8)	4294	3.77 (17)	69.10	35.2	2266	2.92 (8)	66.05	23819.0
1000 (9)	4481	3.82 (14)	68.41	43.4	2393	3.01 (9)	67.72	26676.5
1000 (10)	4002	3.56 (15)	67.97	26.4	2239	2.92 (9)	67.17	20917.7
1000 (11)	4042	3.62 (20)	67.63	30.0	2311	2.96 (10)	67.16	23504.0
1000 (12)	4690	3.95 (18)	69.75	48.9	2430	3.03 (8)	67.97	28115.9
1000 (13)	3876	3.58 (16)	68.72	26.3	2253	2.94 (9)	67.61	19564.7
1000 (14)	4336	3.75 (14)	68.77	32.6	2309	2.97 (10)	66.33	25001.8
1000 (15)	4151	3.63 (14)	68.34	28.3	2331	2.99 (10)	68.62	23183.7

Table 1: FST-generation, *OR-library* instances. Count: Number of FSTs generated, including MST-edges. Size: Average number of terminals spanned in generated FSTs (resp. maximum number of terminals spanned). Incomp: Percentage of FST pairs sharing at least one terminal which are incompatible. CPU: CPU-time (seconds).

Instance	Rectilinear				Euclidean			
	Count	Size		Incomp	CPU	Count	Size		Incomp	CPU
d198	275	2.38	(5)	37.08	0.6	807	4.95	(21)	87.74	4153.9
lin318	998	3.32	(10)	63.28	4.5	1235	3.54	(8)	71.84	10299.9
fl417	1192	2.93	(17)	46.94	4.9	3715	6.84	(24)	82.12	81803.9
pcb442	558	2.27	(7)	39.43	1.2	-	-	-	-	-
att532	2239	3.76	(16)	69.76	19.4	1246	3.02	(9)	67.92	7403.8
u574	1506	3.06	(9)	62.31	4.9	1340	2.98	(10)	68.66	5347.9
p654	933	2.55	(11)	57.84	3.2	-	-	-	-	-
rat783	3899	4.02	(25)	68.31	45.7	1954	3.15	(11)	69.27	21258.2
pr1002	2198	2.90	(11)	59.18	7.6	2322	2.96	(8)	66.92	15447.9
u1060	2818	3.04	(9)	59.04	11.9	3630	4.10	(13)	77.00	59230.9
pcb1173	3001	3.07	(12)	65.67	11.5	3052	4.61	(36)	89.96	72887.3
d1291	1393	2.09	(7)	15.51	5.8	-	-	-	-	-
rl1323	1957	2.49	(9)	52.63	7.7	2351	2.67	(7)	71.72	8569.4
fl1400	5870	3.02	(11)	40.61	32.7	-	-	-	-	-
u1432	1431	2.00	(2)	0.00	4.8	-	-	-	-	-
fl1577	3822	2.71	(7)	47.40	12.5	-	-	-	-	-
d1655	2219	2.34	(7)	44.85	7.7	-	-	-	-	-
vm1748	4015	3.12	(15)	66.59	18.1	3329	2.77	(10)	66.56	28653.0
rl1889	2920	2.56	(11)	58.98	11.3	3579	3.13	(16)	82.81	33193.8
u2152	2173	2.01	(3)	1.53	9.6	-	-	-	-	-
pr2392	4792	2.83	(9)	55.59	16.1	5923	3.12	(10)	71.48	101522.3
pcb3038	9356	3.28	(12)	64.48	44.1	-	-	-	-	-
fl3795	7770	2.62	(7)	36.28	45.7	-	-	-	-	-
fnl4461	27959	4.59	(34)	71.05	543.5	12246	3.19	(12)	67.90	965730.9
rl5934	8168	2.38	(8)	49.15	71.0	-	-	-	-	-
pla7397	10595	2.46	(9)	53.84	100.9	-	-	-	-	-

Table 2: FST-generation, *TSPLIB* instances. See Table 1 for captions.

n	Rectilinear					Euclidean				
	Count	Size		Incomp	CPU	Count	Size		Incomp	CPU
100	363.3	3.42	(9.6)	67.44	1.7	220.0	2.92	(7.0)	66.81	162.3
	±47.7	±0.21	±1.7	±2.53	±0.8	±21.7	±0.16	±1.4	±3.79	±49.6
200	797.4	3.60	(12.3)	68.08	5.0	454.5	2.95	(7.7)	66.61	781.0
	±78.9	±0.18	±2.9	±2.20	±1.8	±33.2	±0.12	±1.2	±2.93	±202.2
300	1196.6	3.62	(13.4)	68.24	7.8	671.8	2.94	(8.0)	66.75	1740.7
	±90.1	±0.15	±2.6	±1.68	±2.5	±36.4	±0.09	±1.1	±1.96	±394.9
400	1599.9	3.61	(13.6)	68.10	10.7	910.3	2.96	(8.3)	67.20	3171.4
	±127.1	±0.15	±2.3	±1.62	±3.7	±36.9	±0.07	±1.1	±1.49	±593.3
500	1997.6	3.62	(13.9)	68.30	13.5	1139.8	2.96	(8.6)	67.16	5194.8
	±125.0	±0.12	±2.3	±1.30	±3.6	±55.7	±0.09	±1.4	±1.86	±985.4

Table 3: FST-generation, randomly generated instances. Averages over 50 instances for each size; standard deviations on the second line of each row. See Table 1 for captions.

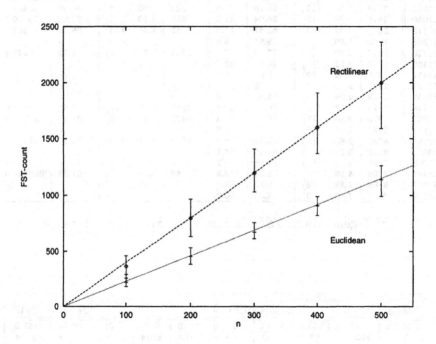

Figure 7: FST generation, randomly generated instances. Total FST-count, including MST-edges; average, minimum and maximum.

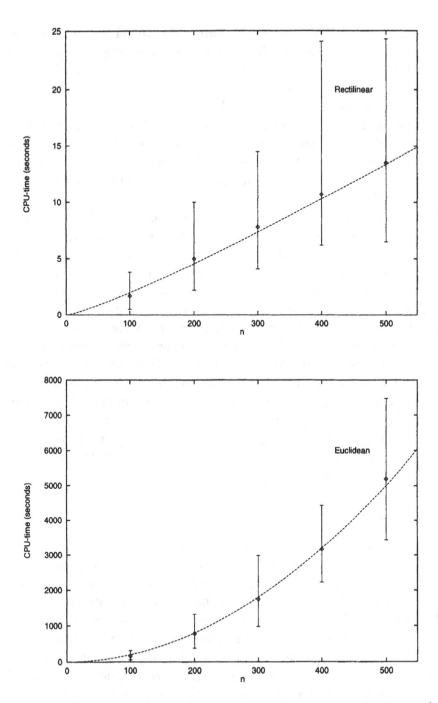

Figure 8: FST generation, randomly generated instances. CPU-time (seconds); average, minimum and maximum. Note different scaling on y-axes.

two-thirds of all FST pairs sharing at least one terminal are incompatible. The effect of adding incompatibility constraints to the integer programming formulation is small compared to the speed-up obtained when using compatibility in conjunction with backtrack search. In some cases it is even negative since larger LPs need to be solved.

Another measure of incompatibility could have been the number of incompatible FST pairs normalized by either n or m (number of FSTs) — or by the square of these numbers. However, this would not have given a good basis for comparing the two problem variants because of the larger number of rectilinear FSTs.

Rectilinear FSTs are generated in a fraction of the CPU-time needed for generating Euclidean FSTs (more than 100 times faster). However, as noted in Section 2, the generation of Euclidean FSTs is much more complicated and involves heavy use of floating point operations and trigonometric functions. When comparing the CPU-times with those given in [26, 27], the following should be noted: First of all a faster workstation has been used in this study (20-30% faster). The code for the Euclidean generator has been optimized to make fewer calls to trigonometric functions and the upper bounding procedure has been improved. These improvements have reduced the running time by 30-40%. The running times reported for the rectilinear generator are *greater* than those reported in [27]. This comes from the fact that incompatibility information has been computed in this study. This increases the running time by a factor of approximately five — thus the computation of incompatibility information completely dominates the generation of rectilinear FSTs.

4.3   Concatenation

Statistics on the performance of the branch-and-cut algorithm for solving the FST concatenation problem are given in Table 4, 5, 6 and Figure 9. We present root LP/optimal IP value gap, number of branch-and-bound nodes and total number of LPs solved, i.e., number of separation iterations.

The strength of the LP relaxation is indicated by a very small LP/IP gap; for many instances the root LP solution is integral and no branching is needed. Only two instances have a gap of more than 0.1% and few instances require more than 10 branch-and-bound nodes. However, the computational effort required to solve the FST concatenation problem using integer programming shows a very large variance compared to the FST generation algorithms. The running time growth is clearly exponential (Figure 9).

One subtle observation regarding Table 4 is that even though the rectilinear FST generator used in the current study generates significantly fewer FSTs than the FST generator by Warme [24], this does not necessarily have a positive effect on the time to solve the concatenation problem. For many instances the branch-and-cut algorithm requires *more* separation iterations (LPs) and branch-and-bound nodes (Nds). The reason for this behaviour is as yet unclear.

Out of 26 instances in the TSPLIB selection, 19 have been solved as rectilinear problems and 13 as Euclidean problems using a cut-off of approximately one week. Instance pr2392 was the largest instance solved, both for the Euclidean and rectilinear problem. It is interesting to note that instances with many co-linear and equidistant terminals are easy to solve as rectilinear problems while instances with a more uniform and less structured distribution are easier to solve as Euclidean problems. Also note that all Euclidean instances for which FSTs have been generated are solved using only a few hours for the concatenation (except for fnl4461 which was not solved within one week).

4.4 Steiner Minimum Tree Properties

In this section we present some structural properties of rectilinear and Euclidean SMTs (Table 7, 8, 9 and Figure 10). Similar statistics have been presented for Euclidean SMTs spanning up to 150 terminals [26]. No such statistics have previously been given for rectilinear SMTs spanning more than 50 terminals.

The number of FSTs in an average RSMT spanning n terminals is approximately $0.5n$ while an ESMT has approximately $0.6n$ FSTs. An ESMT has more MST-edges, $0.3n$, compared to $0.2n$ in an RSMT. This difference is also reflected in the size of the FSTs. RFSTs on average span 2.95 terminals while EFSTs span 2.70 terminals; furthermore, for $n = 500$ the largest RFST spans 7 terminals and the largest EFST 6 terminals (see also Figure 10). The size of the largest FST grows very slowly for both problems, apparently with a growth that is $o(\log n)$. For large instances ($n = 500$) the reduction over the MST is 11.5% for the rectilinear problem and 3.3% for the Euclidean problem.

In order to test the limits of FST based exact algorithms for plane Steiner tree problems, we performed a series of tests on seemingly "difficult" instances. Fößmeier and Kaufmann [9] constructed an infinite series of (rectilinear) instances for which the number of FSTs fulfilling a so-called

n		Rectilinear				Euclidean			
		Gap	Nds	LPs	CPU	Gap	Nds	LPs	CPU
100	(1)	0.000	1	33	5.1	0.000	1	4	0.5
100	(2)	0.000	1	15	2.4	0.000	1	8	0.8
100	(3)	0.000	1	22	2.8	0.000	1	30	1.6
100	(4)	0.000	1	7	1.1	0.000	1	3	0.2
100	(5)	0.000	1	6	0.8	0.000	1	2	0.2
100	(6)	0.000	1	7	6.0	0.000	2	6	0.4
100	(7)	0.000	1	55	6.4	0.000	1	3	0.6
100	(8)	0.001	1	22	3.2	0.000	1	3	0.5
100	(9)	0.000	1	7	4.4	0.000	1	26	1.3
100	(10)	0.000	1	22	2.6	0.000	1	6	0.5
100	(11)	0.109	5	20	2.2	0.000	1	3	0.3
100	(12)	0.000	1	27	1.9	0.000	1	3	0.3
100	(13)	0.000	1	8	3.0	0.000	1	3	0.3
100	(14)	0.046	2	72	5.3	0.000	1	4	0.7
100	(15)	0.000	1	10	4.4	0.000	1	4	0.4
250	(1)	0.000	1	24	7.8	0.000	3	10	2.2
250	(2)	0.009	3	60	18.7	0.000	1	12	2.3
250	(3)	0.000	1	102	30.7	0.000	1	42	4.5
250	(4)	0.013	2	36	22.7	0.000	2	22	3.3
250	(5)	0.000	1	272	37.4	0.000	1	129	23.1
250	(6)	0.003	2	110	44.3	0.003	5	12	2.1
250	(7)	0.000	1	44	17.1	0.000	1	4	1.6
250	(8)	0.053	5	41	26.1	0.000	1	18	4.8
250	(9)	0.000	1	604	242.7	0.000	3	23	2.6
250	(10)	0.060	3	95	63.8	0.000	3	30	4.0
250	(11)	0.017	3	32	18.9	0.000	1	6	1.7
250	(12)	0.005	2	58	38.3	0.000	4	11	3.1
250	(13)	0.028	8	325	330.7	0.000	2	34	6.2
250	(14)	0.000	1	12	9.8	0.001	1	85	7.4
250	(15)	0.000	1	58	21.4	0.000	1	38	4.0
500	(1)	0.004	2	38	65.1	0.000	1	52	17.4
500	(2)	0.009	1	50	63.8	0.004	2	68	28.9
500	(3)	0.000	1	216	966.4	0.000	1	180	138.7
500	(4)	0.000	1	169	367.3	0.000	3	182	52.3
500	(5)	0.034	9	455	2112.7	0.000	2	210	105.0
500	(6)	0.000	1	28	48.6	0.000	1	47	22.0
500	(7)	0.006	1	31	66.1	0.000	3	50	15.2
500	(8)	0.001	1	142	505.8	0.000	2	80	37.6
500	(9)	0.017	4	52	80.9	0.000	4	60	47.9
500	(10)	0.000	1	66	98.9	0.000	2	24	12.9
500	(11)	0.000	2	206	1186.4	0.007	1	215	158.0
500	(12)	0.002	2	70	184.9	0.000	2	284	118.3
500	(13)	0.005	2	52	86.7	0.000	2	52	24.3
500	(14)	0.038	4	112	474.7	0.000	1	75	46.7
500	(15)	0.007	2	44	40.0	0.000	2	615	396.5
1000	(1)	0.047	25	232	2780.9	0.000	4	93	117.2
1000	(2)	0.000	2	124	616.5	0.000	2	299	347.9
1000	(3)	0.001	2	1099	34313.0	0.000	1	27	31.9
1000	(4)	0.011	7	490	1172.3	0.000	1	157	107.8
1000	(5)	0.004	3	1330	34970.7	0.000	1	66	73.0
1000	(6)	0.032	16	5560	351933.9	0.000	2	328	3006.8
1000	(7)	0.006	2	1310	39551.3	0.001	5	56	72.3
1000	(8)	0.005	5	743	28679.8	0.003	5	1160	21956.0
1000	(9)	0.004	5	86	544.4	0.000	2	60	102.4
1000	(10)	0.013	6	1202	48770.7	0.000	2	33	44.4
1000	(11)	0.012	5	485	1341.6	0.000	1	114	72.1
1000	(12)	0.010	4	2107	214567.0	0.000	2	503	513.6
1000	(13)	0.000	1	4300	263917.3	0.000	1	35	35.0
1000	(14)	0.022	17	5416	540376.5	0.000	2	448	1579.3
1000	(15)	0.010	2	285	2701.5	0.000	2	472	569.5

Table 4: FST-concatenation, *OR-library* instances. Gap: Root LP objective value vs. optimal value (gap in percent). Nds: Number of branch-and-bound nodes. LPs: Number of LPs solved. CPU: CPU-time (seconds).

Instance	Rectilinear				Euclidean			
	Gap	Nds	LPs	CPU	Gap	Nds	LPs	CPU
d198	0.000	1	63	4.4	0.000	1	47	10.4
lin318	0.046	7	130	451.7	0.000	1	154	499.8
fl417	0.018	52	331	267.2	0.000	1	75	48.9
pcb442	0.000	1	13	7.0	-	-	-	-
att532	0.014	5	633	16110.9	0.000	1	431	1698.8
u574	0.003	1	91	69.1	0.002	5	36	19.9
p654	0.024	5	74	27.1	-	-	-	-
rat783	0.008	6	126	209.6	0.000	1	131	73.4
pr1002	0.010	14	69	143.1	0.000	1	21	61.3
u1060	0.016	113	660	3807.8	0.005	157	533	1220.9
pcb1173	-	-	-	-	0.004	1	2088	9002.1
d1291	0.000	1	44	43.0	-	-	-	-
rl1323	0.024	3	87	86.9	0.000	1	24	46.9
fl1400	-	-	-	-	-	-	-	-
u1432	0.000	1	1	0.4	-	-	-	-
fl1577	0.002	1	268	701.5	-	-	-	-
d1655	0.000	15	270	513.9	-	-	-	-
vm1748	0.004	3	168	402.0	0.000	1	89	186.7
rl1889	0.011	4	2353	14147.6	0.000	1	2366	11580.1
u2152	0.000	1	7	4.4	-	-	-	-
pr2392	0.000	3	130	693.9	0.000	8	92	1109.2
pcb3038	-	-	-	-	-	-	-	-
fl3795	-	-	-	-	-	-	-	-
fnl4461	-	-	-	-	-	-	-	-
rl5934	-	-	-	-	-	-	-	-
pla7397	-	-	-	-	-	-	-	-

Table 5: FST-concatenation, *TSPLIB* instances. See Table 4 for captions.

n	Rectilinear				Euclidean			
	Gap	Nds	LPs	CPU	Gap	Nds	LPs	CPU
100	0.008	1.1	15.6	2.3	0.001	1.1	5.9	0.5
	±0.028	±0.4	±14.3	±1.4	±0.009	±0.4	±6.4	±0.4
200	0.005	1.4	42.6	15.9	0.001	1.2	23.1	3.6
	±0.010	±0.9	±41.6	±16.2	±0.002	±0.6	±29.8	±3.4
300	0.008	1.7	71.3	73.4	0.000	1.5	35.8	7.3
	±0.010	±1.2	±63.8	±122.9	±0.000	±0.7	±48.6	±11.5
400	0.012	3.0	98.9	161.9	0.000	1.7	88.5	30.1
	±0.013	±2.8	±91.6	±297.3	±0.000	±1.1	±125.6	±42.7
500	0.010	3.3	155.5	504.4	0.000	2.0	89.7	54.7
	±0.012	±3.7	±153.4	±803.6	±0.001	±1.3	±95.7	±90.9

Table 6: FST-concatenation, randomly generated instances. Averages over 50 instances for each size; standard deviations on the second line of each row. See Table 4 for captions.

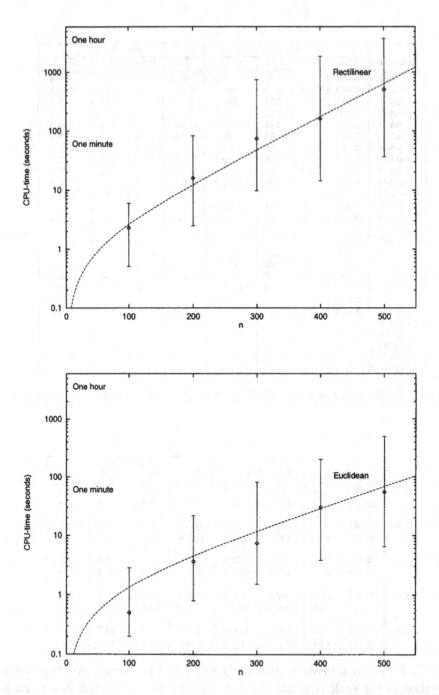

Figure 9: FST concatenation, randomly generated instances. CPU-time (seconds); average, minimum and maximum. Note the logarithmic scaling on *y*-axes.

n		Rectilinear				Euclidean			
		Count	Size	Red	CPU	Count	Size	Red	CPU
100	(1)	48 (18)	3.06 (9)	12.11	7.0	55 (22)	2.80 (5)	3.24	209.4
100	(2)	45 (12)	3.20 (8)	12.45	7.0	56 (26)	2.77 (5)	3.48	265.5
100	(3)	53 (17)	2.87 (4)	11.63	4.2	60 (29)	2.65 (5)	3.42	216.4
100	(4)	51 (19)	2.94 (5)	10.80	2.5	55 (25)	2.80 (5)	3.25	193.4
100	(5)	55 (23)	2.80 (7)	10.21	2.2	62 (34)	2.60 (5)	3.31	177.7
100	(6)	50 (22)	2.98 (7)	12.89	10.8	60 (32)	2.65 (5)	3.40	234.0
100	(7)	48 (12)	3.06 (6)	12.69	10.5	54 (23)	2.83 (5)	3.98	333.7
100	(8)	52 (19)	2.90 (7)	11.27	4.7	61 (34)	2.62 (5)	3.57	177.7
100	(9)	52 (19)	2.90 (6)	13.56	7.0	51 (21)	2.94 (6)	3.50	207.4
100	(10)	54 (22)	2.83 (6)	11.40	4.4	61 (32)	2.62 (6)	3.37	188.8
100	(11)	52 (20)	2.90 (7)	9.97	3.4	59 (29)	2.68 (6)	2.82	154.1
100	(12)	49 (14)	3.02 (5)	12.49	3.3	58 (23)	2.71 (5)	2.69	115.0
100	(13)	46 (13)	3.15 (6)	12.27	6.0	62 (34)	2.60 (5)	2.66	135.6
100	(14)	56 (23)	2.77 (5)	10.63	6.2	55 (27)	2.80 (5)	3.57	180.7
100	(15)	51 (20)	2.94 (9)	11.43	5.8	65 (39)	2.52 (5)	2.78	241.2
250	(1)	120 (36)	3.08 (7)	10.86	14.3	149 (70)	2.67 (5)	3.08	1305.6
250	(2)	127 (43)	2.96 (6)	11.66	22.6	147 (81)	2.69 (6)	2.99	937.6
250	(3)	138 (54)	2.80 (6)	11.23	34.4	153 (80)	2.63 (5)	3.28	1283.7
250	(4)	132 (47)	2.89 (7)	11.35	27.1	145 (68)	2.72 (6)	3.23	1132.7
250	(5)	131 (50)	2.90 (7)	12.00	41.7	148 (78)	2.68 (7)	3.48	1054.3
250	(6)	129 (46)	2.93 (6)	10.58	48.4	149 (74)	2.67 (5)	2.93	977.4
250	(7)	130 (47)	2.92 (7)	10.60	21.1	144 (70)	2.73 (6)	2.80	1055.8
250	(8)	126 (43)	2.98 (6)	12.70	34.2	151 (75)	2.65 (5)	3.65	2051.6
250	(9)	125 (41)	2.99 (6)	11.84	247.0	148 (77)	2.68 (5)	3.09	1231.2
250	(10)	125 (37)	2.99 (6)	12.51	72.6	150 (80)	2.66 (5)	3.40	1522.6
250	(11)	126 (46)	2.98 (7)	12.03	25.9	143 (69)	2.74 (5)	3.31	1282.1
250	(12)	130 (52)	2.92 (6)	11.41	44.0	149 (78)	2.67 (6)	3.45	1088.8
250	(13)	123 (39)	3.02 (7)	12.09	335.8	151 (75)	2.65 (4)	3.29	1571.7
250	(14)	128 (49)	2.95 (7)	11.75	15.2	142 (70)	2.75 (7)	2.99	1159.8
250	(15)	125 (38)	2.99 (6)	12.09	27.3	143 (73)	2.74 (7)	3.15	1305.6
500	(1)	253 (79)	2.97 (6)	11.52	79.1	286 (134)	2.74 (6)	3.42	5479.2
500	(2)	244 (74)	3.05 (7)	12.80	84.5	285 (123)	2.75 (5)	3.51	6649.5
500	(3)	248 (86)	3.01 (7)	12.40	982.1	288 (139)	2.73 (6)	3.37	6711.0
500	(4)	266 (106)	2.88 (8)	11.26	378.4	291 (141)	2.71 (6)	3.50	5820.6
500	(5)	256 (90)	2.95 (7)	11.08	2122.2	300 (157)	2.66 (6)	2.87	4217.7
500	(6)	261 (95)	2.91 (7)	11.69	63.0	287 (133)	2.74 (5)	3.37	6400.0
500	(7)	248 (77)	3.01 (7)	11.74	76.3	290 (139)	2.72 (6)	3.38	4753.8
500	(8)	257 (99)	2.94 (7)	11.50	517.8	304 (158)	2.64 (6)	3.17	5064.7
500	(9)	259 (93)	2.93 (6)	11.15	94.1	294 (146)	2.70 (6)	3.38	5493.6
500	(10)	258 (97)	2.93 (7)	11.53	109.2	284 (134)	2.76 (6)	3.60	5339.4
500	(11)	257 (91)	2.94 (5)	11.67	1198.6	297 (156)	2.68 (5)	3.25	6488.8
500	(12)	260 (99)	2.92 (7)	11.21	197.2	302 (153)	2.65 (6)	3.21	5373.4
500	(13)	259 (93)	2.93 (6)	11.66	96.3	288 (141)	2.73 (6)	3.37	4109.3
500	(14)	247 (84)	3.02 (7)	12.02	489.0	285 (136)	2.75 (6)	3.27	7558.5
500	(15)	261 (97)	2.91 (7)	11.22	52.5	294 (147)	2.70 (7)	3.22	5674.0
1000	(1)	521 (188)	2.92 (7)	11.84	2812.0	580 (278)	2.72 (6)	3.45	21050.3
1000	(2)	496 (164)	3.01 (7)	11.43	642.0	587 (296)	2.70 (6)	3.40	23494.8
1000	(3)	520 (197)	2.92 (6)	11.16	34339.7	609 (318)	2.64 (5)	3.17	22919.3
1000	(4)	499 (184)	3.00 (8)	11.61	1199.1	572 (279)	2.75 (8)	3.30	25040.5
1000	(5)	522 (205)	2.91 (7)	11.34	35003.2	604 (305)	2.65 (6)	3.10	23291.7
1000	(6)	507 (188)	2.97 (9)	11.57	351965.7	582 (279)	2.72 (6)	3.23	27293.6
1000	(7)	523 (206)	2.91 (7)	11.33	39580.6	587 (292)	2.70 (6)	3.26	20515.5
1000	(8)	517 (186)	2.93 (7)	11.80	28715.0	586 (288)	2.70 (5)	3.42	45775.0
1000	(9)	501 (167)	2.99 (7)	12.10	587.8	576 (272)	2.73 (6)	3.37	26778.9
1000	(10)	507 (169)	2.97 (7)	11.81	48797.1	598 (305)	2.67 (6)	3.36	20962.1
1000	(11)	511 (181)	2.95 (8)	11.36	1371.6	582 (294)	2.72 (6)	3.14	23576.2
1000	(12)	492 (164)	3.03 (7)	12.71	214615.9	570 (281)	2.75 (6)	3.58	28629.5
1000	(13)	518 (181)	2.93 (8)	11.43	263943.6	591 (290)	2.69 (6)	3.19	19599.6
1000	(14)	517 (179)	2.93 (8)	11.74	540409.9	587 (287)	2.70 (6)	3.48	26581.2
1000	(15)	512 (184)	2.95 (8)	11.58	2729.8	576 (280)	2.73 (8)	3.24	23753.2

Table 7: SMT-properties, *OR-library* instances. Count: Number of FSTs in SMT (resp. number of MST-edges). Size: Average number of terminals spanned by FSTs in SMT (resp. maximum number of terminals spanned). Red: Reduction over MST in percent. CPU: Total CPU-time (seconds).

Instance	Rectilinear				Euclidean			
	Count	Size	Red	CPU	Opt	Count	Size	CPU
d198	171 (146)	2.15 (4)	3.66	5.0	102 (54)	2.93 (8)	2.91	4164.4
lin318	227 (148)	2.40 (4)	8.90	456.2	208 (129)	2.52 (5)	4.77	10799.7
fl417	237 (111)	2.76 (5)	11.94	272.1	180 (47)	3.31 (7)	3.34	81852.8
pcb442	392 (347)	2.12 (5)	3.99	8.2	-	-	-	-
att532	272 (93)	2.95 (8)	11.44	16130.3	310 (159)	2.71 (6)	3.36	9102.6
u574	371 (212)	2.54 (8)	8.94	74.0	344 (175)	2.67 (7)	3.15	5367.7
p654	584 (526)	2.12 (5)	5.89	30.4	-	-	-	-
rat783	414 (167)	2.89 (9)	12.65	255.3	448 (221)	2.75 (6)	3.52	21331.6
pr1002	688 (438)	2.45 (5)	8.63	150.7	581 (278)	2.72 (6)	3.05	15509.2
u1060	674 (392)	2.57 (8)	11.35	3819.8	599 (274)	2.77 (12)	3.25	60451.8
pcb1173	-	-	-	-	802 (522)	2.46 (10)	3.18	81889.4
d1291	1250 (1213)	2.03 (4)	1.80	48.8	-	-	-	-
rl1323	1154 (1011)	2.15 (6)	5.45	94.5	1075 (899)	2.23 (5)	1.65	8616.3
fl1400	-	-	-	-	-	-	-	-
u1432	1431 (1431)	2.00 (2)	0.00	5.2	-	-	-	-
fl1577	1189 (879)	2.33 (5)	10.59	714.0	-	-	-	-
d1655	1508 (1368)	2.10 (4)	3.57	521.6	-	-	-	-
vm1748	1320 (1009)	2.32 (8)	8.93	420.1	1258 (888)	2.39 (6)	2.81	28839.6
rl1889	1609 (1387)	2.17 (5)	5.49	14159.0	1485 (1190)	2.27 (8)	2.02	44773.9
u2152	2141 (2131)	2.00 (3)	0.22	14.0	-	-	-	-
pr2392	1712 (1163)	2.40 (7)	7.75	710.1	1490 (869)	2.60 (5)	3.61	102631.5
pcb3038	-	-	-	-	-	-	-	-
fl3795	-	-	-	-	-	-	-	-
fnl4461	-	-	-	-	-	-	-	-
rl5934	-	-	-	-	-	-	-	-
pla7397	-	-	-	-	-	-	-	-

Table 8: SMT-properties, *TSPLIB* instances. See Table 7 for captions.

n	Rectilinear				Euclidean			
	Count	Size	Opt	CPU	Count	Size	Red	CPU
100	51.7 (19.4)	2.93 (5.7)	11.30	4.0	58.2 (29.0)	2.71 (5.3)	3.16	162.8
	±3.8 ±4.6	±0.14 ±1.0	±1.03	±1.8	±3.2 ±4.2	±0.09 ±0.9	±0.43	±49.7
200	102.0 (35.8)	2.95 (6.3)	11.51	20.9	116.2 (57.7)	2.72 (5.5)	3.25	784.6
	±4.3 ±5.1	±0.08 ±0.8	±0.81	±16.2	±4.7 ±6.7	±0.07 ±0.6	±0.34	±202.3
300	153.5 (54.7)	2.95 (6.6)	11.47	81.2	177.2 (88.5)	2.69 (5.9)	3.25	1748.0
	±5.4 ±6.5	±0.07 ±1.4	±0.51	±123.6	±4.8 ±8.2	±0.05 ±0.9	±0.23	±401.2
400	205.6 (73.3)	2.94 (6.8)	11.54	172.6	234.6 (116.8)	2.70 (6.0)	3.20	3201.5
	±6.7 ±8.1	±0.06 ±1.0	±0.45	±298.3	±5.9 ±9.3	±0.04 ±0.9	±0.20	±596.6
500	256.4 (92.2)	2.95 (7.1)	11.53	517.9	293.1 (145.2)	2.70 (6.0)	3.25	5249.5
	±6.6 ±7.5	±0.05 ±1.0	±0.45	±803.9	±6.9 ±10.5	±0.04 ±0.7	±0.20	±1013.3

Table 9: SMT-properties, randomly generated instances. Averages over 50 instances for each size; standard deviations on the second line of each row. See Table 7 for captions.

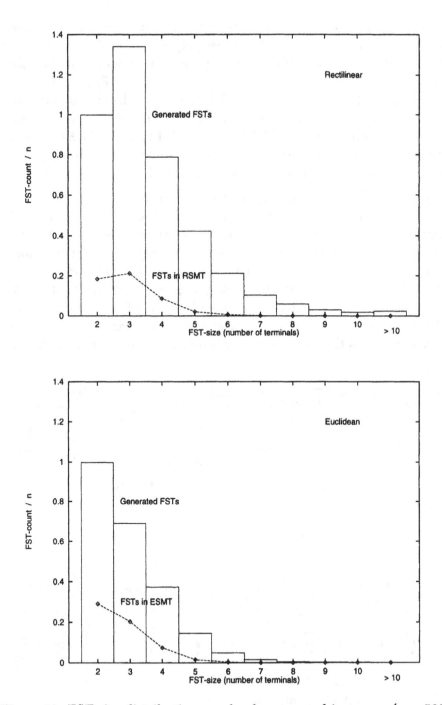

Figure 10: FST-size distribution, randomly generated instances ($n = 500$).

tree star condition is *exponential.* Zachariasen [27] noted that the rectilinear FST generator actually produced a super-polynomial number of FSTs for this series of instances. In Table 10 we present statistics on the first five instances in this series (12-52 terminals). The number of FSTs and the total CPU-time grows rapidly although the structure of the optimal solutions (number of FSTs and average size) does not differ radically from randomly generated instances. The total CPU-time needed to solve the 52 terminal instance is more than 10 minutes; this is almost 200 times the CPU-time needed to solve a randomly generated 100 terminal instance (Table 9).

As previously noted [26], regular lattices are difficult to solve as Euclidean problems using FST based exact algorithms. Polynomial time algorithms for regular lattice problems have recently been given by Brazil et al. [3, 4, 5]. In Table 11 we present data for solving regular lattices spanning 2×7, 3×7, ..., 7×7 terminals. Again the CPU-times are orders of magnitude larger than for solving randomly generated instances of similar size. In particular, it is interesting to note that the LP relaxation is much weaker for these instances and that relatively heavy branching is required. One explanation is the large number of symmetric near-optimal solutions.

	Generation				Concatenation				SMT			
n	Count	Size	Incomp	CPU	Gap	Nds	LPs	CPU	Count	Size	Opt	CPU
12	104	5.44	91.24	0.4	0.000	1	7	0.1	4	3.75	11.90	0.5
22	238	7.87	92.01	3.9	0.000	1	12	0.9	8	3.62	11.47	4.8
32	453	11.16	94.38	21.1	0.000	1	109	10.1	12	3.58	11.43	31.2
42	823	15.46	96.19	100.6	0.000	1	45	30.2	16	3.56	11.42	130.8
52	1509	20.50	97.46	450.5	0.000	1	68	87.5	20	3.55	11.42	538.0

Table 10: Difficult rectilinear instances. See text for instance descriptions and Tables 1, 4 and 7 for captions.

	Generation				Concatenation				SMT			
n	Count	Size	Incomp	CPU	Gap	Nds	LPs	CPU	Count	Size	Opt	CPU
14	96	5.50	90.30	52.2	0.000	1	1	0.1	1	14.00	7.13	52.3
21	216	5.67	79.93	236.6	0.000	1	1	0.1	8	3.50	8.04	236.7
28	376	5.83	76.89	679.3	0.071	7	11	0.7	9	4.00	8.34	680.0
35	565	6.04	75.14	1697.8	0.230	26	72	7.6	12	3.83	8.20	1705.4
42	798	6.42	75.06	4049.6	0.094	25	126	18.5	15	3.73	8.50	4068.0
49	1057	6.76	75.24	8747.7	0.203	92	383	144.8	18	3.67	8.37	8892.4

Table 11: Difficult Euclidean instances. See text for instance descriptions and Tables 1, 4 and 7 for captions.

5 Conclusion

This paper presented an extensive computational study on exact algorithms for the Euclidean and rectilinear Steiner tree problems in the plane. Optimal solutions to problem instances with more than 2000 terminals have been obtained. The branch-and-cut FST concatenation algorithm is orders of magnitude faster than algorithms based on backtrack search or dynamic programming. However, for the rectilinear problem, the concatenation phase is still the bottleneck.

The Euclidean FST generator is remarkably effective on (uniformly) randomly generated problem instances. For structured (real-world) instances, such as those in *TSPLIB*, the generator is less efficient. One reason is simply that fewer FSTs can be pruned away by using current tests. In order to solve larger, structured Euclidean problems, the Euclidean FST generator must be improved. Often such instances contain terminal subsets which can be mapped into other subsets by translation and rotation (and sometimes scaling). If such congruent subsets could be identified, FSTs generated for one subset could essentially be copied to all congruent subsets. However, whether this is a practical option for improving the FST generator is still an open question.

The FST concatenator can, on the other hand, most likely be improved. We are currently investigating three options: Firstly, FST pruning can be applied to the list of generated FSTs. If faster and perhaps even more powerful tests than those already suggested [9, 26] can be applied, it may be worthwhile to reduce the list of FSTs before entering the concatenation phase. But as noted in Section 4.3 the effect of reducing the FST list need not have a positive effect on the concatenation.

Another option is early branching, i.e., branching in the root node before an LP optimal solution has been obtained. The problems that require the most time to concatenate spend the vast majority of their time generating constraints that improve the lower bound only minutely before LP-optimum is reached. Orders of magnitude reduction in total concatenation time are possible when good heuristics are used to detect such *tailing off* in the convergence rate and choosing a good branching variable instead.

Thirdly, other types of valid inequalities may be added to the formulation. For other well-studied problems (e.g., TSP) the inclusion of new types of valid inequalities has formed the basis for substantially improved exact algorithms.

FST concatenation can be avoided altogether by enumerating all so-called full topologies for Z (see, e.g., [15]). A shortest tree for a given full topology or any of its degenerate topologies can be determined in $O(n^2)$ time using the *luminary algorithm* of Hwang and Weng [16] or by using a numerical algorithm suggested by Smith [21]. Since the number of full topologies is superexponential in the number of terminals, these methods can only be effective if the number of full topologies can be reduced by using pruning tests similar to those applied in this paper.

References

[1] J. E. Beasley. OR-Library: Distributing Test Problems by Electronic Mail. *Journal of the Operational Research Society*, 41:1069–1072, 1990.

[2] M. W. Bern and R. L. Graham. The Shortest-Network Problem. *Scientific American*, pages 66–71, January 1989.

[3] M. Brazil, T. Cole, J. H. Rubinstein, D. A. Thomas, J. F. Weng, and N. C. Wormald. Minimal Steiner Trees for 2^k x 2^k Square Lattices. *Journal of Combinatorial Theory, Series A*, 73:91–110, 1996.

[4] M. Brazil, J. H. Rubinstein, D. A. Thomas, J. F. Weng, and N. C. Wormald. Full Minimal Steiner Trees on Lattice Sets. *Journal of Combinatorial Theory, Series A*, 78:51–91, 1997.

[5] M. Brazil, J. H. Rubinstein, D. A. Thomas, J. F. Weng, and N. C. Wormald. Minimal Steiner Trees for Rectangular Arrays of Lattice Points. *Journal of Combinatorial Theory, Series A*, 79:181–208, 1997.

[6] E. J. Cockayne and D. E. Hewgill. Exact Computation of Steiner Minimal Trees in the Plane. *Information Processing Letters*, 22:151–156, 1986.

[7] E. J. Cockayne and D. E. Hewgill. Improved Computation of Plane Steiner Minimal Trees. *Algorithmica*, 7(2/3):219–229, 1992.

[8] S. E. Dreyfus and R. A. Wagner. The Steiner Problem in Graphs. *Networks*, 1:195–207, 1971.

[9] U. Fößmeier and M. Kaufmann. On Exact Solutions for the Rectilinear Steiner Tree Problem. Technical Report WSI-96-09, Universität Tübingen, 1996.

[10] J. L. Ganley and J. P. Cohoon. Improved Computation of Optimal Rectilinear Steiner Minimal Trees. *International Journal of Computational Geometry and Applications*, 7(5):457–472, 1997.

[11] M. R. Garey, R. L. Graham, and D. S. Johnson. The Complexity of Computing Steiner Minimal Trees. *SIAM Journal on Applied Mathematics*, 32(4):835–859, 1977.

[12] M. R. Garey and D. S. Johnson. The Rectilinear Steiner Tree Problem is *NP*-Complete. *SIAM Journal on Applied Mathematics*, 32(4):826–834, 1977.

[13] M. Hanan. On Steiner's Problem with Rectilinear Distance. *SIAM Journal on Applied Mathematics*, 14(2):255–265, 1966.

[14] F. K. Hwang. On Steiner Minimal Trees with Rectilinear Distance. *SIAM Journal on Applied Mathematics*, 30:104–114, 1976.

[15] F. K. Hwang, D. S. Richards, and P. Winter. *The Steiner Tree Problem*. Annals of Discrete Mathematics 53. Elsevier Science Publishers, Netherlands, 1992.

[16] F. K. Hwang and J. F. Weng. The Shortest Network under a Given Topology. *Journal of Algorithms*, 13:468–488, 1992.

[17] K. Mehlhorn and S. Näher. LEDA - A Platform for Combinatorial and Geometric Computing. Max Planck Institute for Computer Science http://www.mpi-sb.mpg.de/LEDA/leda.html, 1996.

[18] F.P. Preparata and M. I. Shamos. *Computational Geometry: An Introduction*. Springer-Verlag, New York, second edition, 1988.

[19] G. Reinelt. TSPLIB - A Traveling Salesman Problem Library. *ORSA Journal on Computing*, 3(4):376–384, 1991.

[20] J. S. Salowe and D. M. Warme. Thirty-Five-Point Rectilinear Steiner Minimal Trees in a Day. *Networks*, 25(2):69–87, 1995.

[21] W. D. Smith. How to Find Steiner Minimal Trees in Euclidean *d*-Space. *Algorithmica*, 7(2/3):137–177, 1992.

[22] I. Tomescu and M. Zimand. Minimum Spanning Hypertrees. *Discrete Applied Mathematics*, 54:67–76, 1994.

[23] D. M. Warme. Practical Exact Algorithms for Geometric Steiner Problems. Technical report, System Simulation Solutions, Inc., Alexandria, VA 22314, USA, 1996.

[24] D. M. Warme. A New Exact Algorithm for Rectilinear Steiner Minimal Trees. Technical report, System Simulation Solutions, Inc., Alexandria, VA 22314, USA, 1997.

[25] P. Winter. An Algorithm for the Steiner Problem in the Euclidean Plane. *Networks*, 15:323–345, 1985.

[26] P. Winter and M. Zachariasen. Euclidean Steiner Minimum Trees: An Improved Exact Algorithm. *Networks*, 30:149–166, 1997.

[27] M. Zachariasen. Rectilinear Full Steiner Tree Generation. *Networks*, to appear.

On Approximation of the Power-p and Bottleneck Steiner Trees

Piotr Berman [1]
Department of Computer Science and Engineering,
Penn State University, University Park, PA 16802.
E-mail: berman@cse.psu.edu.

Alexander Zelikovsky [2]
Department of Computer Science,
Georgia State University, University Plaza,
Atlanta, GA 30303-3083,
E-mail: alexz@cs.ucla.edu.

Abstract

Many VLSI routing applications, as well as the facility location problem involve computation of Steiner trees with non-linear cost measures. We consider two most frequent versions of this problem. In the power-p Steiner problem the cost is defined as the sum of the edge lengths where each length is raised to the power $p > 1$. In the bottleneck Steiner problem the objective cost is the maximum of the edge lengths. We show that the power-p Steiner problem is MAX SNP-hard and that one cannot guarantee to find a bottleneck Steiner tree within a factor less than 2, unless P = NP. We prove that in any metric space the minimum spanning tree is at most a constant times worse than the optimal power-p Steiner tree. In particular, for $p = 2$, we show that the minimum spanning tree is at most 23.3 times worse than the optimum and we construct an instance for which it is 17.2 times worse. We also present a better approximation algorithm for the bottleneck Steiner problem with performance guarantee $\log_2 n$, where n is the number of terminals (the minimum spanning tree can be $2\log_2 n$ times worse than the optimum).

[1] Research partially supported by NSF grant CCR-9700053.
[2] Research partially supported by Volkswagen Stiftung and Packard Foundation.

117

D.-Z. Du et al. (eds.), Advances in Steiner Trees, 117-135.
© 2000 *Kluwer Academic Publishers.*

Contents

1   Introduction

A bottleneck Steiner tree (or a min-max Steiner tree) is a Steiner tree (i.e. a tree spanning a distinguished point set) in which the maximum edge weight is minimized. A power-*p* Steiner tree is a Steiner tree in which the sum of edge weights raised in power $p > 1$ is minimized. Several multifacility location and VLSI routing problems ask for bottleneck and power-*p* Steiner trees.

Consider the problem of choosing locations for a number of hospitals serving homes where the goal is either to minimize maximum weighted distance to any home from the hospital that serves it and between hospitals or to minimize the cost of constructing corresponding roads which may nonlinearly depend on distances [16, 5, 11]. The solution is a tree which spans all hospitals and connects each home to the closest hospital. This tree can be seen as a Steiner tree where the homes are terminals and hospitals are Steiner points. Unlike the classical Steiner tree problem where the total length of Steiner tree is minimized, in this problem it is necessary to minimize either maximum edge weight or sum of weights raised in a certain power $p > 1$.

The other instance of the bottleneck and power-*p* Steiner tree problems occurs in electronic physical design automation where nets are routed subject to delay minimization [3, 15, 10, 2]. The terminals of a net are interconnected possibly through intermediate nodes (Steiner points) and for electrical reasons one would like to minimize maximum distance between each pair of interconnected points or to minimize Elmore delay which is a quadratic function of length.

The most popular versions of the bottleneck and power-p Steiner tree problems in the literature are geometric. Note that if the number of Steiner points is not bounded, then any edge can be subdivided into infinitely small segments and the resulting maximum edge length and the sum of edge weights raised in power $p > 1$ become zero. While ordinarily we should be most happy if we can solve a problem with cost 0, one can clearly see that in our examples these solutions make no sense. The reason is that there exists hard limits on the number of new nodes which can be introduced. For example, in electronic physical design the delay grows quadratically between the points where the signal is amplified, and new points will decrease the delay only if they contain amplifiers (transistors). However, each amplifier consumes energy, and there is a hard limit on the amount of energy that can be dissipated. This complication can be tackled in several ways. One is to introduce the allowed number of amplifiers as a new parameter to Steiner-tree like problem. This approach can be indeed precise, but difficult to deal with algorithmically, in particular, we would need to use a much more complicated estimate of Elmore delay. The other approach is to restrict the positions of amplifiers topologically, so their number would be always smaller than the number of terminals (hence, only on the branching point), and the Elmore delay is easy to estimate (thus, on every branching point). This gives rise to the problem formulation suggested by Ganley and Salowe [7, 8].

Problem 1. Given a set of n points in the plane (called *terminals*), find a bottleneck (power-p) Steiner tree spanning all terminals such that the degree of any Steiner point is at least 3.

Another approach is to require each branching to have an amplifier (so Elmore delay is a function of length), limit the edge length and to minimize the number of amplifiers (which now may be placed on nodes of degree 2). The resulting formulation has been proved to be NP-hard by Sarrafzadeh and Wong [15].

Problem 2. Given a set of n terminals in the plane and $\lambda > 0$, find a Steiner tree spanning n terminals with the minimum number of Steiner points such that every edge is not longer than λ.

In this paper we generalize the bottleneck and power-p Steiner problems to arbitrary metric spaces and weighted graphs. The first such formulation is suggested in [4].

Problem 3. Given a graph $G = (V, E, d)$ with nonnegative weight d on edges, and a set of terminals $S \subset V$, find a Steiner tree spanning S with the smallest maximum edge weight.

Problem 3 can be solved efficiently in the optimal time $O(|E|)$ [4]. Unfortunately, the above formulation does not bound the number of Steiner points and therefore does not capture the practical flavor of the problems 1 and 2. Note that the classical Steiner tree problem in an arbitrary discrete metric space and in weighted graphs are equivalent because any edge can be replaced with a shortest path between its endpoints without affecting the sum of weights objective. This is not valid for the bottleneck or power-p objectives and it is necessary to be more careful in generalizing these two problems to graphs. We first generalize problem 1 to an arbitrary discrete metric in the following way.

Let (V, d) be a metric space on a finite set of *points* V with a nonnegative distance function $d : V^2 \rightarrow R^+$ and let $S \subset V$ be a set of *terminals*. An arbitrary tree T in (V, d) which contains S and in which any point from $V \setminus S$ (a *Steiner point*) has degree at least 3 is called a *Steiner tree* of S. Note that the restriction on the degree of a Steiner point bounds the number of Steiner points.

Bottleneck Steiner Problem. Given a metric space (V, d) and a subset $S \subset V$. Find a Steiner tree in which the maximum distance between adjacent points is minimized.

Power-p Steiner Problem. Given a metric space (V, d) and a subset $S \subset V$. Find a Steiner tree in which the sum of distances between adjacent points each raised to the power p is minimized.

The above formulations imply the graph-theoretical generalization of Problem 1 considered in [7, 8]. The metric space (V, d) is considered to be a metric closure of a graph with edge costs over the vertex set V, i.e., the distance between any pair of vertices is the cost of a shortest path between these vertices. We also assume that edge costs already satisfy the triangle inequality. In this paper we consider the following formulation.

Given a graph $G = (V, E, d)$ with costs d on edges and a set of terminals $S \subset V$, find a Steiner tree spanning S in the metric closure (V, d) of G in which each Steiner point has degree at least 3 such that the maximum edge cost (Bottleneck Steiner Problem) or the sum of edge costs each raised to the power p (Power-p Steiner Problem) is minimized.

In the next section, we show that the both problems are hard to approximate. We first prove that the Power-p Steiner Tree Problem is MAX SNP-hard. We next show that the Bottleneck Steiner Tree Problem cannot be solved by any polynomial-time algorithm with the performance ratio less than 2 unless P=NP.

Note that the NP-hardness of the Power-p Steiner Tree Problem is nontrivial. Indeed, the classical NP-hard Steiner Tree Problem in graphs ($p = 1$) cannot be embedded in the Power-p Steiner Tree Problem with all edge weights raised in power $p > 1$ because the triangle inequality becomes invalid.[3]

A well-known approximation algorithm for Steiner problems, the so called *MST-heuristic*, finds the minimum spanning tree of the terminal set S in G. Similarly to the classical Steiner tree problem, if no Steiner points are allowed (i.e. if the Steiner tree may connect only terminals from S), the MST-heuristic outputs the optimal solution for the both problems: the Bottleneck and the Power-p Steiner Tree Problem [7, 8]. Our next goal is to find the performance ratio [4] of the MST-heuristic which is called a Steiner ratio. We consider the *bottleneck* and the *power-p Steiner ratios* $\rho_B(n)$ and ρ_p, respectively. The bottleneck Steiner ratio is also defined as the supremum over all instances with n terminals of the ratio of the maximum edge weight of the minimum spanning tree for S over the maximum edge weight of the bottleneck Steiner tree.

It was conjectured in [6] that the power-p Steiner ratio ρ_p is finite. In Section 3 we prove that this conjecture is true, i.e. that the power-p Steiner ratio ρ_p is constant for any p. In particular, for $p = 2$, we show that the minimum spanning tree is at most 23.3 times worse than the optimum. But we do not think that this bound is tight since the largest Steiner ratio for the Power-2 Steiner Tree Problem found so far equals 17.2.

It has been proved that $\rho_B(n) = 2\lfloor \log_2 n \rfloor - \delta$, where δ is either 0 or 1 depending on whether mantissa of $\log_2 n$ is greater than $\log_2 3/2$ [7]. This means that he minimum spanning tree heuristic for the bottleneck Steiner tree problem has approximation ratio $2 \log_2 |S|$. In Section 4 we present an approximation algorithm with the better performance ratio of $\log_2 |S|$.

---

[3]Indeed, e.g., $1 + 1 \geq 2$ but $1^p + 1^p < 2^p$ for any $p > 1$. Such a modification of edge weights also cannot help for approximating power-p Steiner trees, since all approximations of classical Steiner trees heavily rely on the triangle inequality for the shortest-path weight.

[4]Performance ratio of an approximation algorithm is an upper bound on the ratio of the output cost of this algorithm over the optimum cost.

2 Approximation Complexity of Power-p and Bottleneck Steiner Problems

In this section we construct reduction from the maximum independent set problem in 3-regular graphs and the exact cover by 3-sets to the power-p and bottleneck Steiner problems, respectively. The both reductions use similar gadgets (see Figure 1).

Theorem 1 *The power-p Steiner Problem is MAX SNP hard.*

Proof. We will construct an approximation preserving reduction from the problem of finding a maximum independent set in 3-regular graphs (see [14]). Let $G = (V, E)$ be a 3-regular graph. We create an instance of the power-p Steiner problem consisting of a graph $G' = (V', E', d)$ and a terminal set $S \subseteq V'$ as follows:

(1) $S = E \cup \{\theta\}$, where θ is the special, "central" node.

(2) For each node $v \in V$ and its three incident edges e_i, e_j, e_k we insert two nodes to V', namely v_1, v_2 and connect six nodes of V', e_i, e_j, e_k, v_1, v_2 and θ with five length 1 edges, as shown in Figure 1(a).

(3) For each pair of edges $e_i, e_j \in E$ we insert node e_{ij} to V' that is connected with them with length 1 edges, in turn, the new node is connected with θ with length 2 edge (see Figure 1(b)),

   Note that any 5-tuple of edges (2) allows to attach e_i, e_j and e_l to θ with cost 5. Any triple of edges (3) allows to attach e_i and e_j to θ with cost $2 + 2^p = 2\alpha$, where $2 < \alpha < 2^p$.

   To show that this is an approximation preserving reducibility, we need to describe how to translate a Steiner trees T in G' into an independent set in G.

   Assume that G has $2m$ nodes and $3m$ edges. Tree T consists of a Steiner components with 4 leaves, b with 3 leaves and c with 2 leaves, where $3a + 2b + c = 2m - 1$. The only Steiner components with 4 leaves are described in point (2) of our construction; each of them corresponds to a node in V and together they correspond to an independent set. Moreover, each of them has cost 5. One can easily check that every possible component with 3 leaves consists of two length 1 edges and one length 2 edge, so its power-p cost is 2α. Now, suppose that there exist at least 2 components with 2 leaves. One

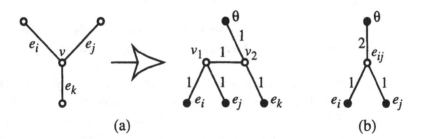

(a)   (b)

Figure 1: Reduction from the independent set problem for 3-regular graphs
to pSTP. The terminals are denoted by filled circles. A node with 3 adjacent
edges is transformed into a 5-tuple (a). Any pair of edges is transformed
into a 3-star (b)

can show that (i) they cost at least 2×2^p; and (ii) they can be replaced
with one component wit h3 leaves, of the kind described in point (3) of
the construction. (After removing these two components, T splits into 3
connected components, one of them contains θ, from the other two we can
select e_i and e_j respectively.) Thus we can assume that $c < 2$, and so (b, c)
is equal to $(2m-1-3a, 0)$ or $(2m-2-3a, 1)$. If we neglect a small constant
additive term, we can express the cost of T as follows

$$5a + (2m - 3a)\alpha = 2m\alpha - (3\alpha - 5)a. \tag{1}$$

Observe that $3\alpha - 5 > 1$. This linear relationship between the cost of T
and a, the size of the corresponding independent set in G, shows that our
reduction preserves the approximability.   □

Theorem 2 *The Bottleneck Steiner Tree Problem cannot be solved by any
polynomial-time algorithm with the performance ratio less than 2 unless P
equals NP.*

Proof. We are going to translate X3C (exact cover by 3-sets, see [12]) into
the Bottleneck Steiner Tree Problem. An instance of X3C is a set with
$3m$ elements, and a collection of 3-element subsets. The task is to find a
collection of m disjoint subsets, i.e. to cover all the elements with a disjoint
collection of subsets. Given an instance with set of elements V and set
of triples E, we create an instance of the Bottleneck Steiner Steiner Tree
Problem consisting of a graph $G' = (V', E', d)$ and a terminal set $S \subseteq V'$
similarly to the previous reduction (see Figure 1(a)).

(1) $S = V \cup \{\theta\}$, where θ is an auxiliary terminal.

(2) For each triple $t = \{u_1, u_2, u_3\}$ we create nodes t_1 and t_2 and a 5-tuple of length 1 edges: $\{u_1, t_1\}, \{u_2, t_1\}, \{t_1, t_2\}, \{u_3, t_2\}$ and $\{t_2, \theta\}$.

It is easy to see that the bottleneck cost is 1 if and only if we can create a Steiner tree that consists of complete 5-tuples corresponding to 3-element subsets of X3C only, i.e. if and only if there exists an exact 3-cover. Otherwise, the bottleneck cost is at least 2. Thus if the Bottleneck Steiner Tree Problem can be approximated with a factor strictly less than 2, then X3C can be solved exactly.   ☐

3   The Power-p Steiner Ratio

In this section we show that the performance ratio of the MST-heuristic is constant bounded.

Theorem 3 *The power-p Steiner ratio ρ_p is finite for any p.*

Proof. We may assume that an instance I of a power-p Steiner problem is defined by a graph (V, E), a length function of edges d and a set of terminals $S \subset V$. In such a case we assume that the distance between two nodes is the length of the respective shortest path. Let $Mst(I)$ and $St(I)$ be the power-p cost of the minimum spanning tree of S and the minimum Steiner tree respectively. For each instance I we also define $ratio(I) = Mst(I)/St(I)$. Our goal is to show how to find the power-p Steiner ratio, i.e. a value ρ_p such that $ratio(I) \leq \rho_p$ for every I. Our strategy is to show that for every instance I there exists an instance I' such that $ratio(I') \geq ratio(I)$, while the form of I' is so special that we can find succinct expressions for $St(I')$ and $Mst(I')$ and compare them using elementary calculus.

Consider now an instance I with minimum Steiner tree T. Our first modification of I is to replace the original graph with the tree T. Obviously, $St(I)$ is unchanged, but $Mst(I)$ could only increase, because for each pair of nodes, say u nd v, the current shortest path (which gives the distance in the modified metric) existed in the old graph, which in turn could contain yet shorter paths; thus the new metric space, induced by T, cannot have smaller distances than the metric space induced by the old graph.

Similarly, if in a Steiner tree T one of the terminals u is an internal node, we may divide T into two parts, T_1 and T_2, that share node u.

Clearly, $St(I(T_1)) + St(I(T_2)) = St(I(T))$, while $Mst(I(T_1)) + Mst(I(T_2))$ $\geq Mst(I(T))$. Therefore we may consider only the trees where the set of terminals and leaves are equal.

We may put further restrictions on the form of T. Replacing a node of degree higher than 3 with a group of nodes connected by 0-length edges changes neither $Mst(I(T))$, nor $St(I(T))$. Similarly, replace a node with a chain of three nodes connected by 0-cost edges. Thus we may assume that all internal nodes of T have degree 3, with the exception of the "root node" r that has degree 2, i.e. that T is a rooted binary tree. We can also replace a leaf with a group of nodes connected by 0-length edges: $Mst(I(T))$ clearly does not change, while $St(I(T))$ can only decrease. Thus we may assume that T is a full binary tree. We say that the *level i* of T consists of nodes that are i edges away from the leaves, and edges that join nodes on levels i and $i - 1$. The final simplification of T is done by the following lemma.

Lemma 1 *For every full binary tree T with function $d(e)$ defining the length of edges, there exists another length function d' such that on every level all edges have the same length, $St(I(T))$ remains unchanged and $Mst(I(T))$ does not decrease.*

Proof. We use the following strategy. We will first define a certain weighted tree $M(T)$ which spans all leaves of T. In our transformations, $St(I(T))$ will be fixed, the cost of $M(T)$ will be nondecreasing, and at the end, $M(T)$ will coincide with $Mst(I(T))$.

Let r be the root of T. For any vertex $u \in T$, let $T(u)$ be the subtree of T with root u. Let $leaf(u)$ be a leaf in $T(u)$ which is closest to u and let $\pi(u)$ be the path from u to $leaf(u)$. Note that for any leaf l, the path $\pi(l)$ is empty. We define $e(u)$ to be an edge between two leaves $leaf(u_1)$ and $leaf(u_2)$, where u_1 and u_2 are the children of u. The cost of $e(u)$ is defined to be sum of costs of two paths $\pi(u_1)$, $\pi(u_2)$ and two edges (u, u_1), (u, u_2). The graph $M(T)$ consists of all edges $e(u)$, $u \in T$ (see Figure 2). The graph $M(T)$ is a tree since it does not contain cycles and the number of edges equals the number of leaves minus one.

After $k - 1$ transformations, the edges on the same level are the same, provided that they are in the same $T(u)$ for some u on level $k - 1$. This claim is trivially true after 0 transformations.

Now we will describe our k-th transformation (see Figure 3). We perform it separately for every node on level k. Let v_k be such a node, with children v_{k-1} and v'_{k-1}, and let $x_k = d(v_k, v_{k-1})$, $x'_k = d(v_k, v'_{k-1})$. Because of the

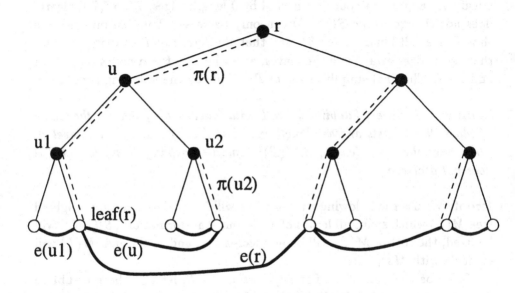

Figure 2: The tree T (solid edges) and the tree $M(T)$ (thick edges) spanning the leaves (empty circles) of the tree T. The paths $\pi(u)$ from vertices of the tree T to the closest leaf $leaf(u)$ are dotted.

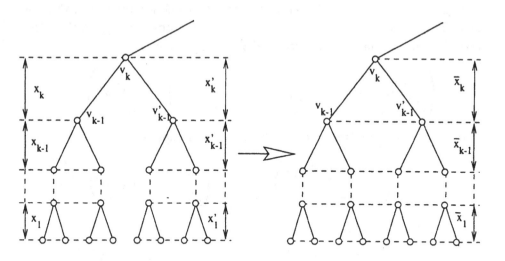

Figure 3: After the k-th transformation, all edges on the same level i, $i \leq k$, have the same length.

previous $k - 1$ transformations, for $1 \leq i < k$, all edges on i-th level of $T(v_{k-1})$ (and $T(v'_{k-1})$) have the same length, x_i (respectively, x'_i).

The transformation replaces the lengths of edges on level i of $T(v_k)$ with

$$\bar{x}_i = \left(\frac{(x_i)^p + (x'_i)^p}{2} \right)^{\frac{1}{p}}$$

It is obvious that this change of edge length does not change $St(T)$. We only need to show the cost of $M(T)$ cannot decrease. Let l be the the length of $\pi(v_{k-1})$ plus x_k, and l' be the the length of $\pi(v'_{k-1})$ plus x'_k. After the transformation, both these lengths become equal to $\bar{l}$. Suppose $l \leq l'$. For an ancestor u of v_k, the length of edge $e(u)$ cannot decrease, because in the respective path we replace l with the greater value $\bar{l}$:

$$l \leq \frac{1}{2} \left(\sum_{i=1}^{k} x_i + \sum_{i=1}^{k} x'_i \right) = \sum_{i=1}^{k} \frac{x_i + x'_i}{2} \leq \sum_{i=1}^{k} \left(\frac{(x_i)^p + (x'_i)^p}{2} \right)^{\frac{1}{p}} = \sum_{i=1}^{k} \bar{x}_i \leq l'$$

The central inequality is a well-known inequality on power-p means (see e.g. [9], page 26, inequality 2.9.1).

Now it remains to show that the sum of costs of $e(u)$ increases for the descendants of v_k. These descendants can be paired, a node u on i-th level from $T(v_{k-1})$ with a node u' on the same level from $T(v'_{k-1})$. Before the transformation, the sum of costs of $e(u)$ and $e(u')$ equals

$$\left(\sum_{j=1}^{i} 2x_j\right)^p + \left(\sum_{j=1}^{i} 2x'_j\right)^p = 2^p\left(\left(\sum_{j=1}^{i} x_j\right)^p + \left(\sum_{j=1}^{i} x'_j\right)^p\right)$$

After the transformation, the corresponding sum equals

$$2\left(\sum_{j=1}^{i} 2\left(\frac{(x_j)^p + (x'_j)^p}{2}\right)^{\frac{1}{p}}\right)^p = 2^p\left(\sum_{j=1}^{i} \left((x_j)^p + (x'_j)^p\right)^{\frac{1}{p}}\right)^p$$

The necessary inequality

$$\left(\sum_{j=1}^{i} x_j\right)^p + \left(\sum_{j=1}^{i} x'_j\right)^p \leq \left(\sum_{j=1}^{i} \left((x_j)^p + (x'_j)^p\right)^{\frac{1}{p}}\right)^p$$

follows from Minkowski's inequality (see e.g. [9], page 32, inequality 2.12.1) after substitution: $(x_j)^p = a_j$, $(x'_j)^p = b_j$ and $\frac{1}{p} = r$. $\square$

According to the last lemma, a tree of the kind that we need to consider has all edges on level i equal to some x_i. Thus a k-level tree T is fully defined by the vector $\vec{x} = (x_1, \ldots, x_k)$. Then $St(T) \leq 2\sum_{i=1}^{k} 2^{k-i} x_i^p = 2S(\vec{x})$, while

$$Mst(T) = \sum_{i=1}^{k} 2^{k-i}\left(2\sum_{j=1}^{i} x_j\right)^p = 2^p\sum_{i=1}^{k} 2^{k-i}\left(\sum_{j=1}^{i} x_j\right)^p = 2^p M(\vec{x})$$

. Thus $ratio(T) \leq 2^{p-1} M(\vec{x})/S(\vec{x})$. Our remaining task is to find an upper bound for $M(\vec{x})/S(\vec{x})$.

To finish the proof, we will show that for suitably chosen α and β the ratio $M(\vec{x})/S(\vec{x})$ is bounded by $\alpha/(1-\beta)$. This appears to be difficult, because the quantity in question is a function of k variables. However, the next lemma reduces this inequality to another one that involves two variables only.

Lemma 2 *Assume that $\alpha > 1$, $1 > \beta > 0$, and*

$$\alpha x^p + 2\beta y^p \geq (x+y)^p \quad \text{for every } x, y \geq 0 \tag{2}$$

Then $M(\vec{x}) \leq \alpha/(1-\beta)S(\vec{x})$ for every vector $\vec{x}$ with nonnegative entries.

Proof. We can rewrite $M(\vec{x})$ as $\sum_{i=1}^{k} \tau_i$, where $\tau_i = 2^{k-i} \left(\sum_{j=1}^{i} x_j \right)^p$; similarly $S(\vec{x}) = \sum_{i=1}^{k} y_i$, where $y_i = 2^{k-i} x_i^p$. Then we can show by induction on i that

$$\tau_i \leq \alpha \sum_{j=1}^{i} \beta^{i-j} y_j = \sigma_i.$$

For $i = 1$, this claim states that $y_i = 2^{k-1} x_1^p < \alpha 2^{k-1} x_1^p$. Now, inductively assume that $\tau_{i-1} \leq \sigma_{i-1}$. Then

$$
\begin{aligned}
\tau_i &= 2^{k-i} \left(y_i + \sum_{j=1}^{i-1} x_j \right)^p \\
&\leq \alpha 2^{k-i} y_i^p + 2\beta 2^{k-i} \left(\sum_{j=1}^{i-1} x_j \right)^p \\
&= \alpha y_i + \beta \tau_{i-1} \\
&\leq \alpha y_i + \beta \sigma_{i-1} \\
&\leq \alpha y_i + \alpha \sum_{j=1}^{i-1} \beta^{1+i-1-j} y_j \\
&= \sigma_i
\end{aligned}
$$

It remains to estimate $\sum_{i=1}^{k} \sigma_i$. Note that this sum is a linear combination of y_js. In particular, y_j appears in σ_j, σ_{j+1}, σ_{j+2}, etc., with coefficients that are $\alpha\beta^0$, $\alpha\beta^1$, $\alpha\beta^2$ etc. Obviously, these coefficients add to less than $\sum_{l=0}^{\infty} \alpha\beta^l = \alpha/(1-\beta)$. Therefore

$$M(\vec{x}) \leq \frac{\alpha}{1-\beta} \cdot \sum_{i=1}^{k} y_i = \frac{\alpha}{1-\beta} S(\vec{x})$$

$\square$

Now it remains to show how to find α and β that satisfy inequality (2). As a preliminary step, we eliminate one of the variables: we can divide both sides by $(x + y)^p$, and substitute $x \leftarrow x/(x + y)$. The range of the new variable is between 0 and 1. Then the existence of the desired α and β follows quickly from the following

Lemma 3 *Assume that δ and x are between 0 and 1 and $p > 1$. Then*

$$\frac{x^p}{\delta^{p-1}} + \frac{(1-x)^p}{(1-\delta)^{p-1}} \geq 1$$

Proof. We will find a necessary and sufficient condition for γ so that

$$\frac{x^p}{\delta^{(p-1)}} + \frac{(1-x)^p}{\gamma^{(p-1)}} \geq 1 \text{ for every } x \in (0,1) \tag{3}$$

Let $f(x)$ be the left hand side of (3), then the derivative of f is

$$f'(x) = p\left(\frac{x^{p-1}}{\delta^{(p-1)}} - \frac{(1-x)^{p-1}}{\gamma^{(p-1)}}\right) = p\left(\left(\frac{x}{\delta}\right)^{p-1} - \left(\frac{1-x}{\gamma}\right)^{p-1}\right).$$

It is easy to see that the minimum of f satisfies

$$f'(x) = 0 \iff \left(\frac{x}{\delta}\right)^{p-1} = \left(\frac{1-x}{\gamma}\right)^{p-1} \iff \frac{x}{\delta} = \frac{1-x}{\gamma}.$$

The last equality has a unique solution $x_0 = \delta/(\delta + \gamma)$ (consequently $1 - x_0 = \gamma/(\delta+\gamma)$). Because x_0 is the minimum, it is sufficient and necessary that $f(x_0) \geq 1$:

$$
\begin{aligned}
f(x_0) &= \frac{x^p}{\delta^{p-1}} + \frac{(1-x)^p}{\gamma^{p-1}} \\
&= \frac{1}{\delta^{p-1}}\frac{\delta^p}{(\delta+\gamma)^p} + \frac{1}{\delta^{p-1}}\frac{\gamma^p}{(\delta+\gamma)^p} \\
&= \frac{\delta}{(\delta+\gamma)^p} + \frac{\gamma}{(\delta+\gamma)^p} \\
&= \frac{1}{(\delta+\gamma)^{p-1}} \geq 1 \\
&\iff (\delta+\gamma)^{p-1} \leq 1 \iff \delta+\gamma \leq 1
\end{aligned}
$$

The last inequality shows that $1 - \delta$ is the maximum value of γ such that condition 3 holds. □

   This concludes the proof of our theorem. □

   To illustrate how this technique allows to estimate the power-p Steiner ratio with a concrete constant, we will review the case of $p = 2$. We know that this ratio is bounded by $2^{p-1}M(\vec{x})/S(\vec{x})$, which in turn, by lemma 2 is bounded by $2\alpha/(1 - \beta)$. By Lemma 3, we can pick α and β by choosing some δ between 0 and 1 and setting $\alpha = (1 - \delta)^{-p+1}$ and $\beta = \delta^{-p+1}/2$; in our case $\alpha = 1/(1 - \delta)$ and $\beta = 1/2\delta$. The resulting ratio is

$$f(\delta) = 2\frac{1}{1-\delta} \cdot \frac{1}{1-1/2\delta}$$

It is easy to check (0 of the derivative etc.) that $f(\sqrt{1/2}) = 8\sqrt{2}+12 \approx 23.314$ yields the minimum value. Thus, we proved the following

Corollary 1 *The power-2 Steiner ratio is at most* $8\sqrt{2} + 12 \approx 23.314$.

This upper bound need not be exact. In the worst case that we have found the edge length x_i form an arithmetical progression with the remainder 1 and the power-2 Steiner ratio is ≈ 17.2.

4 A Better Heuristic for the Bottleneck Steiner Problem

It was proved in [7] that the MST-heuristic for the Bottleneck Steiner Problem has an approximation ratio at most $2\log_2 n$, where $n = |S|$ is the number of terminals. Moreover, it was shown that this bound is tight, i.e. the minimum spanning tree can have an edge as long as $2\log_2 n$ times the length of the longest edge in the optimal bottleneck tree. In this section we suggest a better approximation algorithm for the the Bottleneck Steiner Problem (see Introduction) with the performance ratio half of that of the MST-heuristic.

Let consider a weighted hypergraph $H = (S, U, w)$ in which the vertex set coincides with the terminal set S and the edge set U contains all pairs and triples of terminals. The weight of an edge (s, s') is its length and the weight of a triple τ is the maximum edge length in the minimum bottleneck steiner tree for τ. Let $H_{\leq \alpha}$ be subhypergraph of H which contains only hyperedges of weight at most α. Note that in hypergraphs with the edge size at most 3, we can find efficiently using Lovasz' algorithm [13] the spanning hypertree B (i.e. hypergraph B such that replacing all 3-hyperedges of B with two usual edges makes a usual spanning tree), if it exists. It is easy to see that replacing each 3-hyperedge of B with the corresponding minimum bottleneck Steiner tree converts B to a Steiner tree for S.

Algorithm 1:

(1) For each triple of terminals find the optimal bottleneck tree

(2) Find the minimum α, such that the hypergraph $H_{\leq \alpha}$ contains a spanning hypertree

(3) Output the Steiner tree corresponding to the spanning hypertree of the hypergraph $H_{\leq \alpha}$

Theorem 4 *The performance ratio of Algorithm 1 is at most* $\log_2 n$, *where n is the number of terminals.*

Proof. Let OPT be an optimal bottleneck Steiner tree. We may assume that all internal nodes has degree 3 and all terminals are leaves in OPT. Indeed, if an internal node v has degree at least 4, then we add an auxiliary copy of v connected to v with a zero-cost edge and evenly split vertices adjacent to v between v and and its copy. Similarly, if a terminal s is not a leaf in OPT, then we add a copy of s, say s', which is not a terminal. Then we substitute s with s' in OPT while joining s to s' with a zero-cost edge. These graph transformations do not change the number of terminals n.

Moreover, we may assume that all edges in OPT have the same length 1. Indeed, if there is an edge of length less than the maximum length, then we may increase its length up to the maximum without increasing the bottleneck cost.

We will use the following two lemmas which are well-known graph-theoretical facts (e.g. Lemma 2 can be found in [1]).

Fact 1 *Given a tree with n leaves and all internal nodes of degree 3, except the root r that may have degree 2 or 3. There exists a leaf at most $\log_2 n$ edges away from r.*

Fact 2 *Any tree with n leaves contains a node c (called center) such that removing c splits the tree into connecting components each containing at most $n/2$ nodes.*

In order to proof Theorem, we will span the set of terminals S with a union of Steiner trees for triples (and pairs) of terminals each having the bottleneck cost at most $\log_2 n$.

Let c be a center of the tree OPT and let c_1, c_2, and c_3 be the three nodes adjacent to c. Removing c splits the tree OPT into three connected components, say OPT_1, OPT_2, and OPT_3 such that $c_i \in OPT_i$. Note that each node c_i has degree 2 inside the tree OPT_i, and Lemma 1 implies that there is a path from c_i to a leaf of OPT_i, say l_i, of length at most $\log_2 \frac{n}{2} = \log_2 n - 1$. Thus, $d(c, l_i) = d(c, c_i) + d(c_i, l_i) \leq \log_2 n$, so the tree $\{(c, l_i)|i = 1, 2, 3\}$ shows that one can connect the three different node sets $OPT_i \cap S$ of the hypergraph H by "edge" $\{l_1, l_2, l_3\}$ of cost at most $\log_2 n$. So it suffices to prove that one can also span, separately for $i = 1, 2, 3$, the vertices of $OPT_i \cap S$ by hyperedges of $H_{\log_2 n}$. We recursively produce the necessary links, using the fact that the sets $OPT_i \cap S$ lie in a tree with at most $n/2$ terminals and exactly one internal node of degree 2.

Now we consider a tree T of such form, i.e. T has at most $\frac{n}{2}$ terminals-leaves, the root r of T has degree 2 and all other internal nodes have degree

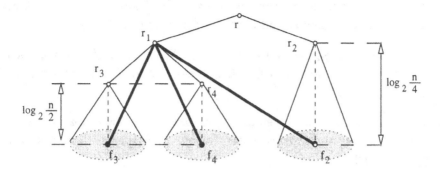

Figure 4: The Steiner tree for terminals f_3, f_4 and f_2 with the additional node r_1 (thick edges) has bottleneck cost at most $\log_2 n$.

3 (see Figure 4). Removing r splits T into two subtrees T_1 and T_2, with roots r_1 and r_2 which are neighbors of r in T. For certainty, we assume that T_1 has as many leaves as T_2, i.e. either $|S \cap T_1| = |S \cap T_2| = 1$ or $|S \cap T_1| \geq 2$ and $|S \cap T_2| \leq \frac{n}{4}$. In the former case we arrive at the basis of the recursion: the cost of the hyperedge connecting the two terminals $S \cap T_1$ and $S \cap T_2$ is 2, i.e., at most $\log_2 n$. Otherwise, the removal r_1 from T_1 splits it into two trees, say T_3 and T_4, with the roots r_3 and r_4, respectively. Let f_i be the leaf closest to r_i in each tree T_i, $i = 2, 3, 4$. Note that $d(r_1, f_2) \leq d(r_2, f_2) + 2 \leq \log_2 \frac{n}{4} + 2 = \log_2 n$. Similarly, $d(r_1, f_3)$ and $d(r_1, f_4)$ are at most $\log_2 n$. Now tree $\{(r, f_1), (r, f_2), (r, f_3)\}$ shows that the cost of the hyperedge $\{f_1, f_2, f_3\}$, which connects the three different terminal sets of T, is at most $log_2 n$. As terminal sets T_2, T_3 and T_4, if cardinality is at least 2, lie again in trees $(T_2, T_3$ and $T_4)$ of the same form, but with fewer terminals, we can continue the process of producing hyperedges of $H_{\log_2 n}$ until all terminals are connected. □

References

[1] C. D. Bateman, C. S. Helvig, G. Robins, and A. Zelikovsky, *Provably-Good Routing Tree Construction with Multi-Port Terminals*, in Proc. International Symposium on Physical Design, Napa Valley, CA, April 1997, pp. 96–102.

[2] K. D. Boese, A. B. Kahng, B. McCoy, and G. Robins, *Near-optimal*

critical sink routing tree constructions, IEEE Trans. on Comput.-Aided Des. of Integr. Circuits and Syst., 14 (1995), pp. 1417–11436.

[3] C. Chiang, M. Sarrafzadeh, and C. K. Wong, *Global Routing Based on Steiner Min-Max Trees*, IEEE Trans. Computer-Aided Design, 9 (1990), pp. 1318–25.

[4] C. W. Duin and A. Volgenant, *The partial sum criterion for Steiner trees in graphs and shortest paths*, Europ. J. of Operat. Res., 97 (1997), pp. 172–182.

[5] J. Elzinga, D. Hearn, and W. D. Randolph, *Minimax multifacility location with Euclidean distances*, Transportation Science, 10 (1976), pp. 321–336.

[6] J. L. Ganley, *Geometric interconnection and placement algorithms*, PhD thesis, Dept of CS, University of Virginia, 1995.

[7] J. L. Ganley and J. S. Salowe, *Optimal and approximate bottleneck Steiner trees*, Oper. Res. Lett., 19 (1996), pp. 217–224.

[8] J. L. Ganley and J. S. Salowe, *The power-P Steiner tree problem*, Nordic Journal of Computing, 5 (1998), pp. 115–127.

[9] G. Hardy, E. Littlewood, and G. Polya, *Inequalities*, Cambridge University Press, 1934.

[10] N. D. Holmes, N. A. Sherwani, and M. Sarrafzadeh, *Utilization of vacant terminals for improved over-the-cell channel routing*, IEEE Trans. Computer-Aided Design, 12 (1993), pp. 780–782.

[11] J. N. Hooker, *Solving nonlinear multiple-facility network, location problem*, Networks, 19 (1989), pp. 117–133.

[12] R. M. Karp, *Reducibility among combinatorial problems*, in R.E. Miller and J.W. Thatcher. Complexity of Computer Computations, 1972, pp. 85–103.

[13] L. Lovasz and M. Plummer, *Matching Theory*, Elsvier Science, 1986.

[14] C. H. Papadimitriou and M. Yannakakis, *Optimization, approximation, and complexity classes*, in Proc. ACM Symp. the Theory of Computing, 1988, pp. 229–234.

[15] M. Sarrafzadeh and C. K. Wong, *Bottleneck Steiner trees in the plane*, IEEE Trans. on Computers, 41 (1992), pp. 370–374.

[16] J. Soukup, *On minimum cost networks with nonlinear costs*, SIAM J. Applied Math., 29 (1975), pp. 571–581.

Exact Steiner Trees in Graphs and Grid Graphs

Siu-Wing Cheng
Department of Computer Science
The Hong Kong University of Science & Technology
Clear Water Bay, Hong Kong

Contents

D.-Z. Du et al. (eds.), Advances in Steiner Trees, 137-162.
© 2000 Kluwer Academic Publishers.

1   Introduction

Given a graph with nonnegative edge lengths and a selected subset of vertices, the *Steiner tree* problem is to find a tree of minimum length that spans the selected vertices. This problem is also commonly called the *graphical Steiner minimal tree* problem or GSMT problem for short. We call the selected vertices *terminals*. In a Steiner tree, any vertex which is not a terminal and has degree at least three is called a *Steiner vertex*.

The GSMT problem was independently posed by Hakimi [21] and Levin [29]. Since the GSMT decision problem is known to be NP-complete [26], it is unlikely that a polynomial time algorithm exists. Despite this, many researchers have been attacking the problems from different angles. Polynomial time approximation algorithms have been developed that construct a Steiner tree of length within a constant factor of optimal [34, 35]. Various exact algorithms with or without performance analyses are also known. Surveys of these results can be found in [24, 33]. There is also a book dedicated to the Steiner tree problem in general [25]. In this paper, we survey several graph-theoretic dynamic programming algorithms that construct exact Steiner trees. We will cover several exponential time algorithms for general graphs and planar graphs, as well as two polynomial time algorithms for special cases in grid graphs.

For general graphs, we will present the algorithm by Dreyfus and Wagner [16] in Section 2. The generality of the algorithm allows it to be refined in other situations. In Section 3, we will present results in [3, 4, 17] for two most notable examples, namely, speedup of the general algorithm for sparse graphs and planar graphs. Then in Section 4 we will turn to the problem of constructing rectilinear Steiner trees in two special cases when the terminals lie on the boundary of a grid graph or a rectangle. Section 5 surveys a number of polynomial time algorithms known for related special cases of the GSMT problem.

Our motivation is two-fold. First, the Dreyfus-Wagner algorithm is a simple and elegant solution for general graphs. It is also quite versatile as evidenced by variants developed for the GSMT problem in different contexts. Thus, we want to give a coherent exposition for these variants to be understood as a whole. Second, we want to survey some newer techniques that lead to faster algorithms for finding rectilinear Steiner trees in some special cases that arise in VLSI routing. The improvements are mainly based on newly discovered structural properties which may be useful for related problems.

2   The Dreyfus-Wagner Algorithm

Let $G = (V, E)$ denote a graph with nonnegative edge lengths and let W denote the set of terminals in G, where $|V| = N$ and $|W| = n$. Take a set of terminals $X \subseteq W$ and a vertex $v \in V \setminus X$. Define $C(v, X)$ to be the minimum length of a Steiner tree that spans $X \cup \{v\}$. Define $B(v, X)$ to be the minimum length of a Steiner tree that spans $X \cup \{v\}$ and in which v has degree at least two. Given an optimal Steiner tree spanning $X \cup \{v\}$ in which the degree of v is at least two, we can split the tree at v to obtain two subtrees, one spanning $Y \cup \{v\}$ and the other spanning $(X \setminus Y) \cup \{v\}$ for some nonempty $Y \subset X$. Thus, we obtain the following recurrence to compute $B(v, X)$.

$$B(v, X) = \min_{\emptyset \subset Y \subset X} \{C(v, Y) + C(v, X \setminus Y)\}. \tag{1}$$

Given an optimal Steiner tree spanning $X \cup \{v\}$, if v has degree at least two, then this tree has length $B(v, X)$. Otherwise, the tree path from v leads to a vertex $u \in X$ or a vertex $u \in V \setminus X$ of degree at least three, whichever comes first. Thus we obtain the following recurrence to compute $C(v, X)$.

$$C(v, X) = \min\{\min_{u \in X} \{C(u, X \setminus \{u\}) + d(u, v)\}, \min_{u \in V \setminus X} \{B(u, X) + d(u, v)\}\}, \tag{2}$$

where $d(u, v)$ is the shortest path distance between u and v in G. The boundary conditions are $B(v, \emptyset) = C(v, \emptyset) = 0$ for all $v \in V$. We assume that there is a preprocessing time of $O(n^3)$ to compute all-pairs shortest paths in G [15].

Recurrences 1 and 2 are evaluated in increasing cardinality of X. At the end, the length of an optimal Steiner tree spanning all terminals is equal to $C(v, W \setminus \{v\})$ for any terminal $v \in W$. To reconstruct an optimal Steiner tree, it suffices to record back pointers during the computation. This will not increase the asymptotic time or space complexities.

Theorem 2.1 *Given a graph $G = (V, E)$ with N vertices where n of them are terminals, an optimal Steiner tree can be computed in $O(N3^n + N^2 2^n)$ time and $O(N 2^n)$ space.*

Proof. Correctness follows from two induction invariants that are clearly maintained throughout the algorithm. The first invariant is: if there is an optimal Steiner tree spanning $X \cup \{v\}$ in which v has degree at least two, then $B(v, X)$ is the length of this tree. The second invariant is that $C(v, X)$ is the length of an optimal Steiner tree spanning $X \cup \{v\}$.

We analyze the time complexity. In recurrence 1, each terminal may belong to Y, or $X \setminus Y$, or $V \setminus X$. Therefore, there are 3^n possible triples of X, Y, and $V \setminus X$. Thus, evaluating all instances of recurrence 1 takes $O(N3^n)$ time. Since there are 2^n possible subsets $X \subseteq W$, there are $O(N2^n)$ instances of recurrence 2 and each takes $O(N)$ computing time. This proves the $O(N3^n + N^2 2^n)$ time bound. Only the $B(v, X)$'s and $C(v, X)$'s need to be stored and there are $O(N2^n)$ of them. This proves the space bound. $\square$

3   Improving the Dreyfus-Wagner Algorithm

In [17], Erickson, Monma, and Veinott studied the minimum-additive-concave-cost flow problems in graphs. They proposed a generic dynamic programming algorithm for solving this class of problems, which finds applications in areas including inventory, production and capacity planning. They also applied it to finding optimal Steiner trees in graphs. In essence, it can be viewed as a method to speed up the evaluations of recurrences in the case of sparse graphs and when the locations of terminals are restricted.

3.1   Speedup for Sparse Graphs

The key idea is to batch the computation of $C(v, X)$ for all vertices $v \in V \setminus X$ for a fixed X in a single-source shortest paths computation. First, replace each edge in G by two oppositely directed arcs with weights equal to the weight of the original edge. Second, create a new vertex s. Third, for each vertex $u \in X$, add an arc (s, u) with weight $C(u, X \setminus \{u\})$, and for each vertex $u \in V \setminus X$, add an arc (s, u) with weight $B(u, X)$. Now, if we invoke the Dijkstra's algorithm with s as the source, then the cost of the shortest path from s to $v \in V \setminus X$ is exactly $\min\{ \min_{u \in X}\{ C(u, X \setminus \{u\}) + d(u, v) \}, \min_{u \in V \setminus X}\{ B(u, X) + d(u, v) \} \} = C(v, X)$. Hence, one invocation of Dijkstra's algorithm evaluates $C(v, X)$ for all $v \in V \setminus X$. Using Fibonacci heaps [18], Dijkstra's algorithm runs in $O(N \log N + |E|)$ time. For dense graphs, $|E| = \Theta(N^2)$ and so there is no gain. But for sparse graphs like planar graphs, this speeds up the running time to $O(N3^n + (N \log N)2^n)$.

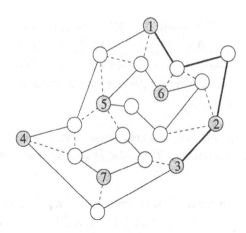

Figure 1: The solid edges are face boundaries and dashed edges are other graph edges. The shaded vertices are terminals. Two examples of intervals on the exterior face are $[1,3] = \{1,2,3\}$ and $[3,1] = \{3,4,1\}$. bd$[1,3]$ is the chain of boundary edges from 1 to 3 which is shown as bold line segments.

3.2   Terminals on a Few Faces in a Planar Graph

Further improvement can be made if G is planar and the terminals lie on the boundaries of a few faces F_i of G, $1 \leq i \leq f$. (Note that G may contain more than f faces.) To simplify the discussion, we assume that the boundary of the exterior face of G is a simple polygon. When this is not true, G can be partitioned into smaller graphs to satisfy this condition and it suffices to solve the problem on each smaller graph obtained. The readers are referred to [3, 17] for details. We also assume that the faces F_i are mutually vertex disjoint.

The speedup comes from restricting the possibilities of X, Y, and $X \setminus Y$ in recurrences 1 and 2. Let S be a set of terminals on F_i. We say that S is *an interval* if there are two terminals u and w in S such that if we traverse from u to w in clockwise order around F_i, we will encounter all terminals in S and no other terminal not in S. We denote S by $[u, w]$. We denote the boundary of F_i on $[u, w]$ by bd$[u, w]$. See Figure 1.

In defining $B(v, X)$ and $C(v, X)$, we require further that $X \cap F_i$ or $(X \cup \{v\}) \cap F_i$ is an interval, for all $1 \leq i \leq f$. We call this the *interval requirement*. The idea of using intervals to speed up the dynamic programming was first proposed and proved in [17]. We present a proof of correctness below which generalizes an argument in [4].

Let T be a Steiner tree that spans $X \cup \{v\}$ for some subset X of terminals and for some vertex $v \notin X$ such that the interval requirement is satisfied. For $1 \leq i \leq f$, let $[u_i, w_i]$ denote $(X \cup \{v\}) \cap F_i$ if it is an interval or $X \cap F_i$ otherwise. Let ρ_i be the path in T from u_i to w_i. We say that a vertex v is *exterior* in T if for all $1 \leq i \leq f$, v lies on ρ_i or outside the finite region bounded by ρ_i and $\mathrm{bd}[u_i, w_i]$.

Lemma 3.1 *Suppose that the interval requirement is enforced in executing the Dreyfus-Wagner algorithm. Then the following invariants hold throughout the execution.*

1. *If there is an optimal Steiner tree spanning $X \cup \{v\}$ in which v is exterior and has degree at least two, then $B(v, X)$ is the length of this tree.*

2. *If there exists an optimal Steiner tree spanning $X \cup \{v\}$ in which v is exterior, then $C(v, X)$ is the length of this tree.*

Proof. This proof generalizes the argument in [4] for requiring terminals on the exterior face to be intervals. We prove by induction on $|X|$. The invariants clearly hold for $X = \emptyset$.

We prove invariant (1) first. Let $T_B(v, X)$ be an optimal Steiner tree spanning $X \cup \{v\}$ in which v is exterior and the degree of v is at least two. If v does not lie on ρ_i for all i, then $X \cap F_i$ is an interval for all i as v is exterior. Then an arbitrary split at v yields an optimal subtree T spanning $Y \cup \{v\}$ and an optimal subtree T' spanning $(X \setminus Y) \cup \{v\}$ for some nonempty $Y \subset X$. Since v is exterior in $T_B(v, X)$, v is exterior in T and T'. Since there is no tree edge between T and T', $Y \cap F_i$ is either empty or equal to $X \cap F_i$ for all i. See Figure 2. Thus, by induction assumption, $C(v, Y)$ and $C(v, X \setminus Y)$ are available (since the interval requirement is satisfied) as the lengths of T and T' respectively. This means that $B(v, X)$ is correctly computed when evaluating recurrence 1. Suppose that v lies on some path ρ_k. We try to choose k such that v lies on F_k as well. If such a k does not exist, then we choose any k such that v lies on ρ_k. We denote the path in $T_B(v, X)$ from from u_k to v by $\rho_k(u_k, v)$. Root $T_B(v, X)$ at v. Let T be the subtree that includes v and the vertices in $T_B(v, X)$ descending from the edge on $\rho_k(u_k, v)$ incident to v. Let T' denote the rest of $T_B(v, X)$. Observe that v is exterior in both T and T'. Let Y be the subset of X spanned by T. Since there is no tree edge between T and T', for all $i \neq k$, $Y \cap F_i$ is either empty or equal to $X \cap F_i$ which is the same $(X \cup \{v\}) \cap F_i$ as v does not lie on F_i. Thus, for all $i \neq k$, both $Y \cap F_i$ and $(X \setminus Y) \cap F_i$

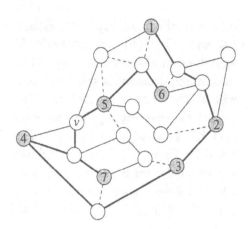

Figure 2: The bold line segments form $T_B(v, X)$ in which v is exterior. We can split the tree at v to yield one subtree spanning $\{v\} \cup [1, 4] \cup [7]$ and another subtree spanning $\{v\} \cup [5, 6]$. Hence, each subtree either does not span any terminal in X on a face or a complete interval in X on a face.

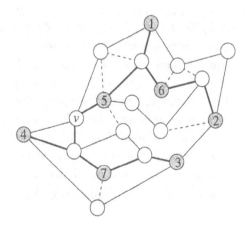

Figure 3: The bold line segments form $T_B(v, X)$ in which v is exterior. The vertex v lies on path from 1 to 4 for the interval $[1, 4]$ spanned on the exterior face. The subtree that includes v and vertices descending from the edge connecting v and 5 spans terminals 1, 2, 5, and 6. The rest of $T_B(v, X)$ spans terminals 3, 4, and 7. Thus the interval $[1, 4]$ on the exterior face are split into two intervals $[1, 2]$ and $[3, 4]$.

are intervals. $(Y \cup \{v\}) \cap F_k$ consists of terminals in $[u_k, w_k]$ such that the paths in $T_B(v, X)$ from v to them overlap with $\rho_k(u_k, v)$. We claim that $(Y \cup \{v\}) \cap F_k$ and $[u_k, w_k] \setminus (Y \cup \{v\})$ are intervals. See Figure 3. If this claim is true, then by induction assumption, $C(v, Y)$ and $C(v, X \setminus Y)$ are available (since the interval requirement is satisfied) as the lengths of T and T' and so $B(v, X)$ will be correctly computed as $C(v, Y) + C(v, X \setminus Y)$. Clearly, $u_k \in (Y \cup \{v\}) \cap F_k$. Let $[u_k, w'_k]$ be the smallest interval on F_k that contains $(Y \cup v) \cap F_k$. Thus $w'_k \in (Y \cup \{v\}) \cap F_k$. If any vertex z in $(Y \cup \{v\}) \cap F_k$ is spanned by $T_B(v, X)$, then z is enclosed by $\mathrm{bd}[u_k, w'_k]$ and two paths in $T_B(v, X)$: one between v and u_k and one between v and w'_k. Thus, the path in $T_B(v, X)$ between v and z must overlap with $\rho_k(u_k, v)$ and so z belongs to $(Y \cup \{v\}) \cap F_k$. This proves the claim that both $(Y \cup \{v\}) \cap F_k$ and $[u_k, w_k] \setminus (Y \cup \{v\})$ are intervals.

We now prove invariant (2). Let $T_C(v, X)$ be an optimal Steiner tree spanning $X \cup \{v\}$ and v is exterior in $T_C(v, X)$. If the degree of v is at least two, then $T_C(v, X)$ has length $B(v, X)$ which is available. Thus, $C(v, X)$ will be computed correctly as $B(v, X) + d(v, v)$ when evaluating recurrence 2. Suppose that the degree of v is one. Then we trace the path in $T_C(v, X)$ leading from v. If we come to a vertex u of degree at least 3 and $u \notin X$, then u is exterior in $T_C(v, X)$ as v is exterior and all vertices between v and u have degree two. So $T_C(v, X)$ has length $B(u, X) + d(u, v)$. If we come to a vertex $u \in X$, then since we have only been encountering degree-2 vertices, $T_C(v, X)$ is equal to the union of an optimal Steiner tree T spanning X and a shortest path from v to u. Since v is exterior in $T_C(v, X)$, u is also exterior in T. Moreover, since v is exterior and has degree one, for all i, if $(X \cup \{v\}) \cap F_i$ is an interval, then this interval can be specified with v as an end vertex. This implies that $X \cap F_i$ is an interval for all i. Thus, $C(u, X \setminus \{u\})$ is available (since the interval requirement is satisfied) as the length of T by induction assumption. Hence, the length of $T_C(v, X)$ is $C(u, X \setminus \{u\}) + d(u, v)$. In all, $C(v, X)$ is computed correctly by recurrence 2. $\qquad\square$

Corollary 3.2 *Given a planar graph G of N vertices where n of them are terminals lying on f faces, an optimal Steiner tree can be computed in $O(N(\frac{n+f}{f})^{3f} + (N \log N)(\frac{n+f}{f})^{2f})$ time and space $O(N(\frac{n+f}{f})^{2f})$.*

Proof. Let W be the set of terminals. In an optimal Steiner tree, it is clear that $W \cap F_i$ is an interval for $1 \leq i \leq f$. One of the terminals on a face must be exterior. Therefore, by Lemma 3.1, it suffices to return $\min\{C(v, W \setminus \{v\}) : v \in W\}$.

We analyze the time and space complexities. Let n_i be the number of terminals on the face F_i. In recurrence 1, for a fixed v there are at most $n_i^3 + 1$ combinations of intervals $X \cap F_i$ or $(X \cup \{v\}) \cap F_i$, $Y \cap F_i$ or $(Y \cup \{v\}) \cap F_i$, and $(X \setminus Y) \cap F_i$. There is an extra 1 to include the case $X \cap F_i = \emptyset$. Thus, for a fixed v, there is a total of at most $\Pi_{i=1}^{f}(n_i^3 + 1) \leq \Pi_{i=1}^{f}(n_i + 1)^3$ combinations of X, Y, and $X \setminus Y$. Observe that $\Pi_{i=1}^{f}(n_i + 1) \leq (\frac{n+f}{f})^f$ by the fact that geometric mean is at most arithmetic mean. Substituting this gives a total time of $O(N(\frac{n+f}{f})^{3f})$ for evaluating all instances of recurrence 1. The same trick applies for bounding the time for evaluating all instances of recurrence 2 and the space. Thus, the algorithm runs in $O(N(\frac{n+f}{f})^{3f} + (N \log N)(\frac{n+f}{f})^{2f})$ time and $O(N(\frac{n+f}{f})^{2f})$ space. $\quad\square$

Remark. Corollary 3.2 was first proved in [17]. The original time and space bounds stated are $O(N 2^{-f} n^{3f} + (N \log N) n^{2f})$ and $O(N n^{2f})$ respectively. This is obtained mainly by overestimating the number of terminals on F_i to be n and hence the terms n^{3f} and n^{2f}. We use the relationship between arithmetic and geometric means to get tighter bounds when f is not too small compared with n.

Corollary 3.3 *Let $G = (V, E)$ be a planar graph of N vertices. If there are n terminals and m of them lies on the exterior face, then an optimal Steiner tree can be computed in $O(N m^3 3^{n-m} + (N \log N) m^2 2^{n-m})$ time and space $O(N m^2 2^{n-m})$.*

Proof. Let F be the exterior face. For a fixed v, there are at most $m^3 + 1$ possible intervals $X \cap F$ or $(X \cup \{v\}) \cap F$, $Y \cap F$ or $(Y \cup \{v\}) \cap F$, and $(X \setminus Y) \cap F$. There are at most 3^{n-m} ways of distributing terminals not on F into $X \setminus Y$, Y, and $V \setminus X$. Thus, evaluating all instances of recurrence 1 takes $O(N m^3 3^{n-m})$ time. The time in evaluating all instances of recurrence 2 and the space bound can be similarly analyzed. $\quad\square$

Corollary 3.3 was first proved in [4]. In the same paper, Bern improved the result in Corollary 3.2 further. The first key idea is that given any optimal Steiner tree T spanning the terminals in G, there is a way to cut G open by removing edges and vertices not in T such that all the f faces are "linked" together to become the exterior face. So the strategy is to enumerate all possible cuts, invoke Corollary 3.2 for each such cut, and return the tree of minimum length among all such cuts. Note that the running time of each invocation is now $O(Nn^3 + (N \log N)n^2)$. The second key idea is that it is not necessary to actually remove edges and vertices from G to make a cut. Each such cut must leave a face between two consecutive terminals (on that face) and then enter another face between two consecutive terminals (on that face). This implies that there are at most n^2 possible ways to choose how a cut should leave and enter two faces successively. Therefore, there are at most n^{2f-2} orderings of terminals on the resulting exterior face after applying some cut. Multiplying n^{2f-2} with $O(Nn^3 + (N \log N)n^2)$ yields a running time of $O(Nn^{2f+1} + (N \log N)n^{2f})$.

Theorem 3.4 *Given a planar graph of N vertices where n of them are terminals lying on f faces, an optimal Steiner tree can be computed in $O(Nn^{2f+1} + (N \log N)n^{2f})$ time and $O(Nn^{2f})$ space.*

Other than an optimal Steiner tree spanning the terminals, recurrences 1 and 2 also yield an optimal Steiner tree spanning all terminals plus one other fixed vertex. The running time and space bound are the same. Thus, one can treat one terminal as not a terminal when computing an optimal Steiner tree. This gives some saving in Corollary 3.2, Corollary 3.3, and Theorem 3.4, since n decreases by one and f may also decrease. The special case where all terminals lie on the exterior face of a planar graph has also been studied by Provan [31] and an $O(N^2n^2)$-time algorithm was given.

4   Rectilinear Steiner Trees

Given a set of terminals in the plane, a tree that spans all terminals and consists of vertical and horizontal edges is called a rectilinear Steiner tree or RST. Finding a RST of minimum length is known as the rectilinear Steiner minimal tree problem or RSMT problem. Despite the restriction on the orientations of tree edges, the RSMT problem is NP-hard [19]. However, some special cases can be solved in polynomial time. All of them restrict the locations of the terminals. The first such results are by Aho, Garey and Hwang [2]. They presented three algorithms. The first algorithm works for n terminals lying on two parallel lines and it runs in $O(n)$ time. The second algorithm works for n terminals lying on $m > 2$ parallel lines. Its running time is linear in n but exponential in m. The third algorithm works for n terminals lying on the boundary of a rectangle. Its running time is $O(n^3)$.

Subsequently, Cohoon, Richards and Salowe [14] and independently Agarwal and Shing [1] presented linear time algorithms for terminals on the boundary of a rectangle. Later, Richards and Salowe [32] extended the algorithm to work for terminals on the boundary of a *rectilinear convex hull*. A rectilinear convex hull of a set of terminals is a *smallest* enclosing simple rectilinear polygon that contains a rectilinear shortest path between every two terminals. The running time is $O(k^5 + k^4 n)$ where n is the number of terminals and k is the number of sides of the rectilinear convex hull. The main result in [32] is that there is a set of structural properties satisfied by a subset of optimal RSTs. These structural properties motivate subsequent work on the same or related problems.

Improved algorithms for terminals on a rectilinear convex hull are outlined in [10] and [11]. In [27], an algorithm is presented for the case when the terminals lie on the exterior face of a *grid graph*. A grid graph is a planar graph with vertical and horizontal edges and each face, except the exterior face, consists of exactly four vertices. The length of an edge is the Euclidean distance between its endpoints. The running time is $O(\min\{n^4, N n^2\})$. This is a significant improvement of the $O(N n^3 + (N \log N) n^2)$-time algorithm implied by Theorem 3.4 for terminals lying on the exterior face of a general planar graph. Recently, an algorithm is presented in [9] for the same problem. The running time is $O(k^4 + k^3 n)$, where k is the number of sides bounding the exterior face. Thus, for terminals lying on the boundary of a simple rectilinear polygon of constant size, there is a linear time algorithm for computing an optimal RST that lie inside the polygon. This relates back to the linear time algorithm in [14] for rectangles and extends the results in [10, 11, 32] for rectilinear convex hulls.

In Section 4.1, we outline the algorithm in [27] for computing an optimal RST spanning terminals on the exterior face of a grid graph. In Section 4.2, we describe structural properties of optimal RSTs proved in [9]. Then we present a linear time algorithm for computing an optimal RST spanning terminals on the boundary of a rectangle. The algorithm is different from the first linear time algorithms obtained in [1, 14]. We opt for presenting this algorithm because the basic strategy can be extended when P is a simple rectilinear polygon. More details can be found in [9].

4.1 Terminals on the boundary of a grid graph

The notations used are slightly different from those in [27]. Let G be the grid graph. The exterior face of G bounds a simple rectilinear polygon P.

All the n given terminals lie on the boundary of P. We can further assume that there is a terminal at each convex corner of P. Otherwise, we can remove this corner by removing the vertex at it and the two incident edges from G as they are not needed in forming the optimal RST. Thus, with an $O(N)$-time preprocessing step, this assumption can be enforced.

In a RST, we call a maximal collinear sequence of tree edges in the interior of P an *interior line* and a maximal collinear sequence of tree edges on the boundary of P a *boundary line*. If there is a line through a vertex v, then a *line from v* is one of the two line segments obtained by splitting the line at v. Vertices on P are called *boundary vertices*. We call the vertex at a corner of P a *boundary corner vertex*. A *grid line* is a maximal sequence of collinear edges in G in the interior of P.

Let G' be a subgraph of G obtained by removing all grid lines of G that are not incident to any terminal or boundary corner vertex. Note that the exterior face of G' is still P. G' can be produced in $O(N)$ time and the size of G' is $O(\min\{n^2, N\})$. By Hanan's results [22], there is an optimal RST which is a subgraph of G'. Thus, G' is our underlying graph from now on.

4.1.1 Subproblems and Recurrences

We denote the directions upward, rightward, downward, and leftward by integers 1, 2, 3, and 4 respectively. Let T be a RST spanning a subset of terminals of G'. Let ℓ be any collinear sequence of tree edges. We say that a tree edge is *pinned to ℓ* at a tree vertex u if the tree edge is perpendicular to ℓ and incident to u on ℓ. We call each connected component of T obtained by removing from T edges on P an *interior component*. By Hwang's results in [23], each interior component can be transformed without increasing total length into one of two types:

Type 1: There is one interior line ℓ with boundary vertices as endpoints. Each tree edge pinned to ℓ has a boundary vertex as an endpoint, and these edges point alternatively in opposite directions.

Type 2: There are two orthogonal interior lines ℓ_1 and ℓ_2 sharing a degree two endpoint in the interior of P, and the other endpoints of ℓ_1 and ℓ_2 are boundary vertices. Each tree edge pinned to ℓ_1 or ℓ_2 has a boundary vertex as an endpoint, and these edges point alternatively in opposite directions.

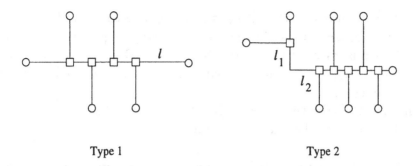

Figure 4: Circles represent boundary vertices and squares represent Steiner vertices.

See Figure 4. Thus, it suffices to focus on optimal RSTs that satisfy the above properties. This motivates the following definitions of subtrees that can exist in such an optimal RST. Given a RST $T(v, [a, b])$ spanning an interval $[a, b]$ and $v \notin [a, b]$, we call $T(v, [a, b])$ a *restricted tree* if:

1. every line contains a boundary vertex and every line from v contains a boundary vertex,

2. for every boundary vertex u in $T(v, [a, b])$ with degree at least two, if u does not lie on $\mathrm{bd}[a, b]$, then the boundary traversed from u to a or from b to u does not contain any terminal.

We call $T(v, [a, b])$ an i-tree if it is restricted and it contains a line from v pointing in direction i. We call $T(v, [a, b])$ an ij-tree if it is restricted and it contains two lines from v pointing in directions i and j. We use $C_i(v, [a, b])$ to denote the minimum length of an i-tree $T(v, [a, b])$. We use $B_{ij}(v, [a, b])$ to denote the minimum length of an ij-tree $T(v, [a, b])$. Define $C(v, [a, b])$ to be $\min_{1 \leq i \leq 4} C_i(v, [a, b])$. Define $B(v, [a, b])$ to be the minimum length of ij-trees $T(u, [a, b])$ in which the line from u in direction i or j contains v. The following recurrence computes $B_{ij}(v, [a, b])$.

$$B_{ij}(v, [a, b]) = \min_{\emptyset \subset [a, x] \subset [a, b]} \{C_i(v, [a, x]) + C_j(v, [a, b] \setminus [a, x])\}. \qquad (3)$$

We set $C_i(v, [a, b])$ to be

$$\min \begin{cases} \min_{u \in V \setminus [a, b]} \{B(u, [a, b]) + d(u, v)\} \\ \min_{u \in [a, b]} \{C(u, [a, b] \setminus [a, u]) + C(u, [a, u] \setminus \{u\}) + d(u, v)\}, \end{cases} \qquad (4)$$

subject to the constraint that the path from v to u consists of up to four parts in this order: an interior line, a consecutive sequence of boundary lines, followed by at most two more interior lines. One or more of the four parts may not appear and in any configuration, the first line in the path points in direction i.

Finally, if W is the set of terminals in G', then we take $\min_{v \in W} C(v, W \setminus \{v\})$ to be the desired length of an optimal RST.

Remark. Recurrence 4 is stated differently in [27] because the vertex u in $B(u, [a, b])$ is not enforced to be outside $[a, b]$ in [27].

It is obvious that recurrence 3 is correct by definition. Recurrence 4 is essentially an adaption of recurrence 2. Its correctness hinges on the following result.

Lemma 4.1 *If an optimal RST contains an i-tree $T(v, [a, b])$, then $T(v, [a, b])$ has the same length as the union of a path ρ from v to u and a tree T, where ρ respects the constraint for recurrence 4 and either T is a jk-tree $T(u', [a, b])$ in which a line from u' in direction j or k contains u, or T is the union of a k-tree and a j-tree spanning $([a, b] \setminus [a, u]) \cup \{u\}$ and $[a, u]$ respectively.*

Lemma 4.1 is almost identical to Lemma 2 in [27] and the same proof applies. The small difference arises from our different formulation of recurrence 4.

Lemma 4.2 *Let W be the set of terminals in G. Then $\min_{v \in W} C(v, W \setminus \{v\})$ is the length of an optimal RST spanning W.*

Proof. Let T be an optimal RST spanning W such that each interior component is of type 1 or 2. Then T is clearly a restricted tree too. Also, there must be at least one terminal v in T with degree one acting as an interval end vertex.   □

4.1.2 Implementing the Recurrences

Consider computing $B_{ij}(v, [a, b])$. The choice of x in recurrence 3 is completely determined by v and the lines from v in directions i and j. For each vertex v, shoot vertical and horizontal rays from v to the boundary of P until they are intercepted. The endpoints of these rays on the boundary of P define a constant number of special intervals. Computing $B_{ij}(v, [a_i, b_j])$ for a special interval $[a_i, b_j]$ takes $O(n)$ time each and hence a total of $O(\min\{n^3, Nn\})$ time for all the vertices v and their special intervals. Then when we compute $B_{ij}(v, [a, b])$, we first identify the special interval $[a_i, b_j]$ for v such that $[a_i, b_j] \subseteq [a, b] \cup \{a_i, b_j\}$. Such a special interval must exist for the following reason. If $\text{bd}[a_i, b_j] \subseteq \text{bd}[a, b]$, then obviously $[a_i, b_j] \subseteq [a, b] \cup \{a_i, b_j\}$. Otherwise, a_i or b_j is a degree-2 boundary vertex outside $\text{bd}[a, b]$ in the ij-tree spanning $[a, b] \cup \{v\}$. By property 2 of restrictedness, $[a_i, b_j] \subseteq [a, b] \cup \{a_i, b_j\}$ or $[b_j, a_i] \subseteq [a, b] \cup \{a_i, b_j\}$ and both $[a_i, b_j]$ and $[b_j, a_i]$ are special intervals. After identifying this special interval in $O(1)$ time for $[a, b]$, we compute $B_{ij}(v, [a, b])$ by adopting the best x found when computing $B_{ij}(v, [a_i, b_j])$. In all, evaluating all instances of recurrence 3 takes a total of $O(\min\{n^2, N\} \cdot n^2) = O(\min\{n^4, Nn^2\})$ time.

To compute $B(v, [a, b])$, we try to consider in one batch all $B_{ij}(u, [a, b])$ in which the line from u in direction i or j contains v. For each direction i, we extract the subgraph G'_i of G' with edges parallel to direction i. Direct each edge in G'_i in direction i and give them zero weight. Add a new source vertex s and for each vertex u of G'_i, add the arc (s, u) with weight $\min_{j \neq i} B_{ij}(u, [a, b])$. Then run a single source shortest path algorithm with s as the source. Repeating this for all directions i determines the shortest path distance from s to v which is $B(v, [a, b])$ for all vertices v of G'. Since G'_i is acyclic, the single source shortest path algorithm can be implemented in $O(\min\{n^2, N\})$ time. Hence, computing $B(v, [a, b])$ for all vertices v and for all $[a, b]$ takes $O(\min\{n^4, Nn^2\})$ time.

To compute $C_i(v, [a, b])$, we formulate four phases of single source shortest path computations based on the structure of the path from v to u: first two phases for the last two possible interior lines on the path, third for the possible consecutive sequence of boundary lines, and fourth for the first interior line in the path. In each phase, extract subgraphs of G' for all four possible directions and run the single-source shortest path computations on them. At the beginning of each phase, we use the result of the previous phase as the current estimate of shortest path distances from s to other vertices. The phases for interior lines are similar to that in computing $B(v, [a, b])$. In the phase for the consecutive sequence of boundary lines, we only extract edges on the boundary of P and we need try two times, once for orienting the edges in the clockwise direction and once for the counter-clockwise direction.

Theorem 4.3 *Given a grid graph of N vertices where n of them are ter-*

minals lying on the exterior face, an optimal RST can be computed in $O(\min\{n^4, Nn^2\} + N)$ *time.*

4.2 A Different Approach

The notations used are slightly different from those in [9] in order to conform with notations used previously. Let P be the boundary of a grid graph G. We denote the boundary of P by ∂P and the interior of P by $\text{int}(P)$. For any two boundary vertices a and b on P, we denote by $[a, b]$ the union of $\{a, b\}$ and all terminals encountered while traversing ∂P from a to b in clockwise order. This extends the earlier definition of intervals. The difference is that the two end vertices of an interval may no longer be terminals. A *complete interior line* is an interior line whose two endpoints are on ∂P. We assume that the two tree edges incident to a degree-2 tree vertex are not collinear, otherwise the degree-2 tree vertex can be removed. Thus, a degree-2 tree vertex in $\text{int}(P)$ is an *interior corner-vertex* at an *interior corner*. An interior line incident to an interior corner-vertex is a *leg*. A degree-3 or degree-4 tree vertex in $\text{int}(P)$ is a T-vertex. A *head* is an interior line that contains two collinear tree edges which are incident to a T-vertex. Given a T-vertex v and a head h through v, a *body* is an interior line from v perpendicular to h. Define the *interior degree* of a tree vertex to be the number of incident tree edges that lie in $\text{int}(P)$. Define a *non-alternating edge pair* to be a pair of tree edges pinned to the same side of an interior line at two adjacent tree vertices in $\text{int}(P)$.

Let V' be the set of endpoints of grid lines incident to concave boundary corner vertices. Let V'' be the union of V' and boundary vertices u such that $\{u, v\} = [u, v]$ or $[v, u]$ for some $v \in V'$ or boundary corner vertex v. A *purple grid line* is a grid line incident to a vertex in V''. Each boundary corner vertex and each purple grid line endpoint is a *purple boundary vertex*. Thus, if P has k sides, then there are $O(k)$ purple grid lines and $O(k)$ purple boundary vertices. We call the intersections of purple grid lines and purple boundary vertices *purple vertices*. There are $O(k^2)$ of these.

4.2.1 Purple Properties

There is a subset of optimal RSTs that have relatively simple geometry as stated in the following result.

Theorem 4.4 *Given a grid graph G with terminals lying on its boundary, there is an optimal RST such that*

1. *it is a subgraph of G,*

2. *there is no non-alternating edge pair,*

3. *for each T-vertex, each body is a tree edge incident to some boundary vertex, and each head is incident to a purple boundary vertex,*

4. *for each interior corner, each leg is incident to a purple boundary vertex.*

We call a RST spanning any subset of terminals on ∂P purple if it satisfies the above properties. An optimal purple RST spanning all terminals can be selected using the following criteria in the listed order of priority: (1) maximize the length of intersection of the tree and ∂P, (2) maximize the sum of interior degrees of purple boundary vertices, (3) minimize the number of interior corner-vertices, (4) maximize the sum of interior degrees of terminals, (5) minimize the number of non-alternating edge pairs. Proofs can be found in [9].

4.2.2   A Linear Time Algorithm for a Rectangle

We present an $O(n)$-time algorithm for finding an optimal RST when P is a rectangle. This is faster than the $O(n^4)$-time algorithm implied by Theorem 4.3. Note that P consists of four sides and so there is a constant number of purple grid lines and purple vertices. Given a boundary vertex u and a terminal v, we call u and v *nearly opposite* if v is a terminal next to the other endpoint of the grid line incident to u. We say that an interval $[a, d]$ is *restricted* if $[a, d]$ consists of boundary vertices from at most three sides of P.

Subproblems. For each interval $[a, d]$ where ad is a grid line, define $i([a, d])$ to be the minimum length of a purple spanning RST subject to the constraint that it contains ad.

For each purple vertex v and restricted interval $[a, d]$ where a and d are purple, define $\ell(v, [a, d])$ to be the minimum length of a purple spanning RST subject to the constraint that v is an interior corner-vertex with legs va and vd.

For each restricted interval $[a, d]$ where a or d is purple or a and d are nearly opposite, define $\tau([a, d])$ to be the minimum length of a purple spanning RST subject to the constraint that it does not overlap with $\partial P \setminus$ bd$[a, d]$.

For each purple vertex v and restricted interval $[a, d]$ where a is purple, define $j_a(v, [a, d])$ to be the minimum length of a purple spanning RST subject to the constraints that va is an interior line in the tree and the tree does not overlap with $\partial P \setminus (\text{bd}[a, d] \cup \{v\})$. Symmetrically, we can define $j_d(v, [a, d])$ for restricted interval $[a, d]$ where d is purple. The constraints are that vd is an interior line in the tree and the tree does not overlap with $\partial P \setminus (\text{bd}[a, d] \cup \{v\})$.

Lemma 4.5 *If all the subproblems are solved, then an optimal RST can be computed in $O(n)$ time.*

Proof. Let T be a purple optimal RST selected by our criteria in section 4.2.1. T exhibits three possible structures and we will find the minimum in each case and return the overall minimum. Case 1: T does not contain any interior line. Then T has length $|\partial P|$ minus the longest boundary length between two neighboring terminals. Case 2: T contains a complete interior line ad. Then T has length $i([a, d]) + i([d, a]) - |ad|$. Case 3: T contains a purple interior corner-vertex v with legs va and vd. Splitting T along va and vd yields a subtree T_1 spanning $[a, d] \cup \{v\}$ and a subtree T_2 spanning $[d, a] \cup \{v\}$. Either $[a, d]$ or $[d, a]$ is not restricted and let it be $[a, d]$. Then T_2 has length $\ell(v, [d, a])$. Since the interior corner at v cannot be flipped to increase overlapping with ∂P (by our selection criteria), there are tree edges in T_1 pinned to va and vd. Thus, T_1 has length $j_a(v, [a, c]) + j_d(v, [a, d] \setminus [a, c])$ for some nonempty $[a, c] \subset [a, d]$ such that both $[a, c]$ and $[a, d] \setminus [a, c]$ are restricted. These three cases exhaust all possibilities by Theorem 4.4. Minimization in each case takes $O(n)$ time.   □

Recurrences. We present recurrences for solving the subproblems in increasing sizes of intervals. The recurrences should be evaluated in the order shown due to dependency.

Set $i([a, d])$ to be the minimum of

- $\min_{\emptyset \subset [a,c] \subset [a,d]} \{j_a(d, [a, c]) + \tau([a, d] \setminus [a, c])\}$. This choice is considered only when ad is a purple grid line.

- $|ad| + |au| + |dv| + \min_{\emptyset \subset [u,c] \subset [u,v]} \{\tau([u, c]) + \tau([u, v] \setminus [u, c])\}$ where u and v are the boundary vertices on $\mathrm{bd}[a, d]$ next to a and d respectively. This choice is considered only when ad is non-purple and uv is purple.

- $|au| + |dv| + i([u, v])$ where u and v are the boundary vertices on $\mathrm{bd}[a, d]$ next to a and d respectively. This choice is considered only when ad and uv are non-purple.

- $|ad| + \min\{\tau([a, d] \setminus \{a\}), \tau([a, d] \setminus \{d\})\}$.

Lemma 4.6 *If ad is a purple grid line or ad is next to a parallel purple grid line with endpoints in $\mathrm{bd}[a, d]$, then $i([a, d])$ is computed correctly in $O(n)$ time. Otherwise, $i([a, d])$ is computed correctly in $O(1)$ time.*

Proof. Let $T([a, d])$ be a RST that satisfies the constraints for $i([a, d])$. If ad is purple, then splitting $T([a, d])$ at d yields a subtree T_1 spanning $[a, c] \cup \{d\}$ and a subtree T_2 spanning $[a, d] \setminus [a, c]$ for some nonempty $[a, c] \subset [a, d]$. By making T_1 and T_2 lie inside their rectilinear convex hulls, T_1 and T_2 satisfy the constraints for $j_a(d, [a, c])$ and $\tau([a, d] \setminus [a, c])$ respectively. In particular, T_1 and T_2 do not overlap with $\partial P \setminus (\mathrm{bd}[a, c] \cup \{d\})$ and $\partial P \setminus (\mathrm{bd}[a, d] \setminus [a, c])$ respectively. So $T([a, d])$ has length $j_a(d, [a, c]) + \tau([a, d] \setminus [a, c])$. If ad is non-purple, then ad does not contain any Steiner vertex as $T([a, d])$ is a purple RST. Let u and v be the boundary vertices on $\mathrm{bd}[a, d]$ next to a and d respectively. So uv is a grid line parallel to ad. Suppose that $T([a, d])$ contains the line segment au and dv. If uv is purple, then $T([a, d])$ clearly has length $|ad| + |au| + |dv| + \min_{\emptyset \subset [u,c] \subset [u,v]} \{\tau([u, c]) + \tau([u, v] \setminus [u, c])\}$. Otherwise, we can sweep ad so that it becomes incident to u and v. The modified tree has length $|au| + |dv| + i([u, v])$ as uv cannot contain any Steiner vertex in a purple RST that satisfies the constraints for $i([u, v])$. Suppose that $T([a, d])$ does not contain au or av. Then $T([a, d])$ is the union of ad and a subtree that spans either $[a, d] \setminus \{a\}$ or $[a, d] \setminus \{d\}$. So $T([a, d])$ has length $|ad| + \min\{\tau([a, d] \setminus \{a\}), \tau([a, d] \setminus \{d\})\}$. $\square$

Set $\ell(v, [a, d]) = \min_{\emptyset \subset [a,c] \subset [a,d]} \{j_a(v, [a, c]) + j_d(v, [a, d] \setminus [a, c])\}$.

Lemma 4.7 $\ell(v, [a, d])$ *is computed correctly in* $O(n)$ *time.*

Proof. Let $T(v, [a, d])$ be a RST that satisfies the constraint for $\ell(v, [a, d])$. Similar to the proof of Lemma 4.6, splitting $T(v, [a, d])$ at v shows that it has length $j_a(v, [a, c]) + j_d(v, [a, d] \setminus [a, c])$ for some nonempty $[a, c] \subset [a, d]$. □

Set $\tau([a, d])$ to be the minimum of

- $|\mathrm{bd}[a, d]|$,

- $\min\{i([b, c]) + |\mathrm{bd}[a, b]| + |\mathrm{bd}[c, d]|\}$ among all b and c on $\mathrm{bd}[a, d]$ where bc is a grid line such that bc is purple or $b = a$ or $c = d$.

- $\min\{\ell(v, [b, c]) + j_b(v, [a, b]) + j_c(v, [c, d]) - |vb| - |vc|\}$ among all purple interior corner-vertices v with legs vb and vc, where b and c lie on $\mathrm{bd}[a, d]$.

Lemma 4.8 $\tau([a, d])$ *is computed correctly in* $O(1)$ *time.*

Proof. Let $T([a, d])$ be a RST that satisfies the constraint for $\tau([a, d])$. The analysis is similar to the proof of Lemma 4.5 except when $T([a, d])$ contains a complete interior line bc. In this case, since $[a, d]$ is restricted and $T([a, d])$ does not overlap with $\partial P \setminus \mathrm{bd}[a, d]$, a and d lie on the same sides of P as b and c respectively. We can also assume that there is no complete interior line in $T([a, d])$ closer to a and d than bc, otherwise we could have chosen it instead. So $T([a, d])$ has length $i([b, c]) + |\mathrm{bd}[a, b]| + |\mathrm{bd}[c, d]|$ if bc is purple. Otherwise, since $T([a, d])$ is a purple RST, bc does not contain any Steiner vertex and it can be swept until $b = a$ or $c = d$. □

Set $j_a(v, [a, d])$ to be the minimum of

- $j_a(v, [a, d] \setminus \{d\}) + |ud|$ if there is a line segment ud in $\mathrm{int}(P)$ where u is on va and $ud \perp va$,

- $j_a(v, [a, c]) + |cd|$ where c is the boundary vertex on $\mathrm{bd}[a, d]$ next to d,

- $\tau([a, d]) + |va|$.

The computation of $j_d(v, [a, d])$ is symmetric.

Lemma 4.9 $j_a(v, [a, d])$ *and* $j_d(v, [a, d])$ *are computed correctly in* $O(1)$ *time.*

Proof. Let $T_a(v, [a, d])$ be a RST that satisfies the constraints for $j_a(v, [a, d])$. Let c be the boundary vertex next to d on $\mathrm{bd}[a, d]$. We assume that no leg of any interior corner is incident to d. If not, since $T_a(v, [a, d])$ does not overlap with $\partial P \setminus (\mathrm{bd}[a, d] \cup \{v\})$, we can first flip this interior corner to overlap with cd. If there is a complete interior line ℓ incident to d, then ℓ must be parallel to va and so there cannot be any tree edge pinned to va (they would cross ℓ otherwise). So $T_a(v, [a, d])$ has length $\tau([a, d]) + |va|$. Otherwise, d is connected to the point u on va by the line segment ud or d is connected to c by the line segment cd. If $T_a(v, [a, d])$ contains both ud and cd, then we can sweep ud so that it becomes incident to c. Thus, $T_a(v, [a, d])$ has length $j_a(v, [a, d] \setminus \{d\}) + |ud|$ or $j_a(v, [a, c]) + |cd|$. The analysis for $j_d(v, [a, d])$ is symmetric. $\square$

From Lemmas 4.5–4.9, we conclude that

Theorem 4.10 *Given n terminals on the boundary of a rectangle, an optimal RST can be computed in $O(n)$ time.*

5   Related Results

There are other results concerning constructing exact Steiner trees in graphs. We briefly survey them below.

Homotopic routing. In VLSI routing, one popular goal is to interconnect terminals with minimum length wiring. The given terminals lie on the boundaries of *modules* which are simple rectilinear polygons. Usually, only vertical and horizontal wires are allowed. The interconnection is based on an infinite underlying grid graph (with vertical and horizontal edges covered by modules removed). Terminals are partitioned into groups and only terminals in the same group need to be interconnected. In the homotopic routing problem, the topologies of the connections in the groups are specified (either manually or by the output of other automatic processes) and the problem is to find a "consistent" minimum length Steiner forest with one RST per group.

The topology of each group is given by a *net*. A net is a simple closed curve that passes through all the terminals in the group and does not enclose any module. The segment between consecutive terminals in a net is called a *subnet*. For each net η, the desired RST T_η must satisfy the constraint that each subnet in η can be deformed continuously to the corresponding tree path in T_η. Steiner trees for different nets should not intersect or touch each other.

In [27], it is described how to combine the result of Theorem 4.3 and the routing results in [20, 28, 30] to solve the homotopic routing problem. Essentially, the routing results in [20, 28, 30] can be applied to turn each net η into a simple rectilinear polygon that contains the terminals in η on its boundary and the desired T_η lies inside the polygon. Thus, Theorem 4.3 can be applied to construct T_η. The total running time is $O((\alpha + n)\beta \log(\alpha + n)\beta + Nn^2)$, where α is the complexity of the modules, β is the complexity of the nets, N is the size of the underlying graph, and n is the number of terminals.

Channel and Switchbox Routing.   Another VLSI routing problem is addressed in [12]. The given terminals lie on two parallel lines which enclose some rectangular modules between them. The region between the two parallel lines is called a *channel*. The goal is to construct a minimum length RST that spans the terminals and avoids the modules. Note that the solution is allowed to touch the module boundaries though. By laying an appropriate grid graph underneath the modules, this problem is solvable using the results in Theorem 3.4. The running time is $O(Nn^3 + (N \log N)n^2)$, where n is the number of terminals and N is the size of the underlying graph. In [12], an algorithm is presented that runs in time linear in n and exponential in the square of the number of modules α^2. This is an improvement when n is large and α is a small constant. In [13], the same result is extended to the case where terminals lie on the boundary of a rectangle (called switchbox) which contains rectangular modules as obstacles.

Convex Layers.   In [5], polynomial time algorithms are presented for two special cases. In first case, the underlying graph is planar and all terminals lie within a constant number of *layers* of the exterior face. The first layer consists of vertices on the exterior face. Remove these vertices and incident edges. The second layer consists of vertices on the exterior face of the resulting graph. Subsequent layers are defined the same way. The running time of the algorithm is $O(N^4)$, where N is number of vertices in the graph. In the second case, the problem is to find an optimal RST for n points in the plane that form an "onion" of constant number of layers. The outermost layer is the boundary P_0 of the rectilinear convex hull. Remove the points on P_0. The next layer is the boundary P_1 of the rectilinear convex hull of the remaining points. Subsequent layers are defined the same way. The running time of the algorithm is $O(n^{6l-1})$, where l is the number of layers.

Parallel Lines and Smooth Curve. The problem of finding an optimal RST spanning terminals on l parallel lines was first solved in [2]. The time and space complexities of this algorithm are reported in [7] to be $O(n16^l)$ and $O(n8^l)$ respectively. An improvement has been obtained in [7] which reduces the running time to $O(nl^3 10^l)$ and space to $O(nl5^l)$. A polynomial time algorithm is recently obtained in [8] for a generalization where the terminals lie on a constant number of disjoint compact curves.

6 Conclusion

We surveyed several closely related algorithmic techniques and structural properties that have been developed for finding exact Steiner trees in graphs and grid graphs. The aim is to give a coherent exposition of these materials so as to inspire future work on the GSMT problem.

References

[1] P.K. Agarwal and M.T. Shing, Algorithms for special cases of rectilinear Steiner trees: I. points on the boundary of a rectilinear rectangle, *Networks*, Vol.20 (1990) pp. 453-485.

[2] A.V. Aho, M.R. Garey, and F.K. Hwang, Rectilinear Steiner trees: Efficient special case algorithms, *Networks*, Vol.7 (1977) pp. 37-58.

[3] M. Bern, Network design problems: Steiner trees and spanning k-trees, Ph.D. thesis, Computer Science Division, University of California at Berkeley, 1987.

[4] M. Bern, Faster exact algorithms for Steiner trees in planar networks, *Networks*, Vol.20 (1990) pp. 109-120.

[5] M. Bern and D. Bienstock, Polynomially solvable special cases of the Steiner problem in planar networks, *Annals of Operations Research*, Vol.33 (1991) pp. 405-418.

[6] M. Bern and R.L. Graham, The shortest-network problem, *Scientific American*, January (1989) pp. 84-89.

[7] M. Brazil, D.A. Thomas, and J.F. Weng, Rectilinear Steiner minimal trees on parallel lines, in D.Z. Du, J.M. Smith, and J.H. Rubinstein (eds.) *Advances in Steiner trees*, (Kluwer Academic Publishers, 1998).

[8] M. Brazil, D.A. Thomas, and J.F. Weng, A polynomial time algorithm for rectilinear Steiner trees with terminals constrained to curves, manuscript, 1998.

[9] S.W. Cheng, Steiner tree for terminals on the boundary of a rectilinear polygon, *Proceedings of DIMACS workshop on network connectivity and facilities location*, April 28-30, 1997, (DIMACS Series in Discrete Mathematics and Theoretical Computer Science 40, 1998) pp. 39-57.

[10] S.W. Cheng, A. Lim, and C.T. Wu, Optimal Steiner trees for extremal point sets, *Proceedings of International Symposium on Algorithms and Computation*, 1993 pp. 523-532.

[11] S.W. Cheng and C.K. Tang, Fast algorithms for optimal Steiner trees for extremal point sets, *Proceedings of International Symposium on Algorithms and Computation*, 1995 pp. 322-331.

[12] C. Chiang, M. Sarrafzadeh, and C.K. Wong, An optimal algorithm for rectilinear Steiner trees for channels with obstacles, *International Journal of Circuit Theory and Application*, Vol.19 (1991) pp. 551-563.

[13] C. Chiang, M. Sarrafzadeh, and C.K. Wong, An algorithm for exact rectilinear Steiner trees for switchbox with obstacles, *IEEE Transactions on Circuits and Systems-I: Fundamental Theory and Applications*, Vol.39 (1992) pp. 446-454.

[14] J.P. Cohoon, D.S. Richards, and J.S. Salowe, An optimal Steiner tree algorithm for a net whose terminals lie on the perimeter of a rectangle, *IEEE Transactions on Computer-Aided Design*, Vol.9 (1990) pp. 398-407.

[15] T.H. Cormen, C.E. Leiserson, and R.L. Rivest, *Introduction to Algorithms*, (Cambridge, Massachusetts, The MIT Press, 1994).

[16] S.E. Dreyfus and R.A. Wagner, The Steiner problem in graphs, *Networks*, Vol.1 (1972) pp. 195-207.

[17] R.E Erickson, C.L. Monma, and A.F. Veinott, Send-and-split method for minimum-concave-cost network flows, *Mathematics of Operations Research*, Vol.12 (1987) pp. 634-664.

[18] M.L. Fredman and R.E. Tarjan, Fibonacci heaps and their uses in improved network optimization algorithms, *Journal of ACM*, Vol.16 (1987) pp. 1004-1023.

[19] M.R. Garey and D.S. Johnson, *Computers and Intractability*, (New York, W.H. Freeman and Company, 1979).

[20] R.I. Greenberg and F.M. Maley, Minimum separation for single-layer channel routing, *Information Processing Letters*, Vol.43 (1992) pp. 201-205.

[21] S.L. Hakimi, Steiner's problem in graphs and its implications, *Networks*, Vol.1 (1971) pp. 113-133.

[22] M. Hanan, On Steiner's problem with rectilinear distance, *SIAM Journal on Applied mathematics*, Vol.14 (1966) pp. 255-265.

[23] F.K. Hwang, On Steiner minimal trees with rectilinear distance, *SIAM Journal on Applied Mathematics*, Vol.30 (1976) pp. 104-114.

[24] F.K. Hwang and D.S. Richards, Steiner tree problems, *Networks*, Vol.22 (1992) pp. 55-89.

[25] F.K. Hwang, D.S. Richards, and P. Winter, *The Steiner Tree Problem*, (Amsterdam, North-Holland, Annals of Discrete Mathematics 53, 1992).

[26] R.M. Karp, Reducibility among combinatorial problems, in R.E. Miller and J.W. Thatcher (eds.), *Complexity of Computer Computations*, (New York, Plenum Press, 1972) pp. 85-103.

[27] M. Kaufmann, S. Gao, and K. Thulasiraman, An algorithm for Steiner trees in grid graphs and its application to homotopic routing, *Journal of Circuits, Systems, and Computers*, Vol.6 (1996) pp. 1-13.

[28] C.E. Leiserson and F.M. Maley, Algorithms for routing and testing routability of planar VLSI layouts, *Proceedings of the 17th Annual ACM Symposium on the Theory of Computing*, 1985 pp. 69-78.

[29] A.J. Levin, Algorithm for the shortest connection of a group of graph vertices, *Soviet Math. Doklady*, Vol.12 (1971) pp. 1477-1481.

[30] F.M. Maley, *Single-Layer Wire Routing and Compaction*, (Cambridge, Massachusetts, The MIT Press, 1990).

[31] J.S. Provan, Convexity and the Steiner tree problem. *Networks*, Vol.18 (1988) pp. 55-72.

[32] D.S. Richards and J.S. Salowe, A linear-time algorithm to construct a rectilinear Steiner tree for k-extremal points, *Algorithmica*, Vol.7 (1992) pp. 246-276.

[33] P. Winter, Steiner problem in networks: A survey. *Networks*, Vol.17 (1987) pp. 129-167.

[34] A.Z. Zelikovsky, An 11/6-approximation algorithm for the network Steiner problem, *Algorithmica*, Vol.9 (1993) pp. 463-470.

[35] A.Z. Zelikovsky, A faster approximation algorithm for the Steiner tree problem in graphs, *Information Processing Letters*, Vol.46 (1993) pp. 79-83.

Grade of Service Steiner Trees in Series-Parallel Networks

Charles J. Colbourn and Guoliang Xue
Department of Computer Science
University of Vermont, Burlington, VT 05405
E-mail: {colbourn, xue}@cs.uvm.edu

Abstract

The grade of service Steiner tree problem is to determine the minimum total cost of an assignment of a grade of service to each link in a network, so that between each pair of nodes there is a path whose minimum grade of service is at least as large as the grade required at each of the end nodes. This problem has important applications in communication networks and in transportation. It generalizes the Steiner tree problem, in which there are two grades of service. When the network has n nodes and there are r grades of service, an algorithm to determine the cost of a grade of service Steiner tree is given which runs in $O(r^3 n)$ time on series-parallel networks.

Contents

D.-Z. Du et al. (eds.), Advances in Steiner Trees, 163-174.

1   The Problem

A communication network is often modeled by an undirected graph $G = (V, E)$ where V is the set of vertices and E is the set of edges. With each edge $e \in E$, there is an associated cost $c(e) \geq 0$. The traditional Steiner tree problem asks for a minimum cost subgraph of G which spans a given subset of vertices. Such problems find important applications in communication networks and have been studied by many researchers. We refer the readers to the survey papers [8, 13] and the book [9] for details.

In this traditional network model, there is only one cost function defined on the edges of the network and there are only two kinds of service requested by the vertices of the network: connected, or not (necessarily) connected. In real-life problems, the vertices may have more than two kinds of service request and the edges may have service-dependent cost functions. For example, all major universities are to be connected to the Internet via a $T3$-line, while other universities and colleges are to be connected to the Internet via a $T1$-line. Linking two points with a $T3$-line is more expensive than linking the two points with a $T1$-line, but the latter has nonzero cost and therefore is more expensive than not linking at all. As an example in transportation, all major cities are to be connected via interstate highways, while small cities and towns must be connected via local highways. Again, the cost of building an interstate highway is higher than the cost of building a local highway.

In this paper, we study the *grade of service Steiner tree (GOSST) problem* formally defined as follows.

Input:

1. A graph $G = (V, E)$ and an integer r.

2. A function $g : V \mapsto \{0, 1, \ldots, r-1\}$. The integer $g(x)$ for $x \in V$ is the *grade of service* required by vertex v.

3. A function $c : E \times \{0, \ldots, r-1\} \mapsto \mathbf{R} \cup \{\infty\}$ (the real numbers adjoined with the number infinity). The value $c(e, g)$ is the *cost* of providing grade of service at least g on the edge e. The function c satisfies $c(e, i) \geq c(e, i-1)$ for $1 \leq i < r$ and $c(e, 1) > c(e, 0) = 0$ for each edge $e \in E$.

Problem: Determine the minimum total cost of an assignment, if one exists, of one grade of service to each edge, so that for every two vertices x and y in G, there is a path connecting x and y on which the minimum grade of service assigned is at least $\min\{g(x), g(y)\}$. If no such assignment exists, report ∞ as the solution.

When $r = 2$, the *GOSST* problem has only two grades of service. Grade 0 offers no service at all, while grade 1 offers full service. Hence the well known Steiner tree problem is a special case of the *GOSST* problem. Since the Steiner tree problem is NP-hard [3], the *GOSST* problem is also NP-hard. Therefore the existence of efficient optimal algorithms for solving either problem is very unlikely. In this paper, we are interested in solving the *GOSST* problem on a class of sparse networks known as *series-parallel* graphs which are subgraphs of *2-trees* defined below.

Following [11], a *2-tree* can be defined recursively as follows, and all 2-trees may be obtained in this way. A triangle is a 2-tree. Given a 2-tree and an edge $\{x, y\}$ of the 2-tree, we can add a new vertex z adjacent to both x and y; the result is a 2-tree. A series-parallel graph (also known as a partial 2-tree) is a subgraph of a 2-tree. With this definition, one can see that a 2-tree on n vertices has $2n - 3$ edges and $n - 2$ triangles.

Farley [6] demonstrated that 2-trees are isolated failure immune (IFI) networks. Wald and Colbourn [11] showed that a minimum IFI network is a 2-tree. This fact made 2-trees an important class of fault tolerant networks. Wald and Colbourn also showed that the Steiner tree problem on a series-parallel graph is tractable, presenting a linear time algorithm.

In this paper, we show that the *GOSST* problem on a series-parallel graph is also tractable. Using the partial 2-tree representation of series-parallel graphs, we present an algorithm for solving the *GOSST* problem on series-parallel graphs whose running time is linear in n, the number of vertices of G, and polynomial in r, the number of grades of service.

The literature on the *GOSST* problem is relatively recent and most research has focused on the cases where $r = 2$ or 3. Current *et al.* [2] formulated this problem as an integer linear programming problem and developed a heuristic algorithm which employs a K shortest paths algorithm and a minimum spanning tree algorithm. Duin and Volgenant [5] proposed two heuristic algorithms for the special case where $r = 3$. Balakrishnan, Magnanti and Mirchandani [1] proposed an approximation algorithm with performance ratio of $\frac{4}{3}$ for the special case where $r = 3$ and that $c(e, 3)/c(e, 2)$ is a constant for all edges. For general values of r, Mirchandani [10] proposed an approximation algorithm whose performance ratio is $r\rho + 1$, where ρ is the best performance ratio of a Steiner tree heuristic. Related work can also be found in [4, 14].

The grade-of-service Steiner tree problem is also closely related to the design of survivable networks. In that case, again various grades of service are required, but multiple paths can be employed to achieve the grade. See [7] for an excellent survey of this topic, and [12] for an efficient algorithm on series-parallel networks in the biconnected case.

2   The Algorithm

First, to solve GOSST on series-parallel graphs, it suffices to solve it for 2-trees. To see this, use the Wald-Colbourn method [11] to embed a series-parallel graph in a 2-tree in $O(n)$ time. Whenever an edge e is added, simply set $c(e, 0) = 0$ and $c(e, i) = \infty$ for $1 \leq i < r$. The cost of the GOSST does not change.

Our basic strategy is standard. We reduce the 2-tree by suppressing degree two nodes, and by merging parallel edges, until only a single edge remains. At every point, an edge serves to represent the subgraph thus far reduced onto it, and a number of measures are associated with each edge.

To facilitate the presentation and implementation, we arbitrarily assign a direction to each edge $e = \{x, y\}$, replacing it with directed arcs $e = (x, y)$ and $\bar{e} = (y, x)$. This is a notational convenience only, to permit us to refer to the first vertex x and the second vertex y of edge e.

With every arc e, a total of $2r^2$ measures are maintained. Since an n-vertex 2-tree has $2n - 3$ edges and hence $4n - 6$ arcs are represented, $(4n - 6)2r^2$ measures are to be stored. The $2r^2$ measures for an arc e have the following representation and meaning. The measures for arc $e = (x, y)$ are of the form $(\alpha, \beta, \kappa, e)$ where $\alpha, \beta \in \{0, 1, \ldots, r - 1\}$ and $\kappa \in \{C, D\}$. The interpretation of each measure depends upon the subgraph S_e thus far reduced onto the arc e. In each case, α indicates the grade of service expected at x (when the cost is finite, this is always at least $g(x)$, but may be higher if x is assumed to be needed to connect a vertex within S_e to a vertex not in S_e). Similarly, β indicates the grade of service expected at y. Depending upon the selection of κ, various assignments are permitted to the arcs of S_e in which each arc and its reversal receive the same grade. Let S_e be the graph thus far reduced onto the arc $e = (x, y)$. If $\kappa = C$, let T_e be S_e;

if $\kappa = D$, let T_e be S_e together with the *auxiliary arc* (x, y), whose cost for grade of service $\min\{\alpha, \beta\}$ is 0. Then we define a *κ-acceptable assignment* to be an assignment of grades of service to the edges of S_e for which

1. For every two vertices w, z of S_e, there is a path from w to z in T_e whose minimum grade of service is at least $\min\{g(w), g(z)\}$; and

2. There is a path from x to y in T_e whose minimum grade of service is at least $\min\{\alpha, \beta\}$.

3. If w is a vertex of S_e, then there is a path in T_e from w to x whose minimum grade of service is at least $\min\{\alpha, g(w)\}$, and a path from w to y whose minimum grade of service is at least $\min\{\beta, g(w)\}$.

When $\kappa = D$, the second condition is met automatically by the presence of the auxiliary arc in T_e. Then $(\alpha, \beta, \kappa, e)$ represents the minimum cost of a κ-acceptable assignment to the arcs of S_e, and takes on the value ∞ when there is no such acceptable assignment.

The letters 'C' and 'D' represent *connected* and *disconnected*, and serve to remind us whether we have already required the ends of the edge to be connected. We also compute the value $H(e)$, the maximum grade of service required on a vertex of S_e *other than* x and y. This value is needed to determine when a vertex in S_e requires a grade of service greater than that provided by x or y. If a vertex in the remainder of the graph also expects such a higher grade of service, we shall be unable to provide it.

Arc e and its reverse $\bar{e}$ are both retained, but $(\alpha, \beta, \kappa, e)$ and $(\beta, \alpha, \kappa, \bar{e})$ are the same, so we actually only need to store one of the sets of measures.

If we can calculate these measures, we can solve GOSST. To see this, suppose that we have reduced the entire graph onto the arc $e = (x, y)$. Then, by the definition of the measures, we need only examine $(g(x), g(y), C, e)$.

So it remains to establish methods to initialize and update these measures. Initially, we want each arc $e = (x, y)$ to have just the single edge $\{x, y\}$ in S_e. Hence we initialize as shown in Table 1.

Next we describe how to suppress a vertex z of degree two. Supposing that $e' = (x, z)$ and $e'' = (z, y)$ are the two arcs incident at z, we replace these by the arc $e = (x, y)$ and compute measures for e. In the expressions given in Table 2, when a minimum is taken over an empty set, its value is ∞. When a measure is referenced whose expected grade of service is outside the permissible range, its value is taken to be ∞. To understand the reductions, observe that $(H(e') \leq \max\{\alpha, \gamma\})$ or $(H(e'') \leq \max\{\gamma, \beta\}))$ is necessary and

$$(\alpha, \beta, C, e) = \begin{cases} \infty \text{ if } \alpha < g(x) \text{ or } \beta < g(y) \\ c(e, \min\{\alpha, \beta\}) \text{ otherwise} \end{cases}$$

$$(\alpha, \beta, D, e) = \begin{cases} \infty \text{ if } \alpha < g(x) \text{ or } \beta < g(y) \\ 0 \text{ otherwise} \end{cases}$$

$$H(e) = 0$$

Table 1: Initialization

$$(\alpha, \beta, C, e) = \min\{(\alpha, \gamma, C, e') + (\gamma, \beta, C, e'') :$$
$$((H(e') \leq \max\{\alpha, \gamma\}) \text{ or } (H(e'') \leq \max\{\gamma, \beta\})),$$
$$\gamma \geq \min\{\alpha, \beta\},$$
$$\gamma \geq \min\{\alpha, H(e'')\},$$
$$\gamma \geq \min\{H(e'), \beta\}\}$$

$$(\alpha, \beta, D, e) = \min\{(\alpha, \beta, C, e),$$
$$\min\{(\alpha, \gamma, C, e') + (\gamma, \beta, D, e'') :$$
$$((H(e') \leq \max\{\alpha, \gamma\}) \text{ or } (H(e'') \leq \max\{\gamma, \beta\}))$$
$$\alpha \geq \min\{\gamma, \beta\}\},$$
$$\min\{(\alpha, \gamma, D, e') + (\gamma, \beta, C, e'') :$$
$$((H(e') \leq \max\{\alpha, \gamma\}) \text{ or } (H(e'') \leq \max\{\gamma, \beta\}))\}$$
$$\beta \geq \min\{\alpha, \gamma\}\}$$

$$H(e) = \max\{H(e'), g(z), H(e'')\}$$

Table 2: Suppressing a Degree Two Vertex

$$(\alpha, \beta, C, e) = \min\{(\alpha, \beta, C, e') + (\alpha, \beta, D, e''), (\alpha, \beta, D, e') + (\alpha, \beta, C, e'') :$$
$$((H(e') \le \max\{\alpha, \beta\}) \text{ or } (H(e'') \le \max\{\alpha, \beta\}))\}$$
$$(\alpha, \beta, D, e) = \min\{(\alpha, \beta, C, e), (\alpha, \beta, D, e') + (\alpha, \beta, D, e'') :$$
$$((H(e') \le \max\{\alpha, \beta\}) \text{ or } (H(e'') \le \max\{\alpha, \beta\}))\}$$
$$H(e) = \max\{H(e'), H(e'')\}$$

Table 3: Reducing Multiple Edges

sufficient for the existence of a path with service level $\min\{H(e'), H(e'')\}$ between the vertex in $S(e')$ (other than x and z) with service request $H(e')$ and the vertex in $S(e'')$ (other than z and y) with service request $H(e'')$. Other conditions ensure that x can reach the vertex of maximum grade in $S_{e''}$, that y can reach the vertex of maximum grade in $S_{e'}$, and that x can reach y.

Now we describe the reduction of two arcs e' and e'', both of the form (x, y), and their replacement by the single arc e again of the form (x, y). Table 3 gives the corresponding computations.

Obviously updating any particular measure takes $O(r)$ time assuming that arithmetic operations take constant time. It follows that the entire graph can be reduced in $O(r^3 n)$ time, and hence we have proved the following theorem.

Theorem 2.1 *The GOSST problem can be solved in $O(r^3 n)$ time on an n-vertex series-parallel graph.* □

3 An Example

A sample series-parallel network is illustrated in Figure 1. Every vertex is labeled with a capital letter followed by an integer in parentheses which gives the degree of service request of that vertex. For example, vertex A has a service request of grade 2 while vertex C has a service request of grade 0. An edge is labeled with a positive integer followed by the letter g. For example, the edge $\{A, B\}$ is labeled with $2g$ which means that the cost of providing a service of grade g on this edge has a cost of $2 \times g$. The given network is only a *partial* 2-tree. Adding the dotted edge connecting vertices

A and U makes it a 2-tree. We use this example to illustrate the reduction steps of our algorithm.

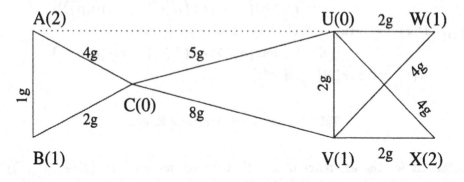

Figure 1: *A series-parallel network:* The number in parenthesis at each vertex represents the grade of service requested by the vertex. For example, the grade of service request at vertex U is 0 while that at vertex V is 1. The edge label represents the edge cost function. For example, the cost to provide grade-g service on edge $\{U, X\}$ is $4 \times g$ for $g = 0, 1, 2$.

We write arcs as pairs of vertices, using AB to denote the arc (A, B), for example. Primes and double primes are used to indicate new measures on the arc after a reduction is completed.

The algorithm is carried out in the following way. After the initialization step using the rules in Table 1, we perform five pairs of reductions. The first consists of suppressing vertex X using the rules in Table 2. A new edge connecting U and V is created as the result of the suppression of vertex X. This new edge is then reduced along with the arc UV using the rules in Table 3, to form UV'. The two steps in eliminating the vertex are tabulated in the two columns headed by UV' in Table 5. The next four pairs of reductions are carried out with the suppression of the vertices W, V, U, C, in the given order, to form arcs UV'', CU', AC' and AB', respectively. The values of the $2r^2(2n-3)$ measures and their updates are shown in Tables 4-5. The $(\alpha\beta\kappa, PQ)$ entry gives the value of $(\alpha, \beta, \kappa, (P, Q))$. For example, the $(00C, VX)$ entry is ∞ which means $(0, 0, C, (V, X)) = \infty$ and the $(21C, AC)$ entry is 4 which means $(2, 1, C, (A, C)) = 4$. The cost of the GOSST for the given example can be read from Table 5; it is 26.

This method for computing the value of the GOSST is a dynamic programming algorithm. Whenever an updated value of a measure is computed, we may set pointers to the items which resulted in this new value (the one

	AB	CB	AC	AU	UC	VC	UV	UW	WV	UX	XV
00C	∞	∞	∞	∞	0	∞	∞	∞	∞	∞	∞
01C	∞	0	∞	∞	0	∞	0	0	∞	∞	∞
02C	∞	0	∞	∞	0	∞	0	0	∞	0	∞
10C	∞	∞	∞	∞	0	0	∞	∞	∞	∞	∞
11C	∞	2	∞	∞	5	8	2	2	4	∞	∞
12C	∞	2	∞	∞	5	8	2	2	4	4	∞
20C	∞	∞	0	∞	0	0	∞	∞	∞	∞	∞
21C	1	2	4	∞	5	8	2	2	4	∞	2
22C	2	4	8	∞	10	16	4	4	8	8	4

Table 4: *Initial Values of C Measures.*

	UV'		UV''		UC'		AC'		AB'	
H	2	2	1	2	2	2	2	2	2	2
00C	∞	∞	∞	∞	6	6	∞	∞	∞	∞
00D	∞	∞	∞	∞	6	6	∞	∞	∞	∞
01C	2	2	4	6	14	11	∞	∞	∞	∞
01D	2	2	4	6	14	11	∞	∞	∞	∞
02C	4	4	4	8	24	20	∞	∞	∞	∞
02D	4	4	4	8	24	20	∞	∞	∞	∞
10C	∞	∞	∞	∞	6	6	∞	∞	∞	∞
10D	∞	∞	∞	∞	6	6	∞	∞	∞	∞
11C	6	4	6	6	14	11	∞	∞	∞	∞
11D	2	2	2	4	6	6	∞	∞	∞	∞
12C	8	6	6	8	24	20	∞	∞	∞	∞
12D	4	4	2	6	22	20	∞	∞	∞	∞
20C	∞	∞	∞	∞	10	10	∞	28	∞	∞
20D	∞	∞	∞	∞	10	10	∞	28	∞	∞
21C	10	8	6	10	18	15	∞	28	30	26
21D	8	8	2	10	10	10	∞	28	28	26
22C	12	8	6	10	26	20	∞	28	32	26
22D	4	4	2	6	10	10	20	20	24	24

Table 5: *Computed Values of Measures after Reductions.*

which attains the minimum). Once the optimal value of the GOSST is computed, we may find the optimal tree by tracing out the pointers which we set during the computation. For our example, the optimal tree is illustrated in Figure 2. Vertices B, C, U and V all receive service of grade 2, which is higher than being requested.

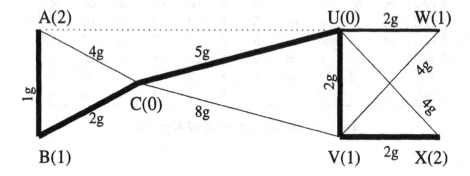

Figure 2: *The optimal solution:* The thickest edges have service of grade 2. The edge connecting U and W has service of grade 1. All other edges have service of grade 0.

4   Conclusions

In this paper, we have studied the grade of service Steiner tree problem on a series-parallel network and have shown that the optimal solution can be found in $O(r^3 n)$ time where n is the number of vertices in the network and r is the number of grades of service. Possible future research directions include generalizing the techniques used in this paper to other sparse networks and to problems where each link has an operational probability.

Acknowledgments

The research was supported in part by the U.S. Army Research Office under Grants DAAG55-98-1-0272 (Colbourn) and DAAH04-96-1-0233 (Xue). Research of G.L. Xue is also supported by the National Science Foundation grants ASC-9409285 and OSR-9350540.

References

[1] A. Balakrishnan, T.L. Magnanti, and P. Mirchandani, Modeling and heuristic worst-case performance analysis of the two-level network design problem, *Management Science*, Vol. 40(1994), pp. 846–867.

[2] J.R. Current, C.S. Revelle, and J.L. Cohon, The hierarchical network design problem, *European Journal of Operational Research*, Vol. 27(1986), pp. 57–66.

[3] M.R. Garey, R.L. Graham and D.S. Johnson, The complexity of computing Steiner minimal trees, *SIAM Journal on Applied Mathematics*, Vol. 32(1977), pp. 835–859.

[4] C. Duin and A. Volgenant, Reducing the hierarchical network design problem *European Journal of Operational Research*, Vol. 39(1989), pp. 332–344.

[5] C. Duin and T. Volgenant, The multi-weighted Steiner tree problem, *Annals of Operations Research*, Vol. 33(1991), pp. 451–469.

[6] A.M. Farley, Networks immune to isolated failures, *Networks*, Vol. 11(1981), pp. 255–268.

[7] M. Grötschel, C.L. Monma, and M. Stoer, Design of survivable networks, in: *Network Models* (M. Ball et al., eds.) Elsevier North-Holland, 1995, pp. 617-672.

[8] F. Hwang and D. Richards, Steiner tree problems, *Networks*, Vol. 22(1992), pp. 55–89.

[9] F.K. Hwang, D.S. Richards and Pawel Winter, *The Steiner tree problem*, *Annals of Discrete Mathematics*, Vol. 53, North-Holland, 1992.

[10] P. Mirchandani, The multi-tier tree problem, *INFORMS Journal on Computing*, Vol. 8(1996), pp. 202–218.

[11] J.A. Wald and C.J. Colbourn, Steiner trees, partial 2-trees, and minimum IFI networks, *Networks*, Vol. 13(1983), pp. 159–167.

[12] P. Winter, Generalized Steiner problem in series-parallel networks, *J. Algorithms*, Vol. 7(1986), pp. 549-566.

[13] P. Winter, The Steiner problem in graphs: a survey, *Networks*, Vol. 17(1987), pp. 129–167.

[14] G.L. Xue, G.-H. Lin and D.-Z. Du, Grade of service Euclidean Steiner minimum trees, *1999 IEEE International Conference on Circuits and Systems*, to appear.

Preprocessing the Steiner Problem in Graphs

Cees Duin
AKE, Faculteit Economische Wetenschappen en Econometrie
Universiteit van Amsterdam
Roeterstraat 11, 1018 WB Amsterdam
E-mail: ceesd@fee.uva.nl

Contents

D.-Z. Du et al. (eds.), Advances in Steiner Trees, 175-233.
© 2000 *Kluwer Academic Publishers.*

1 Introduction

For combinatorial optimization problems that are NP-hard, it is important, before running a time consuming algorithm, to try to reduce the input size of the problem. This is the objective of a so-called preprocessing algorithm. A renowned NP-hard problem is the Steiner Problem in Graphs (SPG). It considers a weighted graph, denoted here as $G = (V, K, E, c)$, with V the set of vertices, K a subset of so-called special vertices, E the set of undirected edges, and $c : E \to Z^+$ a positive integer weight function on E. The problem is to find a tree S of minimum total edge weight that spans the vertices of K.

Two special cases of the SPG are well known and well solvable: the Shortest Path Problem for $|K| = 2$, and for $K = V$ the Minimum Spanning Tree Problem. Generally an optimal solution, called a Steiner tree, is also a minimum spanning tree (MST), but spanning a subset of vertices $K \cup W$, where the set W containing the so-called Steiner vertices is to be determined. The SPG is NP-hard and many special cases remain so, e.g., the Rectilinear Steiner problem with applications in the VLSI design of microchips.

For the SPG there are ample opportunities with regard to preprocessing. Once these opportunities were revealed in [2] as so-called reduction tests, many other tests were proposed in literature. One can find a review of them in [14]. In [6] the use of so-called bottleneck distances is introduced; in [7] such distances are exploited in three tests that dominate other tests. In [20] the bottleneck tests are specialized for use on the Rectilinear Steiner Problem.

The exact branch and cut algorithm of [4] owes part of its limited success to a now outdated preprocessing phase. A more sophisticated implementation of this approach given in [15], can solve much larger problem instances and its preprocessing phase applies the three bottleneck tests. In [9] additional reduction is realized by a sensitivity test; good results are attained by an exact branch and bound algorithm based on reduction only.

This contribution reviews the reduction tests mentioned above and reconsiders their implementation. It also presents stronger tests that could lead to a further reduction of the SPG.

In section 1.1, we first discuss the concept of preprocessing. Next, the reader is familiarized with terminology and opportunities of reduction for the SPG, when we give four simple tests each introducing a basic idea of reduction.

In separate sections −section 2, 3 and 4− we further develop three of these ideas. First we describe more efficient implementations of the three bottleneck tests of [7]. Next we formulate stronger test versions. In section 5 we discuss the sensitivity test of [9], which relates to the fourth reduction idea. By exploiting vertex implied Steiner edges, a concept introduced in section 6, it is possible to further improve each of the mentioned reduction tests. In section 7 we give some computational results to illustrate the impact of preprocessing.

1.1 Preprocessing

Generally speaking, the idea of preprocessing is to transform an instance I of problem P by means of a forerunner algorithm, here the preprocessing algorithm π, into a simplified instance I' of problem P. One expects that the new instance I' is easier as input for the solver, the main algorithm α.

In order for preprocessing to be valid the instance I' should be equivalent to I in the following sense. Any feasible solution T' for instance I' must translate back, say by means of a backtracking algorithm, τ, into a feasible solution T, such that τ preserves optimality: any optimal solution for I' is

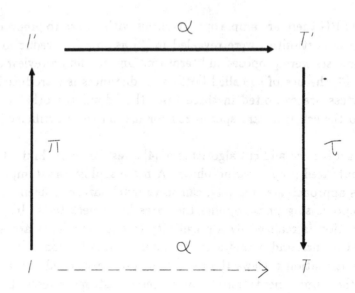

Figure 1: Preprocessing as a composition of algorithms

mapped by τ to an optimal solution of I, see Figure 1. Normally, algorithm τ will be much simpler and less time consuming than π. Besides T' the input of τ additionally consists of records of actions of π; by a reversal of the relevant part of the actions of π, algorithm τ transforms solution T' back to a solution T of the original instance I. When the preprocessing actions of algorithm π are of erasive type only, i.e. $I' \subset I$, then algorithm τ acts trivially, without operations whatsoever, comparable to an identity mapping.

In order for preprocessing to be worth while, it should be expected that the entire trajectory, algorithm $\tau \cdot \alpha \cdot \pi$ (first π, then α and finally τ), will run faster than a straightforward application of main algorithm α. This is the sole objective if algorithm α is an exact algorithm, otherwise, where the main algorithm α is an approximate algorithm, a second motivation can be a likelihood for algorithm $\tau \cdot \alpha \cdot \pi$ to produce a better solution than α. The first objective requires algorithm π to run faster than algorithm α and such that the generated new instance I' is significantly easier for algorithm α. When α is an approximate algorithm, it may also happen that due to the preprocessing trajectory we get returned a worse solution. The extent of such a negative influence, if any, and whether or not the attained speed up remains attractive, can normally be decided only by numerical experiments.

For the Steiner problem in graphs the instance I is given as $G = (V, K, E, c)$

and after preprocessing the instance I' is a so-called reduced graph $G' = (V', K', E', c')$. As will become clear in the next sections an effective preprocessing algorithm for the SPG can consist of a number of subroutines, called reduction tests. One of the tests, an exclusion test, may recognize that an edge is redundant; if so, it simply removes the edge. Another type of test, an inclusion test, may recognize an edge as an optimal edge; if so, it contracts such an edge, thereby changing the graph and its edge costs. Formally, it is not enough to output G' plus the total costs of all optimal edges found during the process. Algorithm π should additionally produce records for each non-erasive graph modification, for instance for each edge contraction. These are necessary as input for the backtracking algorithm τ, in order to retrace from T' the solution T in the original graph G.

As mentioned earlier, if π's actions are of erasive type only, then it is redundant to run algorithm τ. For the SPG it often happens, that not algorithm τ, but in stead the 'main' algorithm α is without work. Frequently, the reduced instance will consist of a 'graph' with just one (special) vertex and no edges, and algorithm α is to transfer this vertex (the obvious optimal solution T') to algorithm τ for backward translation.

On the other hand, for a polynomial time preprocessing algorithm π one cannot always have $|I'| < |I|$, unless $P = NP$. If π could guarantee a minimal reduction, then by repetition of π one would obtain an exact algorithm of polynomial time.

1.2 Reduction Ideas and Terminology

First, we list below some definitions and notations for weighted graphs that are used throughout this paper.

- For a subgraph H of (V, E) the set $V(H)$ contains the vertices of H and $E(H)$ the edges of H. Often for short, we identify H with its set of vertices or its set of edges, writing e.g., $k \in H$ instead of $k \in V(H)$.

- Given a tree T and a subset of its vertices A we denote by T_A the minimal subtree in T that spans A; in particular $T_{\{i,j\}}$ is the fundamental path in T between the vertices i and j.

- Sets are denoted by uppercase and set-cardinality by the same character in lowercase. Thus $k = |K|$ in an arithmetic context such as a time complexity order. In other contexts we write $k \in K$ for a vertex of K.

- For a finite set A and a real valued function $b : A \to \Re$, we write $b(A)$ for the sum $\Sigma_{a \in A} b(a)$.

- By c_{ij} or $c(i,j)$ we denote the weight (length or cost) of an edge; for convenience $c_{ij} = \infty$ if edge (i,j) does not exist. The length of a shortest path between vertices i and j in G is denoted by d_{ij} or $d(i,j)$. By $[i..j]$ we denote a shortest $i - j$ path.

- The distance graph (A, d) for a subset of vertices $A \subset V$ is defined as the complete graph with vertex set A and for any pair $a, b \in A$ an edge $[a..b]$ of weight d_{ab}. Note that, depending on the context, $[a..b]$ is an edge (in distance graphs) or a set of edges (in G).

Next, as an introduction, we present four simple tests that can be seen as forerunners of more sophisticated tests to be developed in later sections. While doing so, we familiarize the reader with other preprocessing terminology.

The first idea that comes to mind is to eliminate edges for which there exists a more attractive path.

Least Cost (Lc) An edge $(i,j) \in E$ can be eliminated if $d_{ij} < c_{ij}$.

Here the *test's object* is an edge and the *test's condition* is $d_{ij} < c_{ij}$; when true the *test's action* is elimination from E.

Observe that all 'Lc-edges' (edges for which the Lc-condition, $d_{ij} < c_{ij}$ holds) and possibly more edges are eliminated when the test condition covers also certain cases of equal cost, such as in the new formulation: Eliminate edge (i,j) if there exists a shortest $i - j$ path that is non-trivial ($\neq \{(i,j)\}$).

In the above paragraph we exemplified the technique of *relaxing a test condition*, reformulating a condition such that it is again valid and more generally fulfilled by the tested objects. If the relaxed condition can still be checked efficiently, then the result is a more effective reduction test, which is said to *dominate* the former test.

If part of a Steiner tree, a vertex $i \notin K$ must be adjacent to at least two other vertices. This implies:

Degree 1-2 ($D_{\leq 2}$) A vertex $i \notin K$ of degree 1 or 2 can be eliminated.

If the degree is two, say i is adjacent to vertices p and q, then the test action is *pseudo-elimination*, that is: i and its edges (p, i), (i, q) are substituted by a single edge (p, q) of weight

$$c_{pq} := \min\{c_{pq}, c_{pi} + c_{iq}\}.$$

The next test is reminiscent of a well known property of minimum spanning trees and can be verified analogously.

Minimum Adjacency (MA) Edge (i, j) with $i, j \in K$ is part of a Steiner tree if its weight is minimal among edges incident to i (or j).

Unlike the other tests, MA is constructive and can be called an *inclusion* test. Trivially, *exclusion* tests, where edges and/or vertices are eliminated, decrease the size of the input. To attain a reduction of the problem graph the test action of MA, can be edge contraction, that is, merge the end vertices of the detected optimal edge. One could however imagine other scenarios. For example, if removal of the optimal edge separates the vertices of K into two components (a so-called K-articulation edge), then the preferred test action is problem decomposition, into two smaller problems defined on these components.

Suppose, we have available an upper bound solution U of cost $c(U)$. Then we can eliminate vertices i that are non-reachable, in the sense that a lower bound on the cost of a Steiner tree containing vertex i exceeds the value $c(U)$.

Reachability (R$_1$) Eliminate vertex i if $\max_{k \in K}\{d_{ik}\} > c(U)$.

The first three tests given here operate −from a topological point of view− locally, processing only a part of the input graph. For example test LC, when applied on (i, j), needs only to scan edges in a neighborhood $\varepsilon = c_{ij}$ of i. The last test R$_1$ however, uses global information on the entire graph. Tests that dominate respectively LC, D$_{\leq 2}$ and MA will also use global information by means of so-called bottleneck distances. A specific type of bottleneck distance (between any two vertices i and j) is the so-called special distance s_{ij}; it can be shorter than the shortest path distance d_{ij}. This will lead to a stronger version of test LC, the so-called special distance test in section 2.

1.3　Steiner Trees and Bottleneck Distances

'Bottleneck distances' were introduced and exploited for the SPG in [6]. We review in this section their definitions and implications. A parallel between minimum spanning trees and Steiner trees is given.

- For a path $P \subset G$, the *bottleneck length* $bl(P)$ is the maximum edge weight on P.

- The *bottleneck distance* b_{ij} is a minimum bottleneck length, i.e.,

$$b_{ij} = \min\{bl(P)| \ P \text{ is an } i-j \text{ path}\}.$$

If edge (i,j) exists then the most simple path between i and j is $\{(i,j)\}$, so we have $\forall i, j \in V : b_{ij} \leq c_{ij}$. To calculate b_{ij} one need not check every possible path from i to j. Given a tree S we let $S_{\{i,j\}}$ denote the fundamental path in S between two vertices i, j in S. Because of the next theorem, first observed in [11], all bottleneck distances can be found in a MST T on $G = (V,E,c)$ in $O(v^2)$ time.

Theorem 1.1 b_{ij} *is equal to* $bl(T_{\{i,j\}})$, *where* T *denotes the MST on* G.

Proof. Let $T_{\{i,j\}}$ have (s,t) as maximum weighted edge and assume that $b_{ij} < c_{st}$ (by definition $b_{ij} \leq c_{st}$). Let

$$\{(i = i_0, \ i_1),(i_1, \ i_2), ...,(i_n, i_{n+1} = j)\}$$

be a path of bottleneck length b_{ij} and consider the two components of $T\backslash\{(s,t)\}$, say T_i, T_j, with $i \in T_i$ and $j \in T_j$. Considering the largest index m with $i_m \in T_i$, we can form a feasible tree $T' = \{(i_m, i_{m+1})\} \cup T\backslash\{(s,t)\}$ that contradicts the optimality of T, because $c(i_m, i_{m+1}) \leq b_{ij} < c_{st}$.　□

Corollary 1 *Edge* $(i,j) \in E$ *is part of a minimum spanning tree iff* $b_{ij} = c_{ij}$.

For the SPG we examine two types of bottleneck distances: special distances and Steiner distances; the first by restriction to so-called special paths, the other by considering so-called Steiner paths. A *special* $i-j$ *path* is a path between i and j in the distance graph of $K \cup \{i,j\}$. A *Steiner path* is a path that is part of some Steiner tree.

- The *special distance* s_{ij} between $i, j \in V$, is the minimum bottleneck length over special $i - j$ paths, i.e.,

$$s_{ij} = \min\{bl(Q_{ij}), \text{ where } Q_{ij} \text{ is an } i - j \text{ path in } (K \cup \{i, j\}, d)\}.$$

- The *Steiner distance* σ_{ij} is the maximum edge weight on a Steiner path from i to j (if any), i.e.,

$$\sigma_{ij} = \max\{c_{vw}|(v, w) \in S_{\{i,j\}}, \text{ where } S \text{ is a Steiner tree with } i, j \in S\}.$$

In particular $\sigma_{ij} = -\infty$ if i and j cannot be simultaneously in a Steiner tree. A set of vertices that can be spanned by a Steiner tree is called a *Steiner set*.

The next theorem reviews all distances and their relations.

Theorem 1.2 *Generally:* $b_{ij} \leq s_{ij}$ *and* $\sigma_{ij} \leq s_{ij} \leq d_{ij} \leq c_{ij}$. *If* $\{i, j\}$ *is a Steiner set:* $b_{ij} \leq \sigma_{ij} \leq s_{ij} \leq d_{ij} \leq c_{ij}$.

Proof. The first inequality follows from a concatenation of the shortest paths of the special path of bottleneck length s_{ij}. The inequality $s_{ij} \leq d_{ij}$ holds, because $\{[i..j]\}$ is a special $i-j$ path of bottleneck length d_{ij}. Further, only $\sigma_{ij} \leq s_{ij}$ is new and needs a proof. Suppose the contrary: S is a Steiner tree containing i and j such that the maximum weighted edge (v, w) on $S_{\{i,j\}}$ has weight $c_{vw} = \sigma_{ij}$ larger than s_{ij}. Let s_{ij} be attained for the special path

$$Q_{ij} = \{[i = k_0..k_1], [k_1..k_2], ..., [k_n..k_{n+1} = j]\},$$

i.e., $s_{ij} = \max\{d_{xy}| [x..y] \in Q_{ij}\}$. In $S\backslash\{(v, w)\}$, let m be the maximum index with k_m in the same component as i. We deduce from $c_{vw} = \sigma_{ij} > s_{ij} \geq d(k_m, k_{m+1})$ the contradiction that feasible solution $S' = [k_m..k_{m+1}] \cup S\backslash\{(v, w)\}$ is of smaller cost than S. $\square$

Also, let us call an edge a *Steiner edge* if it can be part of a Steiner tree. By means of the Steiner distances we can translate corollary 1 to the SPG.

Theorem 1.3 *Edge* $(i, j) \in E$ *is a Steiner edge iff* $\sigma_{ij} = c_{ij}$.

Proof. For the forward implication note first that $\sigma_{ij} \leq c_{ij}$ by theorem 1.2. Next, suppose that $(i, j) \in E$ is a Steiner edge, say S is a Steiner tree containing (i, j). As the path $S_{\{i,j\}} = \{(i, j)\}$ is a Steiner path, we have here additionally $\sigma_{ij} \geq c_{ij}$, so $\sigma_{ij} = c_{ij}$.

For the backward implication, suppose $\sigma_{ij} = c_{ij}$ for an edge $(i,j) \in E$. With the Steiner distance σ_{ij} being equal to $c_{ij}(> -\infty)$, there exists a Steiner tree S, containing i and j, such that the bottleneck length of Steiner path $S_{\{i,j\}}$ is c_{ij}, i.e., this path has a maximum weighted edge, say (v,w) of cost $c_{vw} = c_{ij}$. Then, with the same optimal cost, $S' = \{(i,j)\} \cup S\backslash\{(v,w)\}$ is feasible and contains (i,j). Thus (i,j) is a Steiner edge. $\qquad\square$

2 Edge Exclusion

As a corollary of theorem 1.3: one may eliminate $(i,j) \in E$ if $\sigma_{ij} < c_{ij}$. As it is NP-hard, to determine the Steiner distance σ_{ij} exactly, one should exploit upper bounds. The removal of a Steiner edge from a Steiner tree S, will disconnect at least two special vertices, say k and k'. Consequently, for any Steiner edge (i,j):

$$c_{ij} = \sigma_{ij} \leq \max\{\sigma_{kk'}|k, k' \in K\} \leq \max\{s_{kk'}|k, k' \in K\}$$

So, as observed in [17], one can delete the edges (i,j) of cost higher than $\Delta := \max\{s_{kk'}|k, k' \in K\}$. This reduction −let us call it 'the excessive cost test'− can be executed fast. The upper bound Δ is equal to the largest weight of a MST on distance graph (K, d); it can be determined in $O(v \log v + e)$ time, see [16].

2.1 The Special Distance Test

Besides Δ, also the special distance s_{ij} is an upper bound on σ_{ij} (theorem 1.2), and also this bound, can be computed in polynomial time. So, as another condition for edge elimination we have:

Special distance test (SD) Eliminate edge (i,j) if $s_{ij} < c_{ij}$.

To compute the special distances, it is most straightforward to use an algorithm similar to Dijkstra's labelling algorithm for shortest paths [5]; it is given in Figure 2. With input the shortest distances d, procedure SPECIAL(i) determines for a given vertex i the special distances s_{iv} to all vertices $v \in V$. As in Dijkstra's algorithm, whenever a vertex k enters the set L of labelled vertices, the so far tentative 'special distance' t_{ik} is accurate: $s_{ik} = t_{ik}$,

procedure SPECIAL-V(i);
begin
$L := \{i\}$
$\forall v \in V : t_{iv} := \min\{\Delta, d_{iv}\}$
repeat
   $l := \arg\min\ \{t_{ik}|\ k \in K \backslash L\}$
   $L := L \cup \{l\}$
   $\forall v \notin L :$ **if** $\max\{t_{il}, d_{lv}\} < t_{iv}$ **then** $t_{iv} := \max\{t_{il}, d_{lv}\}$
until $L = K$
$\forall v \in V : s_{iv} := t_{iv}$
end

Figure 2: Special distances from i to all vertices of V in time $O(kv)$

and this vertex is inspected to update tentative distances from i to non-labelled vertices. Since only special vertices need to enter L for inspection, the procedure takes $O(kv)$ time per vertex i.

Thus, after the computation of all shortest distances, all special distances are generated in total time $O(v^2 \log v + ve + kv^2)$ by running SPECIAL-V(i) for every $i \in V$; where the $O(v^2 \log v + ve)$ part arises from the computation of all pair shortest paths, see [12].

As a consequence of the edge eliminations by test SD, some shortest distances d_{vw} in the reduced graph G' may increase, but, as one can easily check, this cannot happen for special distances s_{vw}. So one can repeat the test (in the reduced graph) without updating the special distances s.

One can indeed eliminate more edges (i, j) by test SD under the relaxed condition that the bottleneck length of a non-trivial special path ($\neq \{(i, j)\}$) is smaller *or equal* to c_{ij}. But then, before re-applying the test, one should update distances d and s. (E.g., when given three edges of equal weight between three special nodes, one would otherwise wrongly eliminate all three edges.) Such a repeated application would require much computational effort. We rather apply the inequality check only, eliminating (original) cases of equality by means of a perturbation of the edge weights. One possible perturbation c' of c is:

$$c'_{ij} = 2v^2 c_{ij} - \delta(number(i) + number(j)),$$

procedure SPECIAL-K(i);
begin
Change MST S into MST S^i on $(K \cup \{i\}, d)$
Traverse S^i once to find s_{ik} as $bl(S^i_{\{i,k\}})$ for all $k \in K$
end

Figure 3: Special distances from i to all vertices of K in time $O(k)$

where δ is a small enough number and $number(i)$ is a natural number representing node i, and chosen in the set $\{1...v(= |V|)\}$ in such a way that nodes of K receive higher numbers than nodes of $V \backslash K$. The number δ must not exceed the least possible difference between the cost $c(S^*)$ of an optimal tree S^* and the cost $c(S)$ of a non-optimal solution S (i.e. 1 in case of integer edge costs). To validate this edge cost transformation we show that any non-optimal tree S remains worse than an optimal tree S^*. Since S^* contains at least one and S contains at most $v - 1$ edges, we have:

$$c'(S^*) < 2v^2 c(S^*) \leq 2v^2(c(S) - \delta) < 2v^2 c(S) - 2v\delta(v - 1) < c'(S).$$

2.2   Alternative Implementations

It is easy to show that the SD test dominates other tests given in literature, we already mentioned the 'least cost' test of [2] (see section 1.2), the tests 'R-R deletion', 'R-S deletion' and 'S-S deletion' of [1] −presented unifiedly as test MST in [7]− and 'reduction test (f)' in [19] −recorded as VNK test in [7]. In order to dominate also the 'excessive length' test, one should redefine s_{ij} as $\min\{s_{ij}, \Delta\}$.

With regard to an application of the SD test on all edges $(i, j) \in E$, we now show that the computation of all shortest distances dominates, eliminating the $O(kv^2)$ term from the time complexity given earlier. This is accomplished in a procedure, called MST-SCAN. In the context of already available shortest distances, MST-SCAN uses three steps to return the special distance s_{ij}, for each edge $(i, j) \in E$, in $O(k^2 + kv + ke) = O(ke)$ time.

Term $O(k^2)$ comes from the first step, which is to calculate a MST S on (K, d). In the second step of MST-SCAN, subroutine SPECIAL-K(i), summarized in Figure 3, determines for a vertex i the special distances s_{ik} to all $k \in K$, on the basis of available shortest distances d and MST S of (K, d). According to theorem 1.1, the distance s_{ik} is equal to the bottleneck length of the fundamental $i - k$ path in a MST of distance graph $(K \cup \{i\}, d)$. For a given vertex $i \in V$ the MST S on (K, d) can be changed into a MST on $(K \cup \{i\}, d)$ efficiently in $O(k)$ time, by means of the algorithm of [18]. Thereafter, in this new MST, the values s_{ik} are found for every $k \in K$, again in $O(k)$ time, by a postorder traversal. Subroutine SPECIAL-K(i) is run for all $i \in V$, determining in $O(vk)$ time the special distances between vertices of V and vertices of K.

In its final step, MST-SCAN obtains the special distance s_{ij} as the minimum value in the set

$$\{\Delta, d_{ij}\} \cup \{\max\{s_{ik}, s_{kj}\}|\ k \in K\}$$

this is done, per edge $(i, j) \in E$, in $O(k)$ time.

The time complexity of SD can be further decreased to $O(kv \log v + ke)$ for Euclidean weighted graphs or, more generally, metric graphs, where the edge weights satisfy the triangle inequalities:

$$\forall p, q, r : c_{pq} + c_{qr} \geq c_{pr}.$$

Under these inequalities one cannot have $d_{ij} < c_{ij}$; so now for each edge (i, j) the special distance s_{ij} can be computed in $O(k)$ time as the minimum value in the set

$$\{\Delta, c_{ij}\} \cup \{\max\{s_{ik}, s_{kj}\}|\ k \in K\}.$$

Further the procedure MST-SCAN needs to exploit for its MST's only shortest distances between vertices in K and vertices in V. So, when providing as input only distances d_{ki} for $k \in K$ and $i \in V$, one can obtain distances s_{ik} for $i \in V$ and $k \in K$ in total time $O(kv \log v + ke)$.

In graphs that do not satisfy the triangle inequalities one could follow —for time saving reasons— the same approach. Then, for each edge $(i, j) \in E$, there is returned an upper bound on s_{ij}, say s'_{ij}. Checking this value is also a valid test.

(Special distance)' test (SD') Eliminate edge (i, j) if $s'_{ij} < c_{ij}$.

As another interesting application: the use of MST-SCAN with d–weights everywhere replaced by corresponding c–weights, leads to an upper bound on s_{ij} for edge (i, j), one that is higher than s'_{ij} (if different). The test associated with this other value is equivalent to test MST, which shows that one can apply test MST in $O(ke)$ time. For the edge elimination tests mentioned so far, as well as for a test discussed in the next section, we summarize

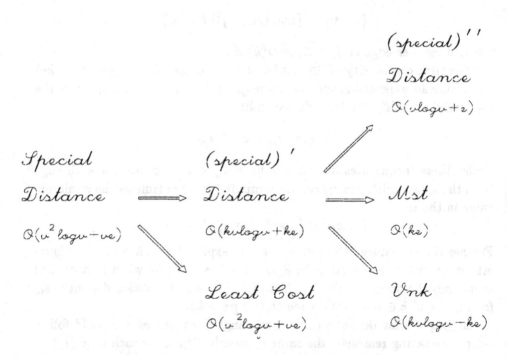

Figure 4: Complexity and domination among edge elimination tests

problem				reduction method			
$|V|$	$|K|$	$|E|$	SD″	SD	ML	MST	VNK
50	5	150	86	83	105	144	89
		500	90	87	112	331	95
		1225	91	87	111	350	95
	10	150	76	75	105	134	85
		500	81	79	112	202	89
		1225	72	71	105	212	79
	25	150	60	60	81	90	73
		500	61	60	82	93	73
		1225	61	60	82	92	76
100	10	300	182	178	235	292	197
		1000	184	181	264	740	200
		4950	188	183	264	820	204
	20	300	158	157	232	281	181
		1000	162	160	257	459	189
		4950	159	157	255	467	183
	50	300	125	124	174	187	163
		1000	124	124	177	194	167
		4950	123	123	173	191	165

Table 1: Test results in the number of remaining edges

in Figure 4 the dominance relations and the time complexity orders. The performance of these tests on uniform random graphs is tabulated in Table 1.

2.3 An Efficient Approximation

All edge elimination tests discussed so far, have a time complexity order that is cubic in the number of vertices. In order to cut down rigorously the computation time, while retaining most of the reduction, we recommend the so-called SD″ test. It is based on other upper bound values s''_{ij} $(\geq s'_{ij})$ that can be computed for all edges $(i,j) \in E$ in time $O(v \log v + k^2 + e)$. Before discussing these values, let us first see what they can do. Table 1 gives

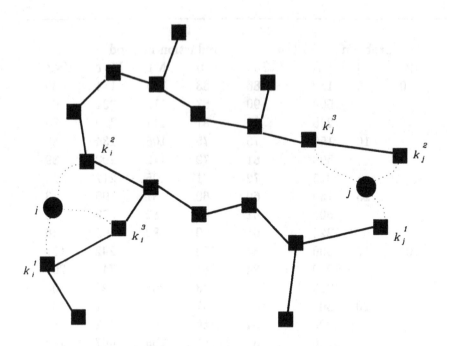

Figure 5: Approximating s_{ij} with paths (...) to MST S of (K, d)

a computational comparison of test SD'' with the other elimination tests. The test problem instances are random graphs with edge weights drawn as uniform random integers in the interval $[1..500]$. All results are averages over 25 instances, rounded to integer values. The column ML gives the results after an exhaustive application of both the MST and the LC test.

The SD'' test establishes a good trade off between computation time and reduction; on the test problems it nearly matches the results of SD. The SD test sort of merges the ideas of LC and MST, but it is stronger than LC and MST together.(Being of little interest, there is no column for SD' in Table 1.)

As already mentioned the idea of the SD'' test is to use a good upper bound s_{ij}'' on s_{ij}; these upper bounds become available after the next three steps.

   1: Compute a MST, S, on (K, d);

   2: In S determine s_{kl} for $k, l \in K$ as $bl(S_{\{k,l\}})$;

   3: Compute from $i \in V \backslash K$ three shortest paths ending in different end vertices of K: k_i^1, k_i^2 and k_i^3.

Step 1 takes $O(v \log v + e)$ time with Mehlhorn's method [16]. Step 2 takes $O(k^2)$ time with $|K|$ postorder tree traversals of S. For step 3 we use an extension of Mehlhorn's method called SPATHS, detailedly given in [10]. SPATHS determines in $O(v \log v + e)$ time, paths from i to vertices of K in which intermediary vertices (if any) are non-special, computing the shortest such path, say to $k_i^1 \in K$ of length $d(i, k_i^1)$, similarly a second such shortest path to $k_i^2 \in K \backslash \{k_i^1\}$ and a third to $k_i^3 \in K \backslash \{k_i^1, k_i^2\}$. (For convenience of presentation we further define for $i \in K : k_i^1 = k_i^2 = k_i^3 = i$). After step 3 we have at hand for each edge (i, j) nine upper bounds s_{ij}^{tr} on s_{ij}; one for each pair of nodes $\{k_i^t, k_j^r\}$, where $t, r = 1, 2, 3$. The value

$$s_{ij}^{tr} = \max\{d(i, k_i^t), s(k_i^t, k_j^r), d(k_j^r, j)\}$$

represents the bottleneck length of the special $i - j$ path

$$\{[i..k_i^t]\} \cup S_{\{k_i^t, k_j^r\}} \cup \{[k_j^r..j]\}$$

in $(K \cup \{i, j\}, d)$; see Figure 5. With the results of step 2 and 3 in computer memory, one can produce, in $O(1)$ time, for any pair $i, j \in V$, the minimum value in the set $\{\Delta, c_{ij}\} \cup \{s_{ij}^{tr} | t, r = 1, 2, 3\}$, which is, by definition, the upper bound s_{ij}'' on s_{ij}. Thus one can apply the following test on all edges in $O(v \log v + k^2 + e)$ time.

(Special distance)'' test (SD'') Eliminate edge (i, j) if $s_{ij}'' < c_{ij}$.

With a more sophisticated implementation, see [17], one can apply test SD'' on all edges in time $O(e + v \log v)$, removing the term k^2.

3 Edge Inclusion

3.1 The Bottleneck Condition

For a constructive test on the basis of theorem 1.3 the relation $\sigma_{ij} = c_{ij}$ is hard to check. For every $i, j \in V : \sigma_{ij} \leq c_{ij}$, and for $i, j \in K$ the relation $b_{ij} = c_{ij}$ is sufficient for equality; then $\{i, j\}$ is a Steiner set and the bottleneck distance b_{ij} is a lower bound on σ_{ij} (see theorem 1.2). The bottleneck test in [7] also considers edges with non-special ends. For an edge (i, j) to be a Steiner edge a sufficient condition is the so-called bottleneck condition:

Theorem 3.1 *Edge (i, j) is a Steiner edge if there are vertices $k_1, k_2 \in K$ such that*

$$d_{k_1, i} + c_{ij} + d_{j, k_2} \leq b_{k_1, k_2},$$

where $\underline{b}$ is to denote the bottleneck distance in $G\backslash\{(i,j)\}$.

Proof. Let T be a Steiner tree with $(i,j) \notin T$. The path $T_{\{k_1,k_2\}}$ lies in $G\backslash\{(i,j)\}$, so according to the condition there is an edge (v,w) on this path such that

$$c_{vw} \geq \underline{b}_{k_1,k_2} \geq d_{k_1,i} + c_{ij} + d_{j,k_2}$$

Transform T into the solution $T' = T\backslash\{(v,w)\} \cup P$, P being the shortest $k_1 - k_2$ path with (i,j) on it. Then $(i,j) \in T'$ and T' is a Steiner tree, as:
$$c(T') \leq c(T) + d_{k_1,i} + c_{ij} + d_{j,k_2} - c_{vw} \leq c(T). \qquad \square$$

As presented, it at first appears to be of high time complexity order to test this condition for all edges. However, in [7] it was shown that the bottleneck condition can only hold for edges (i,j) that are part of a MST T of G. Suppose that the condition holds for edge (i,j). When deleting (i,j), let MST T break into components T_i and T_j, respectively containing i and j. It can be further shown that the special vertices with which (i,j) satisfies the bottleneck condition are respectively found as k_i a special vertex in component T_i nearest to i and k_j a special vertex in component T_j nearest to j. So a candidate edge (i,j) is part of T_K, the minimal subtree spanning K. The conclusion is that the following test finds all Steiner edges in MST T satisfying the bottleneck condition:

> **Nearest Special Vertices** (Nsv) Let (i,j) be in the interior T_K of MST T with k_i, k_j as defined above; further let the cheapest chord (v,w) replace branch (i,j) in the MST of $G\backslash\{(i,j)\}$. Then (i,j) is a Steiner edge satisfying the bottleneck condition if and only if
>
> $$d_{k_i,i} + c_{ij} + d_{k_j,j} \leq c_{vw}$$

When all shortest path distances are already computed, the remaining effort to execute this test on all MST edges is $O(v^2)$ time. In [8] it is additionally shown that one can find all Steiner edges satisfying the bottleneck condition in $O(v^2)$ time, from scratch, that is, without prior knowledge of shortest paths. Also, a test dominating Nsv, the so-called extended Nsv test has been presented there. We omit this test, since in turn it is improved by a more easy relaxation of the Nsv condition, to be described next.

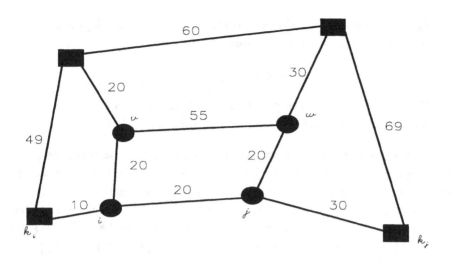

Figure 6: Test Nsv stalls, but Nsc detects Steiner edge (i,j)

3.2 Relaxing the Condition

The basic idea of the Nsv test on edge (i, j) is that the fundamental path in a Steiner tree S between the vertices k_i, k_j must contain at least one edge (v, w) of the cutset of edges that run between T_i and T_j; we denote this cutset as E_{ij}. So, if the Nsv condition is true for edge (i, j) while $(i, j) \notin S$, then $c(S)$ will not deteriorate if one replaces (v, w) by the path

$$[k_i..i] \cup \{(i, j)\} \cup [j..k_j]$$

between k_i and k_j.

If the edge $(v, w) \in E_{ij} \cap S$ has, say, $v \in T_i$ and $w \in T_j$, then one may reconnect the components of $S \backslash \{(v, w)\}$ with available alternative paths via (i, j) : an alternative for $[k_i..i]$ is $[v..i]$ and an alternative for $[j..k_j]$ is $[j..w]$; note that these alternative paths can have zero length, if $v = i$ or if $w = j$. This leads to a stronger form of the Nsv test. For (i, j) in the MST T of G we consider for each edge $e = (v, w)$ of cutset E_{ij}, the smallest alternative path length value

$$\min\{d_{k_i,i}, d_{v,i}\} + c_{ij} + \min\{d_{j,k_j}, d_{j,w}\};$$

let us denote the first and third term as respectively $d_e(i)$ and $d_e(j)$. Then a test that dominates the Nsv test is Nsc (nearest special *or* chord vertices):

Nearest Special or Chord vertices (Nsc) Edge $(i, j) \in T_K$ is a Steiner edge if for each chord $e = (v, w)$ in E_{ij} :

$$c_{vw} \geq d_e(i) + c_{ij} + d_e(j)$$

If $E_{ij} = \emptyset$ then (i, j) is a K-articulation edge, whereas i (or j) is a K-articulation vertex if $\forall e = (v, w) \in E_{ij} : v = i \ (w = j)$.

A vertex i is called a K-articulation vertex if its removal from the graph would generate two components for which the vertex sets, V_1 and V_2 say, contain both special vertices. If a K-articulation vertex is found, then an appropriate reduction is to solve two smaller subproblems, one subproblem considers the induced subgraph of $V_1 \cup \{i\}$ with special set $(K \cap V_1) \cup \{i\}$, the other likewise considers only vertices $V_2 \cup \{i\}$ with special set $(K \cap V_2) \cup \{i\}$ —the main solution is found afterwards by joining the two subsolutions at i. Other inclusion tests found in the literature ([1],[2]) are dominated by the Nsc test (also by Nsv).

To appreciate the Nsc test version, consider Figure 6. Here edge (i, j) is not detected by Nsv because of $e = (v, w)$ of weight 55. For chord e we have $d_e(i) = 10$ and $d_e(j) = 20$, so Nsc succeeds on (i, j).

As shown in [8] one can accurately apply test Nsv on all test objects in time $O(v^2)$. We describe here an alternative implementation, which is also suited for Nsc; it uses expected time $O(e \log v)$. This implementation can even be faster, when the problem graph is sparse (or has become sparse after application of test SD″).

As in the first part of the $O(v^2)$ implementation for Nsv, one can determine for all test objects, edges (i, j) of MST T on G, the accurate values d_{i,k_i} and d_{j,k_j} in time $O(v \log v + e)$, see [8]. Then, for each chord $e = (v, w)$ in turn, one visits only the edges on the fundamental $v - w$ path $T_{\{v,w\}}$ in MST T; note that path $T_{\{v,w\}}$ contains precisely the MST edges (i, j) with $e \in E_{ij}$. For each chord $e = (v, w) \in E \backslash T$ one checks for an expected number of $\log v$ edges (i, j) on the associated fundamental path, each single Nsc inequality itself in constant time.

4   Local Inspection

4.1   Vertex Exclusion

Let $i \in V \backslash K$ be a vertex of degree higher than two. E.g., consider a vertex of degree 3 adjacent to vertices n_1, n_2, n_3, by edges of cost c_1, c_2 and c_3

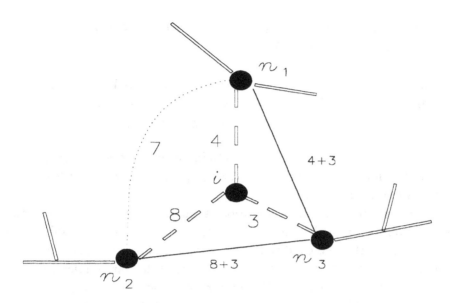

Figure 7: Vertex i can be pseudo-eliminated (dashed: a shortest $n_1 - n_2$ path)

respectively; see Figure 7. When it is shown that i's degree in a Steiner tree need not exceed two, then we may pseudo-eliminate i: remove i's edges, while inserting new edges (n_1, n_2), (n_2, n_3) and (n_1, n_3) of costs respectively $c_1 + c_2$, $c_2 + c_3$ and $c_1 + c_3$ (to account for the case that i is in the optimal solution with degree 2). Clearly, vertex i need not be part of a Steiner tree with degree 3, if for instance

$$d_{n_1, n_2} + d_{n_1, n_3} \leq c_1 + c_2 + c_3.$$

Note that here one of the inequalities $d_{n_1, n_2} < c_1 + c_2$ or $d_{n_1, n_3} < c_1 + c_3$ must hold; so by the least cost test, no more than two new edges need to be inserted.

The given example leads us to an effective test for vertices $i \in V \backslash K$ of general (but small) degree. One interprets the left hand side, $d_{n_1, n_2} + d_{n_1, n_3}$, as the MST weight of distance graph $(\{n_1, n_2, n_3\}, d)$, and one further relaxes by replacing distances d for special distances s ($\leq d$). Consider graph $(N(i), s)$, the complete graph on vertex set

$$N(i) = \{j \in V \mid j \text{ is adjacent to } i\}$$

weighted with special distances s_{pq} for $p, q \in N(i)$. Of all subgraphs of adjacent vertices, the 'local special distance' test considers the MST weight.

The test condition holds trivially for sets N' with $|N'| \leq 2$, so the next test comprises test $D_{\leq 2}$ of section 1.2.

> **Local Special Distance test** (LSD) Vertex $i \in V \backslash K$ can be pseudo-eliminated if for every subset $N' \subset N(i)$, the weight z_{MST} of a MST on (N', s) satisfies $z_{MST} \leq \Sigma_{j \in N'} c_{ij}$.

The test-action of pseudo-elimination is to insert edges (p, q) of cost $c_{pi} + c_{iq}$, but only for those $p, q \in N(i)$ with $s_{pq} = c_{pi} + c_{iq}$ (in view of test SD, other insertions are not necessary).

A verification of the LSD condition is implicitly given in section 4.3, where we validate a further relaxed condition. As the number of subsets to be considered in the LSD condition grows exponentially with the degree m of i, one should apply LSD for small m only. A straightforward implementation leads to the complexity $O(m^2 2^m)$ per vertex, because the computation of a minimum spanning tree on a complete graph with m vertices requires $\Omega(m^2)$ steps. We can here, however, calculate MST's on increasing subsets of vertices and use the linear time MST update procedure of [18], then the test takes $O(m 2^m)$ time per vertex.

Sometimes partial results of test LSD can turn into full results of graph reduction: by changing the test's object to an edge. Suppose that test LSD fails on vertex $i \in V \backslash K$, then it need not have failed with regard to one or more of the edges incident to i.

> LSD(*edge*) Edge (q, i) with $i \in V \backslash K$, can be pseudo-eliminated if the LSD test on i succeeds for all subsets $N' \subset N(i)$ with $q \in N'$.

Here the test action of pseudo-elimination replaces (q, i) by edges (q, r) of cost $\min\{c_{qr}, c_{qi} + c_{ir}\}$ for $r \in N(i) \backslash \{q\}$ (but again (q, r) is only inserted if $s_{qr} = c_{qi} + c_{ir}$). As a consequence $|E|$ can increase instead of decrease?! Still a pseudo-elimination of edge (q, i) can be of value as the edge is replaced by longer edges, which may retrigger other tests, e.g., the NSC test or test LSD on other nodes. A posterior comment on the action of pseudo-elimination is made in section 7.

4.2   Changing the Test's Objective

By a reversal of the arguments in the LSD test condition, an inclusion test can be formulated with the objective of showing, for a non-special vertex,

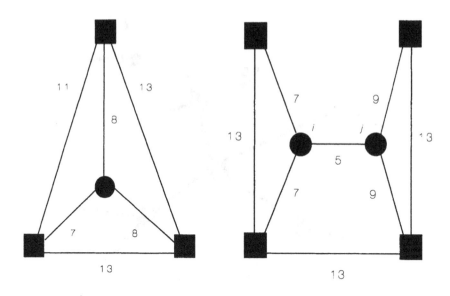

Figure 8: Detecting Steiner vertices

that this vertex can be part of an optimal solution. To show that a non-special vertex can be a Steiner vertex, it is sufficient to show for (just) one set of adjacent vertices $N = \{n_1, n_2, ..., n_p\}$, that the star $\{(i, n_t)| \ t = 1, 2, ..., p\}$ can be extended to a vertex set $L = \{k_1, k_2, ..., k_p\} \subset K$ (taking $k_t = n_t$ if $n_t \in K$), such that the connection cost of L via Steiner vertex i and edges $(i, n_1), ..., (i, n_p)$ is less than this cost would be without vertex i. The latter cost is hard to compute, but we can assess a lower bound by means of the bottleneck distances $\underline{b}$. More precisely:

> **Local Bottleneck Distance test** (LBD) $i \notin K$ is a Steiner vertex if there is a subset $N \subset N(i)$, mapping one-one onto a subset of special vertices $L = \{k(n) \mid n \in N\}$, such that
>
> $$\Sigma_{n \in N}(c_{in} + d_{n,k(n)}) \leq \text{MST weight of } (L, \underline{b})$$
>
> where $\underline{b}$ denotes the bottleneck distance in $G \backslash \{i\}$.

Proof. Let S be a feasible tree without vertex i. As long as a pair of vertices of L is connected in S, choose a pair say m and n. Remove a longest edge from the connecting path; its weight will be at least $\underline{b}_{mn}$. After $|L| - 1$ such deletions, S consists of $|L|$ components each containing a single vertex of L. The cost of S has decreased by at least the MST weight $(L, \underline{b})$. Since

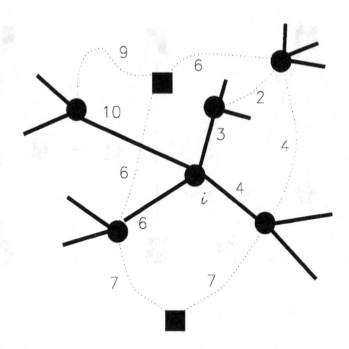

Figure 9: Without the paths of length 7 test Lsc still holds

the test condition is satisfied, we can restore connectedness at no larger a
cost by bringing in, for $n \in N$: edges (i, n) plus a shortest path from n to
$k(n)$, if $n \neq k(n)$. Thus it is possible to incorporate vertex i into any feasible
tree S without increasing its cost; so i is a Steiner vertex.   □

An appropriate test action is to add i to set K; later, a re-application of
test Nsc will find the appropriate Steiner edges. To arrive at a polynomial
time test one could try out only a few heuristic sets N, e.g. the vertices adjacent to i in the MST on (V, E). Further one could try as L the assignment
of N in K that minimizes $\Sigma_{n \in N} d_{nk(n)}$. In the left hand side example of Figure 8 test LBD detects the Steiner vertex; here N, L and K coincide, the $\underline{b}$
-MST weight is 24 and $\Sigma_{n \in N}(c_{in} + d_{n,k(n)}) = \Sigma_{n \in N} c_{in} = 23$. In Figure 6 one
detects vertex i as a Steiner vertex for $N = \{k_i, j, v\}$ and $L = \{k_i, k_j, k_v\}$,
where k_v is the top-left special vertex. Test LBD does not succeed on the
right-hand side graph in Figure 8; we return to this example in section 6.3

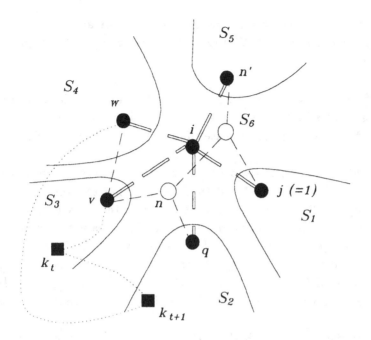

Figure 10: The connectednees can be restored for a lower cost

4.3 Relaxing the LSD Condition

Here we relax the test condition of LSD for vertices: by replacing the left hand side, the minimum spanning tree cost Z_{MST}, by yet a lower connection cost. To motivate it, let us apply test LSD on vertex i in Figure 9, where only a relevant part of the graph is given. The degree 4 option $10+3+6+4$ can be discarded as the MST weight on $(N(i), s)$ is $6+7+9$, also each of the degree 3 topologies for vertex i, with associated weights: $10+3+6$, $10+3+4$, $6+4+3$, is refuted by lower MST weights, respectively: $7+9$, $9+6$, $6+7$. However if the shortest paths of length 7 would have been longer, say of length 8, then LSD would not any more eliminate ($6+8 > 6+4+3$). The next test, which locally considers the Steiner costs would then still succeed: with the upper-right vertex as Steiner vertex, the Steiner tree connection costs are $6+2+4 < 6+4+3$.

> **Local Steiner Cost test (LSC)** Vertex $i \notin K$ can be pseudo-eliminated if for every $N' \subset N(i)$ with $|N'| \leq |K|$
>
> $$Z_{SPG} \leq \Sigma_{j \in N'} c_{ij}$$

where Z_{SPG} is the optimal value for the SPG-instance $(V\backslash\{i\},N',s)$, which has special set N' and a complete set of edges weighted by s.

Proof. We will show that under the condition of the theorem, the minimum degree of i in a Steiner tree is 0 or 2. The insertion of edges (p,q) for $p,q \in N(i)$ accounts for the case that vertex i is in the Steiner tree with degree 2.

So let S^* be a Steiner tree of (V, K, c) in which vertex i has degree 3 or more. Define $N' = N(i) \cap S^*$ and let T be a Steiner tree on $(V\backslash\{i\}, N', s)$ with cost $Z_{spg} \leq \Sigma_{j \in N'} c_{ij}$. We will construct an alternative optimal solution S that does not contain i as a crucial vertex. The idea is to exploit shortest paths induced by 's-edges' of MST T on $(V(T), s)$ to join the components of $S^*\backslash\{i\}$ for a cost of at most Z_{spg}. As the cost of $S^*\backslash\{i\}$ is $c(S^*) - \Sigma_{j \in N'} c_{ij}$ this will contradict the optimality of S^* in case $Z_{spg} < \Sigma_{j \in N'} c_{ij}$. For the equality case we will prove that i, if in Steiner tree S, must have degree 2 in S.

Number the vertices such that $N' = \{1, 2, ..., n'\}$ and let graph $S^*\backslash\{i\}$ consist of components $S_1, ..., S_{n'}$ with $j \in S_j$. First we reduce the size of tree T by merging redundant vertices into vertices of N': any vertex $r \in V(T)\backslash N'$ that is also present in $S^*\backslash\{i\}$, say in S_v, is merged into vertex $v \in N'$, but to prevent a cycle we first delete an arbitrary s-branch from $T_{\{v,r\}}$. By these operations (if any) T and $V(T)$ shrink; at the end $(V(T)\backslash N') \cap S^*\backslash\{i\} = \emptyset$ and the weight of T is $Z'_{spg} \leq Z_{spg}$. Now we renumber $V(T)\backslash N'$ as $\{n'+1, n'+2, ..., n\}$ and we initialize the (not yet feasible) solution S as

$$(S_1 \cup ... \cup S_{n'}) \cup (S_{n'+1} \cup .. \cup S_n);$$

i.e., we add to $S^*\backslash\{i\}$, each of the remaining Steiner vertices $t \in V(T)\backslash N'$ as a single vertex component $S_t = (\{t\}, \emptyset)$. The situation is sketched in Figure 10: the candidate optimal edges incident to i are double-lined; their deletion produces components S_j in which the non-deleted edges of S^* are not pictured to avoid an overload; an irreducable Steiner tree T is depicted (dashes), with for an s-edge (v, w) a corresponding special path (small dashes). Now we describe a number of steps that are to be performed for each s-edge of T on the basis of its associated special path.

Arbitrarily select an s-branch (v, w) from T. This edge corresponds to an original s-edge (r, p) of the unreduced Steiner tree T with $r \in S_v$ and $p \in S_w$; normally $v = r$ and $w = p$ but r or p may have been merged into v

or w. So there is a special path

$$[k_0(=p)..k_1], [k_1..k_2], ..., [k_{x-1}..k_x(=r)]$$

starting in S_v, ending in S_w and of bottleneck length s_{rp}, which is the branch weight of (v, w) in T. Perform two operations.

The first is to add to S an artificial edge: "edge" (k_t, k_{t+1}) of cost $d(k_t, k_{t+1}) \leq s_{rp}$, for the first index t with k_t and k_{t+1} in different components of S, say, $k_t \in S_v$ and $k_{t+1} \in S_q$. This operation artificially links component S_q to S_v and we say that S_q is merged into S_v; in this stage we prefer to connect by artificial "edges" instead of using directly the shortest path $[k_t..k_{t+1}]$, thus we avoid consideration of anomalies with regard to the intersection of $[k_t..k_{t+1}]$ and S.

The other operation for selected s-edge (v, w) reduces tree T. An arbitrary s-edge of $T_{\{v,q\}}$ is deleted and vertex q is merged into v.

Continuing in this way each artificial "edge" addition simultaneously reduces the number of branches in T and the number of components in S by one, at a cost not larger than the length of the deleted branch. After $n-1$ "edge" additions, S is a (partially artificial) tree spanning K and T has empty s-edge set. The total cost of the $n-1$ "edges" is at most $s(T) = Z'_{spg}$.

Now replace in S each added artificial "edge" for the associated series of E-edges of a shortest path; here edges that are already there or that would introduce a cycle are not inserted. Let the total cost of the inserted edges be $Z''(\leq Z'_{spg})$. The final result is a feasible tree S spanning $K \cup V(T)$, for a cost of at most $c(S^*) - \Sigma_{j\in N'}c_{ij} + Z''$. So $c(S) \leq c(S^*)$, because $Z'' \leq Z'_{spg} \leq \Sigma_{j\in N'}c_{ij}$. This would contradict the optimality of S^*, if $Z'' < Z'_{spg}$ or $Z'_{spg} < \Sigma_{j\in N'}c_{ij}$.

So assume $Z'_{spg} = \Sigma_{j\in N'}c_{ij}$, $Z'' = Z'_{spg} = Z_{spg}$ and $c(S) = c(S^*)$. S has arisen from $S^*\backslash\{i\}$, by the insertion of $n-1$ shortest paths with end vertices in $K \cup V(T)$, which excludes vertex i in this respect. The positive edge costs imply that not any two of these paths intersect in an interior path vertex, otherwise $Z'' < Z_{spg}$. This means that at most one of these paths can have re-inserted vertex i. We conclude that i can have at most degree 2 in a Steiner tree. □

A complete check of all the inequalities in the Lsc condition seems out of reach: the number of subsets of adjacent vertices is exponential in the degree m of i, while the determination of a Steiner tree for one subset is already NP-hard. However, the situation is less dramatic. For vertex i of degree

m, all subsets N' can be checked in one run of the dynamic programming
algorithm of [11], determining the Steiner cost of $(V\backslash\{i\}, N(i), s)$ and, along
the way, the Steiner costs of all subsets $N' \subset N(i)$, in time $O(3^m v + 2^m v^2)$.
So, if limited to vertices with degree m or lower, where m is fixed, the test
can be applied on all these vertices in $O(v^3)$ time.

4.4   Increasing the Scope of Inspection

So far, we have improved the technique of local inspection by decreasing
the left-hand side of the LSD condition, replacing d distances by s distances
and MST costs by Steiner costs. The other possibility is to increase the
right-hand side. By exploiting the fact that a local subtree of a Steiner tree
can be extended at each non-special leaf, we increase the size, and weight,
of the object under inspection.

4.4.1   Test LSD^p

The LSD test on a vertex i refutes all star trees containing i: $\{(i,n)|n \in N'\}$
for each $N' \subset N(i)$. In a star of i each leaf is one edge hop away from i. Let
us call a tree T a 'h-tree of i', if T contains i, has no special vertices in its
interior and has each leaf at most h edge hops away from i, i.e.,

$$h = \max\{|E(T_{\{i,j\}})| \quad | \; j \text{ a leaf of } T\}$$

An h-tree T is called an 'exact h-tree' if *all* non-special leaves (if any) are
at *precisely* h edge hops from i.

   If test LSD does not succeed there is one (or more) 1-tree of i, say T, for
which the LSD condition is not fulfilled. We can intensify the inspection by
examining all possible 2-tree extensions of T. Considering again the example
graph of Figure 6, we see that LSD fails on the star tree T containing v with
degree 3; it has set of non-special leaves $L_0 = \{i, w\}$. Tree T has eleven
different 2-tree extensions, depicted in Figure 11. One need not inspect all
these trees. The total set of tree extensions can be partitioned in a number
of $2^{|L_0|} - 1$ so-called *full* subsets of trees, each associated with a different
subset L_{00} of L_0. A tree extension T' belongs to the full set of L_{00} if all the
extending edges of $T'\backslash T$ have one end vertex in L_{00}. For elimination of i, it
is sufficient to refute all the trees of just one of these full sets. E.g., all the
trees of the full set of *exact* extensions, extending T at each node of L_0. In
Figure 11 we can choose between three full sets: the five tree extensions in
the top two rows are the exact extensions of T, a middle row of three trees

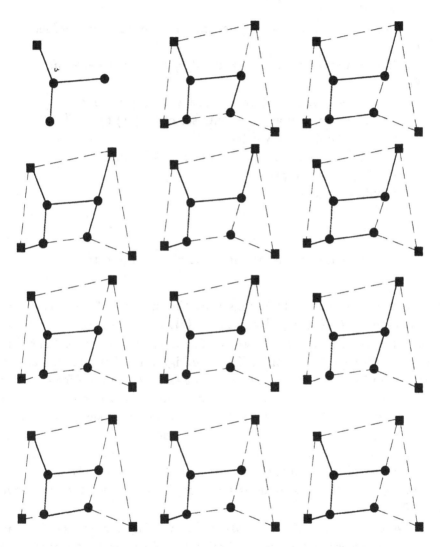

Figure 11: Eleven 2-tree extensions of v's 1-tree (see Fig.6)

LSD^p(vertex i)
BEGIN
Put on a Queue, marked, the 0-tree of i
repeat
Remove a tree T from Queue, say an h-tree of i with leaf set L
   if T is marked **then** *(* replace T, if possible, by extensions *)*
      if $h = p$ or $L \subset K$ **then** Exit, reporting test failure
      else Queue, unmarked, a 'full' set of $(h + 1)$-trees $T' \supset T$
   else *(* check unmarked tree T *)*
      if $Z_L > c(T)$ **then** Put T back on Queue, marked
until Queue is empty
Pseudo-eliminate i
END

Figure 12: Increasing the scope of local inspection

and a bottom row of three trees constitute the full set of 2-tree extensions of v, for respectively $L_{00} = \{w\}$ and $L_{00} = \{i\}$.

Thus, for a given value $p > 1$, we can formulate a test named LSD^p that dominates tests LSD^{p-1},..., LSD$^1 \equiv$ LSD, see Figure 12. The test maintains a first-in-first-out queue of trees. These can be exact h-trees of i for a certain value $h \leq p$ and each tree on queue is either marked (to be extended) or unmarked (to be checked). Initially the queue consists of one marked 'tree': $(\{i\}, \emptyset)$, the 0-tree of i with empty edge set and i as single leaf. At first $h = 0$ and all star trees of i get queued, each unmarked. At any time L denotes the set of leaves of the considered tree T and as a check on T we compare Z_L, the MST cost of s-distance graph (L, s), with the total edge cost of T.

All two-edge 1-trees of i will pass without being marked, which is all right because we will only *pseudo*-eliminate i. As in test LSD any other tree T, containing i with degree three or more, is refuted as a subtree of a Steiner tree, under the condition $Z_L \leq c(T)$. This time, if $Z_L > c(T)$, then we intensify the check, putting T back on the queue with a mark. Each tree T that is not discarded right away (by passing the check $Z_L \leq c(T)$) is to be replaced by a full set of tree extensions; when later all these tree extensions

LsD^p(edge (i,j))
BEGIN
Put on a Queue, unmarked, the 0-tree of (i,j)
repeat
Remove a tree T from Queue, say an h-tree of (i,j)
if T is marked **then** *(* replace T, if possible, by extensions *)*

if $h = p$ or $L \subset K$ **then** Exit, reporting test failure
else Queue, unmarked, a 'full' set of $(h+1)$-trees $T' \supset T$
else *(* check unmarked tree T *)*
if $Z_L \geq c(T)$ **then** Put T back on queue, marked
until Queue is empty
Eliminate (i,j)
END

Figure 13: A similar test on edges dominates test SD

do pass the check, then one has thereby discarded T. In the execution of the second 'if statement' the test fails and is cancelled: we are there stuck with a marked h-tree T of i, which did not pass its check, so we should check a set of $(h+1)$-tree extensions, but we cannot (all leaves of T are special) or we will not (when $h = p$ we prefer to give up for reasons of time complexity).

If LsD would eliminate vertex i, then also LsD$^p(i)$ for $p > 1$ succeeds, as efficiently, without any re-queuing operations. With the FIFO-queue implementation, we plan to enter tests LsD, LsD^2, ..., LsD^p sequentially, as needed. In Figure 6 test LsD fails on vertex v whereas test LsD^2 pseudo-eliminates v, replacing it for an edge of weight 40 from i to the top left vertex. Initially for v only one 1-tree T is queued, the associated MST cost appears too large and T is substituted by five 2-trees of v. Each of these 2-trees T' has a MST cost on the leaf set that is lower than $c(T')$.

Of course the weak aspect of test LsD^p is the threat of processing an exponential number of trees. A first way out may be preprocessing: reduce the number of trees in a full set. After applying two tests discussed in the next section, one seldom needs to inspect a full set wholly; in Figure 12 the apostrophes around *'full'* hint to this fact. E.g., after preprocessing on v's 1-tree T given in Figure 11, the 'full' set of exact tree extensions will become empty. Further we can resort to more heuristic short-cuts (after all the test is already heuristic):

- Cancel the test whenever the number of tree leaves becomes too large.

- Select each time a full set of trees that extend T at just one leaf l, adjacent to only a few vertices of $N(T)$.

Analogously as for vertices, one can consider h-trees for edge objects: T is an h-tree of $e = (i,j)$, if T contains e, has each non-special leaf at edge distance h from e (i.e., the minimum number of edge hops before reaching the object, that is, either end i or end j), has each special leaf at most h edges away from e and has no special vertices in the interior. In our formulation of $\text{LsD}^p(\text{edge})$, see Figure 13, we altered $Z_L > c(T)$ to $Z_L \geq c(T)$, thus the test action can change to complete elimination. We start out with the 0-tree $(\{i,j\},\{e\})$ of e, as an *unmarked* first object, thus $\text{LsD}^p(\text{edge})$ generalizes edge elimination test SD, as the first check compares $Z_L = s_{ij}$ with $c(T) = c_{ij}$.

4.4.2   Preprocessing within Preprocessing

One can imagine that a full set of exact $(h + 1)$-tree extensions is empty: e.g., when it is impossible to extend at all non-special leaves of T without introducing cycles. Like it is unnecessary to inspect a graph extension that is not a tree, it is also unnecessary to inspect a tree that cannot be a subtree of a Steiner tree. Therefore, we can discard extensions for which we know beforehand that the resulting tree cannot be part of a Steiner tree.

Let L_0 again denote the set of non-special leaves of an h-tree T of i, and let us write $N(T)$ for the set of vertices $n \in V \backslash T$ adjacent to one or more leaves of L_0. First, we can restrict the attention to extensions T' that are minimum spanning on $V(T')$. Therefore we may assign each $n \in N(T)$ to just one leaf $l_n \in L_0$ of minimal weight c_{n,l_n}. Now each subset $N' \subset N(T)$ corresponds to an $(h + 1)$-extension T' of T by the addition of edges $\{(l_n, n)|n \in N'\}$. The number of trees in the full set of exact $(h + 1)$ extensions, is equal to the number of subsets N' for which the mapping $l : N' \longrightarrow L_0$ is 'onto', leaving no non-special leaf of T unassigned; whereas the full set of L_{00} corresponds to the collection of those subsets of $N(T)$ that are mapped by l onto L_{00}. Applying these rules to the previous example in figure 14, we see that the number tree extensions decreases: e.g., the full set of $L_{00} = \{w\}$ now contains just one tree.

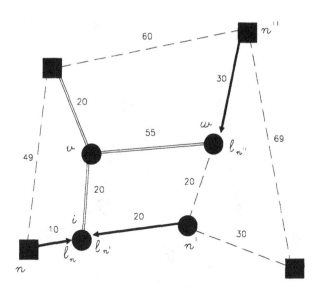

Figure 14: Choosing an assignment $l : N(T) \to L_0$

Normally the number of subsets N' that map onto a specific set L_{00} can be reduced further, by means of two tests that compare special distances with bottleneck distances in tree T.

1. Remove a node n from $N(T)$ if for some $t \in T \setminus \{l_n\}$: $s_{n,t} \leq b_T(n,t)$, where $b_T(n,t)$ is the $n - t$ bottleneck length in T as extended with (n, l_n); for $t \in L_0$ the relation must hold with inequality.

2. N' contains at most one node of a pair $n, n' \in N(T)$ with $s_{n,n'} < b_T(n,n')$, measuring $b_T(n,n')$ in T as extended with (n, l_n) and $(l_{n'}, n')$.

As mentioned we need not inspect those extensions, that are impossible as optimal subtrees. Assuming that tree $\{(n, l_n)\} \cup T \cup \{(l_{n'}, n')\}$ is a subtree of a Steiner tree, we may see all its vertices as special, but then $s_{n,n'} < b_T(n,n')$ implies, by test SD, that the $n - n'$ bottleneck edge in T is non-optimal . A similar argument validates the inequality relation of the first test. (Observe in Figure 14 that the other assignment possible for n' : $l_{n'} = w$, would have led to the removal of n' by test 1.)

Furthermore, we can skip the inspection of alternative optimal tree extensions as long as we do inspect, in case i is indispensable as a Steiner vertex of degree ≥ 3, at least one optimal 1-tree, one optimal 2-tree, and so on, up to the inspection of one optimal p-tree. Let us explain one case

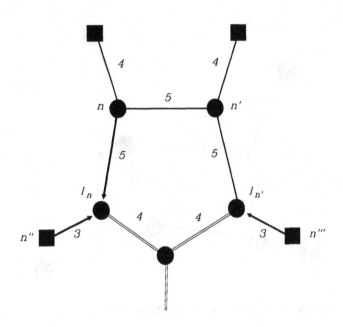

Figure 15: When extending exactly to n, n'', n''' (and not to n'), then for a further extension test 1 will disregard edge (n, n')

of equality in test 1. Assume $s_{n,t} = c_{n,t} = c_{n,l_n} = b_T(n,t)$ and t interior in T. Then an optimal extension T' of T with edge (l_n, n) can be transformed into alternative optimal extension that uses the edge incident to t instead; for a marked h'-tree, with $h' < h$, this alternative, has been queued already. When applying test 1 in Figure 14, node n' gets removed as for the argument $t = w : b_T(n', w) = 55 > 20 = s_{n',w}$. Figure 15 shows, however, that one should be cautious with the equality case.

Finally, there is a nuance by which we can improve the impact of both tests 1 and 2. Consider the exact extension as depicted in Figure 14. In applying test 1 for $n'' \in N(T)$ with as node t the top left vertex the test seems not to succeed, but still the extension of T with $w = l_{n''}$ having degree 2, cannot be optimal: after pseudo-eliminating $l_{n''}$ the replacing edge is longer than edge (t, n''). So test 1 improves if one computes a superior bottleneck length $b'_T(n,t) \geq b_T(n,t)$, simulating the act of pseudo-elimination, where possible between n and t in tree $T \cup \{(l_n, n)\}$; likewise for test 2 with regard to n and n'. With this enhancement there will be no exact extension left for queuing by test LSD^p, so the test succeeds with minimal effort.

Consider now the same example with two weights changed as follows:

the top edge decreases its weight from 60 to 49 and edge (v, w) decreases from 55 to 20. Then the same assignment l applies and the same reduction applies, so again the set of exact extensions is empty; however this time LSD^P would not have succeeded without the preprocessing operations! This illustrates a comment of section 1.1: a preprocessing algorithm π in front of an heuristic algorithm α (here α is test LSD^P), may not only reduce time requirements, but also improve the result.

The suggested tests are to be done for each marked tree in order to prepare the queuing of a reduced number of tree extensions. In a context of available special distances this requires little time per tree T. One first computes in $O(|T|^2)$ time bottleneck distances b'_{vw} between every pair of vertices in tree T. Thereafter, test 1 takes $O(|T|)$ time per vertex n and test 2 will take $O(1)$ time per pair $n, n' \in N(T)$. For sparse graphs with the degree of each vertex of L_0 being small, the total time complexity is $O(|T|^2)$. In general the time complexity for preprocessing is $O(|V(T) \cup N(T)|^2)$. The number of exact $(h + 1)$-tree extensions has an $O(2^{|N(T)|})$ bound. With each node n removed from $N(T)$ this bound is halved and when one pair of related nodes $\{n, n'\}$ is found then a quarter of the extensions previously possible, becomes unnecessary.

It could be the case that $Z_{L'} \leq c(T')$ is not true for some *exact* $(h + 1)$-tree $T' \supset T$, whereas tree T will be discarded, if we would inspect another full set of $(h + 1)$-trees $T'' \supset T$, associated with a leaf set $L_{00} \subset L_0$. But we do not know in advance whether a subset L_{00} is a better choice than L_0, and if so for which set this is true.

However, with an adaptation, the set L_0 itself, corresponding to the exact extensions, will be a dominating choice. In the adaption we first compute Z_L as the MST cost of (L, s), but if this cost exceeds $c(T)$, then later, before generating the 'full' set of extensions of this marked tree T, we re-check with Z_L computed as the Steiner cost of $(V(T) \backslash \{i\}, L, s)$.

Further, in our implementation, we have provided next to p two other computational parameters f and q. The test is exited when the number of tree markings (leading to local Steiner cost computation and/or the queuing of a 'full' set of tree extensions) exceeds f or when for some tree T either $|N(T)|$ or $|V(T)|$ exceeds q. We recursively generate a 'full' set of $(h+1)$-trees, letting subset L' of new leaves grow, computing $Z_{L'}$ by a linear MST update. Under the restrictive parameters f and q this implementation of test LSD^P takes $O(fq2^q)$ time per vertex (independent of p). This includes the computation of the Steiner costs of $(V(T) \backslash \{i\}, L, s)$, which per tree T takes $O(|T|^2 + |T|2^{|T \backslash L|}) = O(q2^q)$ time with an exact enumerative SPG algorithm (of time complexity $O(|V|^2 + |V|2^{|V \setminus K|})$ for instances (V, K, E, c), see section 2.1 of [8]). Under the same restrictive parameters, and with the same arguments as given for test Lsc, it is again in theory possible to run on all vertices a test like Lsc^p in time $O(v^3)$.

5   Global Inspection

In this section we discuss reduction tests that exploit global lower and upper bounds on the cost of a Steiner tree. A classical technique is to analyze the sensitivity of a linear programming lower bound in relation to a known upper bound, thus pricing out variables or fixing variables in. We show how one can reinforce such a sensitivity analysis with the idea of 'reachability', see section 1.2.

First we improve the test condition of reachability. The lower bound on a Steiner tree S containing i can be increased by restricting the attention to Steiner trees S in which i has degree 3, accounting for the other case by means of pseudo-elimination. Graph $S \setminus \{i\}$ consists of at least three components, with in one of them a path to the vertex that is farthest from i. Let again U be a heuristic solution with cost $c(U)$. For vertex $i \notin V(U)$, define $l, k_1, k_2 \in K$ as respectively $\arg\max_{k \in K} d_{ik}$, $\arg\min_{k \in K} d_{ik}$, and $\arg\min_{k \in K \setminus \{k_1\}} d_{ik}$.

3Reachability (R_3) (Pseudo-)eliminate $i \notin U$ if

$$d_{il} + d_{i,k_1} + d_{i,k_2} \geq c(U).$$

If $d_{il} + d_{i,k_1} > c(U)$, then i is eliminated, otherwise i is pseudo-eliminated. Another change in R_3 is, that here we also allow equality.

5.1   Sensitivity

Arguments similar to the reachability test can also be used in the context of a (continuous) linear programming solution. To show this let us consider the multicommodity flow model formulated in [21] for DSPG, the directed Steiner problem on a digraph (V, K, A, c) with costs c_{ij} on arcs $(i, j) \in A$ and a given root vertex r in special set $K \subset V$. The DSPG is to find a least cost directed tree with root vertex r, which contains a directed path from r to all other vertices of K. (By choosing an arbitrary vertex in K as root and substituting each edge for two arcs (i, j) and (j, i) of equal cost, any

SPG instance can be modelled as an instance of DSPG.) The integer linear programming model P below considers binary arc variables y_{ij} and, for each $k \in K \setminus \{r\}$, continuous flow variables $f^k \in R^{|A|}$, as though different special vertices are to receive different commodities from r in the amount of one unit. Let K_0 denote $K \setminus \{r\}$.

(P) Min $z = \Sigma_{(i,j) \in A} c_{ij} y_{ij}$

subject to

$$\Sigma_{(i,j) \in A} f_{ij}^k - \Sigma_{(j,i) \in A} f_{ji}^k \;\; = \;\; \left\{ \begin{array}{ll} 1 & i = r \\ -1 & i = k \\ 0 & \text{otherwise} \end{array} \right\} \;\; i \in V, \; k \in K_0$$

$$f_{ij}^k \;\; \leq \;\; y_{ij} \qquad (i,j) \in A, \; k \in K_0$$
$$f_{ij}^k \;\; \geq \;\; 0 \qquad (i,j) \in A, \; k \in K_0$$

Positive flows f^k from vertex r to special vertices k are only possible if the arc set $\{(i,j) \in A \mid y_{ij} = 1\}$ contains paths from the root r to all $k \in K_0$ and therefore this set constitutes a Steiner solution.

The dual of the linear programming relaxation LP of P is given as:

(D) max $z_D = \Sigma_{k \in K_0} u_r^k$

subject to
$$u_k^k \;\; = \;\; 0 \qquad k \in K_0$$
$$u_i^k - u_j^k - w_{ij}^k \;\; \leq \;\; 0 \qquad k \in K_0, \; (i,j) \in A$$
$$\Sigma_{k \in K_0} w_{ij}^k \;\; \leq \;\; c_{ij} \qquad (i,j) \in A$$
$$w_{ij}^k \;\; \geq \;\; 0 \qquad k \in K_0, \; (i,j) \in A$$

Consider an optimal solution y, f^k of LP, say with optimal value z_{LP}. A linear programming solution can incorporate a corresponding optimal dual optimal solution, giving the reduced costs: $\tilde{c}_{ij} = c_{ij} - \Sigma_{k \in K_0} w_{ij}^k$ of variables y_{ij}. The classical technique of pricing out variables by a sensitivity analysis proceeds as follows: eliminate an edge (i,j) from the problem if $z_{LP} + \tilde{c}_{ij}$ exceeds $c(U)$. Likewise the tentative addition of constraint $\Sigma_{t \neq i} y_{ti} = b$ for right-hand side values $b = 0$ and $b = 1$ might lead to inclusion, respectively exclusion of vertex i.

Generally, the LP lower bound z_{LP} approximates the Steiner cost z^* very well, but an optimal solution of LP is difficult to obtain, as the size of the model is cubic in k and v. Often, the lower bound value of the dual ascent solution (u, w) comes quite close; it can be computed more efficiently and it delivers a reduced costs $c(w)$. For any feasible solution (u, w) of the dual D of LP and any $y \in \{0, 1\}^{|A|}$ that is the incidence vector of a feasible tree spanning K, we have, by Benders' reformulation of P in the so-called master problem, the following lower bound on the optimal Steiner cost:

$$\Sigma_{(i,j)\in A}(c_{ij} - \Sigma_{k\in K_0}w_{ij}^k)y_{ij} + \Sigma_{k\in K_0}u_r^k \leq z^*.$$

Here the last term $\Sigma_{k\in K_0}u_r^k$ is the dual criterion value $z_D(u, w)$, and the first term is at least $\tilde{c}(S^*)$, with S^* denoting a Steiner tree with respect to the reduced arc costs $\tilde{c}$, where $\tilde{c}_{ij} = c_{ij} - \Sigma_{k\in K_0}w_{ij}^k$. This leads to $z^* \geq z_D(u, w) + \tilde{c}(S^*)$, so the lower bound $z_D(u, w)$ may be increased by the Steiner cost of $DSPG(V, K, r, A, \tilde{c})$, or a lower bound on this cost. For a more sophisticated dual solution such as the final dual ascent solution (u, w) of Wong the reduced Steiner cost is zero; however this is not true in a sensitivity analysis.

Let z_i^* denote the optimal solution value under the (extra) restriction that vertex $i(\in V \setminus K)$ must be in the solution tree as a vertex of degree 3 or higher. By restricting incidence vectors y to these types of solutions one obtains analogously lower bounds for z_i^*. Let S_i^* denote a directed Steiner tree with respect to reduced weights $\tilde{c}(w)$ under the additional restriction that this tree contains i as a vertex of degree 3 or higher.

Theorem 5.1 *For dual feasible solutions (u, w) of D : $z_D(u, w) + \tilde{c}(S_i^*) \leq z_i^*$.*

5.2   Reduced Cost Reachability

For an integer linear program model in arc variables y, assume that we have solved the linear program relaxation (or that we have at hand a feasible solution of its dual), giving a lower bound $\underline{z}$ plus reduced arc costs $\tilde{c}$. Further assume that U is a feasible tree of upper bound cost $c(U)$. Then we can exploit theorem 5.1 for vertex exclusion.

Corollary 2 *Pseudo-eliminate $i \notin U$ if $\underline{z}+\tilde{c}(S_i^*) \geq c(U)$.*

The Steiner cost $\tilde{c}(S_i^*)$ may be awkward to obtain; it is given by three paths of least total reduced cost, mutually vertex disjoint (except for i): one

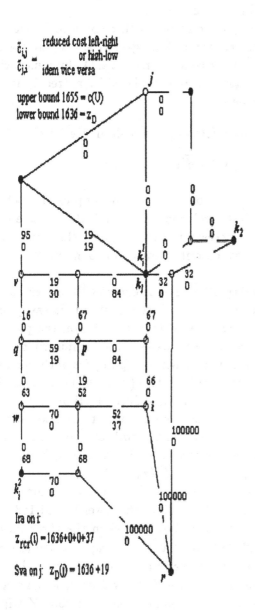

Figure 16: Ira eliminates all non-special vertices (o) to the left of vertex j

path must go from root r to i, the other two must go from i to different nodes in K. In a directed graph the determination of edge disjoint paths is NP-complete; so we will settle for a lower bound.

With $\tilde{d}$ denoting the shortest path distance with respect to $\tilde{c}$, one can recognize in the following test the features of R_3, again k_1 and k_2 are special vertices respectively nearest to i and second nearest to i. A difference with R_3 is that the farthest special vertex l is to be replaced by r; tree S_i^* is not directed from i but from root node r.

Reduced Cost Reachability (RCR) Pseudo-eliminate $i \notin U$ if

$$\underline{z} + \tilde{d}(r,i) + \tilde{d}(i,k_1) + \tilde{d}(i,k_2) \geq c(U).$$

The above formulation is a rough one and can be improved. Let us denote the left hand side of the test condition by $z_{RCR}(i)$. Its variable part $\tilde{d}(r,i) + \tilde{d}(i,k_1) + \tilde{d}(i,k_2)$ might increase if one computes the underlying paths in a more disjoint fashion; we describe how this is done in [8], in the implementation IRA of RCR.

Test IRA (improved reachability after ascent) exploits the lower bound $z_D(u,w)$ of z_{LP}, as obtained by Wong's dual ascent method. It easily prevents the non-node-disjoint case that special vertex k_1 lies on the path to k_2: by exploiting once more procedure SPATHS (see the computation of s'' in section 2.3). With directed input $\tilde{c}$, it will compute, for each vertex i paths without intermediary K-intersection to two (as such) nearest special vertices k_i^1 and k_i^2. Consider as an example the reduced rectilinear problem given in Figure 16. Here we have obtained by dual ascent the lower bound value 1636 and reduced arc costs as specified for both arc directions. The $\tilde{c}$-shortest path from root r to non-special vertex i has zero length, also the two $\tilde{c}$-shortest paths from i to k_1 and k_2 in $K \backslash \{r\}$ 'add' zero length to the lower bound 1636; the path to k_2 runs via k_1. With SPATHS the second path ends in the lower-left special vertex, adding 37 to the lower bound 1636.

Still, especially arcs incident to i may be doubled in the three paths. Consider, for instance, the middle-left boundary vertex q; here the last arc of cost 63 of the path from r to q is retraced in a reverse direction by the first arc of one of the other paths. For vertices i of low degree, we have also prevented this non-edge-disjoint case, by scanning each possible 'Y-triple' of arcs $(p,i), (i,v), (i,w)$ incident to i, determining the value $z_{RCR}(i)$ as the minimum encountered value of

$$z_D(u,w) + \tilde{d}(r,p) + \tilde{c}_{p,i} + \tilde{c}_{i,v} + \tilde{c}_{i,w} + \hat{d}(v,k_v^1) + \hat{d}(w,k_w^1)$$

where $\tilde{d}$ denotes the ordinary shortest path length with respect to $\tilde{c}$, whereas path lengths $\hat{d}$ to vertices k_v^1, k_v^2 are determined for all $v \in V$ by SPATHS; when $k_v^1 = k_w^1$, there is an appropriate substitution by the path length to k_v^2 or k_w^2.

Before the actual testing, one calculates all $\tilde{d}$-paths from r and one determines for every $v \in V \backslash K$, the paths to k_v^1 and k_v^2 by means of modules $1^{st} K$-path and $2^{nd} K$-path of procedure SPATHS (detailedly described in [10]). Essentially, this boils down to running three times a shortest path algorithm with one source and all destinations. So, when one restricts, as we did, the enumeration of Y triplets to vertices of degree smaller than m then after the ascent the application of test IRA on all vertices has time complexity $O(v \log v + e + m^3 v)$. The ascent itself has a worse time bound, though it is relatively fast in comparison with linear programming.

In the IRA implementation of RCR the term $z_D(u, w)$ can be lower than the lower bound z_{LP} of the linear programming relaxation LP of P, but possibly this is compensated by higher path terms $\tilde{d}(r, i), \hat{d}(v, k_v^1), \hat{d}(w, k_v^2)$. Unlike z_{LP}, the dual ascent ascent lower bound may vary with the choice of the root node; one enhances IRA by repeating the test for more than one root node.

5.3 Vertex Inclusion

Consider an optimal solution $\bar{y}$ of a linear programming relaxation in arc variables y. The tentative addition of constraint $\Sigma_{t \neq i} y_{ti} = 0$, for a vertex i with $\Sigma_{t \neq i} \bar{y}_{ti} > 0$, leads to the conclusion that i is a Steiner vertex if the associated linear programming lower bound exceeds $c(U)$. Equivalently, one can update the solution after the change $c_{ti} = \infty$ for all arcs (t, i). However, the optimal solution for a good relaxation such as LP is hard to compute. It is much easier to approximate this reduction method, exploiting once more the results of dual ascent.

We first describe the dual ascent method in more detail. Note that the dual variables w in D are induced by the variables u via the relation $w_{ij}^k = \max\{u_i^k - u_j^k, 0\}$. After initializing all variables u (and w) at zero, one raises each time a set of them, and one updates the set A_0, being the set of arcs with reduced cost

$$\tilde{c}_{ij} = c_{ij} - \Sigma_{k \in K_0} w_{ij}^k$$

at value zero. Initially A_0 is empty and at the conclusion A_0 holds a feasible Steiner arborescence, that is, each special vertex k is reachable from the root vertex r by arcs of zero reduced cost. This property is the driving rule as well as the stopping rule. If not satisfied a subsequent ascent step selects $k \in K$, not reachable from r in A_0, and raises by the same amount, α, the variables u_v^k for the vertices v in

$$N^k = \{v| \text{ in } A_0 \text{ there is no path from } v \text{ to } k\}.$$

Dual feasibility is retained by choosing

$$\alpha \leq \min\{\tilde{c}_{ij}|i \in N^k, j \notin N^k\}$$

and one ensures a finite number of steps by choosing the maximum value for α, as then with each step the set A_0 increases.

Suppose *after* dual ascent, one changes the SPG instance by raising the arc costs to infinity for all arcs incident to a specified vertex of $V(A_0)\backslash K$. For the new dual problem D, the present dual ascent solution remains feasible. But the set A_0 diminishes and it may happen that root r gets separated from some $k \in K$. Thus, under the assumption that i is not a Steiner vertex, it may be possible to continue the ascent:

Steiner Vertex by further Ascent (SVA)

Remove the arcs incident to i from A_0 and $A\backslash A_0$, update sets N^k accordingly and, if possible, continue the ascent, obtaining a (better) bound $z_D(i)$. If $z_D(i) \geq c(U)$ then i is a Steiner vertex.

In Figure 16 we verify that vertex j is a Steiner vertex by SVA: when eliminating j's arcs, the upper-left special vertex gets separated from the root, a continued ascent increments the lower bound with 19 from 1636 to $c(U)$. Also the mid-lowest non-special vertex is an SVA vertex; here we have a luxury problem: this vertex can also be *pseudo*-eliminated by IRA since $1636 + 0 + 52 > c(U)$.

6 Integration of Test Conditions

If a test fails, it may have almost succeeded. If possible, partial results should be exploited. To put it differently, in this section we promote the idea of integrating test conditions. Frequently, one can exploit a partial result of one test condition within the condition of another test. So-called conditionally optimal edges, the partial results of test NSC, are useful in other tests. Finally, we integrate local inspection and reduced-cost-reachability.

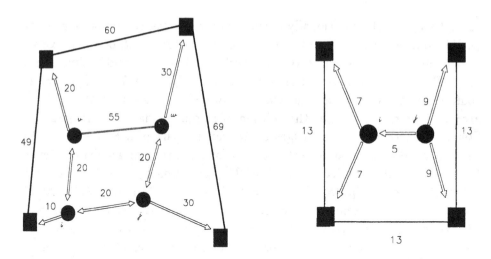

Figure 17: For Figures 6 and 8 the directed graphs of implied edges

6.1   Generating Vertex Implied Edges

A vertex implied edge is 'conditionally optimal'. It is present in a Steiner tree under the condition that i is part of the Steiner tree (while we do not yet know whether i is a Steiner vertex or not).

When checking the NSC test condition for any edge (i, j) of MST T, it takes minor extra computation time to see whether this edge −if not a Steiner edge− is a *vertex implied Steiner edge*, i.e., a Steiner edge under the assumption that either end i or end j can be part of the Steiner tree. Whenever the NSC condition on (i, j) is violated, say for a chord $e = (v, w)$: $d_e(i) + c_{ij} + d_e(j) > c_{vw}$, one can record, in case $i \notin K$, whether the condition $c_{ij} + d_e(j) < c_{vw}$ does hold, and, in case $j \notin K$, whether $d_e(i) + c_{ij} < c_{vw}$ holds. With NSC failing on edge (i, j), it can very well be that $c_{ij} + d_e(j) < c_{vw}$ for every chord $e = (v, w)$ in E_{ij}, which means that (i, j) is a Steiner edge whenever i is to be chosen in the Steiner tree (then effectively i enters K and value $d_e(i)$ becomes zero). The order of time complexity for the NSC test on all edges of MST T is not affected when one additionally detects all such vertex implied Steiner edges; shortly we will give an application in connection with test SD.

The implied Steiner edges give rise to a directed graph: let *arc* (i, j) be part of this graph if edge (i, j) is optimal under the condition that i is in the Steiner tree. While performing the NSC test one constructs this associated directed graph; let us denote it by T^d, as it arises from MST T by removing edges that are not conditionally optimal and replacing the remaining edges (i,j) by one or even two directed arcs. For any vertex let M_i be the maximal subtree in T^d directed out of root i. If test Nsc has not found Steiner edges in MST T, then for i special or a vertex of $T \backslash T_K$, tree M_i is degenerate, containing only vertex i and no arcs. For other vertices i, tree M_i can have arcs, and we refer to M_i as the subtree of i-implied edges: if i is part of a Steiner tree, then i and all edges of M_i can be part of a Steiner tree. In Figure 17 we picture the directed graph for the problem graphs of Figure 6 and 7. For the implied trees M_i in T^d, generated by test Nsc in MST T, we observe some properties. $M_i = \cup\{M_j | \ j \in V(M_i)\}$ and $M_i = M_j$ if i and j are in the same strongly connected component. Secondly, any tree M_i cannot contain special vertices as non-leaf vertices, as these trees are generated in the context that Nsc has not found (unconditional) Steiner edges.

6.2 Implied Exclusion

As a first example, let us enhance with implied edges test SD.

> SD (implied version): Edge (i,j) *not* present in MST T of (V, E, c), can be eliminated if $s^*_{ij} < c_{ij}$, where distance s^* is defined as
>
> $$s^*_{ij} = \min\{s_{vw} | \ v \in V(M_i), w \in V(M_j)\}.$$

In the left-hand side example of Figure 17, the edge (v, w) of weight 55 would be eliminated by the above test; both M_v and M_w coincide with the entire MST T, so $s^*_{vw} = 0$.

Now we describe how implied Steiner edges can be used for vertex exclusion. We do this comprehensively for test LSD only, very similar improvements are possible in the stronger test versions LSC or LSD^P.

Suppose that $i \in V \backslash K$ is tested by LSD, and one or more edges incident i are i-implied, say edges $(i, p_1), ..., (i, p_t)$. The LSD test condition is to refute the possibility that i is chosen in Steiner tree S, in particular the condition on subset N' shows, by contradiction, that it is not a necessity to have i with edge set

$$E_{N'} = \{(i, n) | n \in N'\}$$

in a Steiner tree S. Now as a first step, if i is in S, then edges $(i, p_1), ..., (i, p_t)$ can be chosen in S as well. So, as a first usage of implied Steiner edges, one can relax the LSD condition on i, by considering fewer subsets of adjacent vertices, only those subsets $N' \subset N(i)$ with $\{p_1, ..., p_t\} \subset N'$. This so-called 'lean' LSD test dominates test LSD while consuming less time.

Now, consider a relevant subset N' of $N(i)$ in the lean LSD test on i, as an example suppose $N' = \{p, q, r\}$ with edge (i, p) being i–implied, see Figure 18. The associated edge set, $E_{N'} = \{(p, i), (q, i), (r, i)\}$, represents, in the LSD check on N', a candidate subtree of the Steiner tree. As such, one may extend this tree with the edges of M_i, the set of i-implied Steiner edges; say, in Figure 18, the double lined edges above i. Moreover –only for the purpose of the subtest on N'– one can extend here with the conditional Steiner edges implied by q and r. Thus, using the partial results of the NSC test, one arrives at a maximal implied Steiner tree component

$$M_{N'} = E_{N'} \cup \bigcup_{j \in N'} M_j$$

having the following connotation: if the adjacency set $E_{N'}$ of i is part of a Steiner tree S, then (it is possible to transform S such that) all edges of $M_{N'}$ are part of Steiner tree S.

$M_{N'}$ can be generated efficiently in $O(M_{N'})$ time by starting a search in digraph T^d generated by test NSC. Further note that if it so happens that $M_{N'}$ is not a tree, then the implied version of test LSD has already succeeded for N', since apparently it is not consistent to have $E_{N'}$ in Steiner tree S. For the other (normal) case with $M_{N'}$ containing no cycle, we discuss two ways of continuation.

The first method will not notably increase the complexity of the test. Its idea, pictured in the middle of Figure 18, is to merge each component of $M_{N'} \backslash E_{N'}$ into a single node, say the (generating) vertex of N'; after which we can relax the check of the LSD condition for N' by performing it in the new s-distance graph (N', s). Because of the merge operations this graph may have lower s-weights than in the ordinary LSD.

The second method of continuation aims at increasing the right-hand side $\Sigma_{j \in N'} c_{ij}$ of the LSD condition. We merge all inner (non-leaf) nodes of candidate tree $M_{N'}$ into vertex i, storing the total edge cost of this merged inner tree as $c_{N'}$. Now the former LSD *check* on N', $Z_{MST} \leq \Sigma_{j \in N'} c_{ij}$, is substituted by a new *application* of the LSD test on the merged vertex i, considering the subsets N'' of the new set $N(i)$, with two enhancing features: The right hand sides, $\Sigma_{j \in N''} c_{ij}$, are to be increased with $c_{N'}$, and a larger set $\{p_1, ..., p_t\}$ is compulsory in subsets N'' of $N(i)$, namely all leaves of former tree $M_{N'}$. On the negative side, the time complexity for such a test

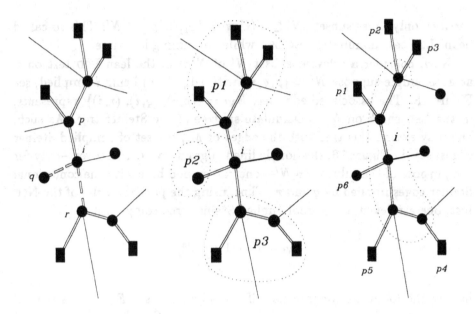

Figure 18: Using "implied" edges in LSD

grows considerably. Because each previous check on N' is substituted by a separate LSD application (on a vertex that is likely to have a higher degree).

In the *lean* LSD^p test one (further) preprocesses the chosen 'full' set of tree extensions by fixing in the edges (l_n, n) that are l_n-implied, each such fix about halves the number of trees remaining in the set. If the hop parameter p is high enough, then in fact, the lean version of LSD^p comprises the above described second method of the implies LSD test.

6.3 Implied Inclusion

Here we describe how inclusion tests such as LBD and SVA can be improved by using vertex implied edges. Consider the vertices from which vertex i can be reached in tree T^d, i.e., the vertices of set $R_i := \{j \in V \backslash K \mid i \in M_j\}$; note that R_i also includes i. To validate the LBD test we assumed the contrary, that i cannot be part of any Steiner tree. Under this assumption any other vertex of R_i can also not be in a Steiner tree. Knowledge of R_i can be used to reformulate test LBD, computing this time the bottleneck distances under deletion of all vertices of R_i.

LBD (implied version) Vertex i (and implied edge set M_i) are optimal, if there is a h-tree T_i of i with leaf set $L \subset K$, such that

$$\text{MST weight of } (L, b') \geq c(T_i),$$

where b' denotes the bottleneck distance in $G \backslash R_i$

Heuristically, when all the leaves of tree M_i are special, one could put to test $T_i = M_i$; otherwise (e.g. i itself may be a non-special leaf of M_i), one could put to test the largest h-tree of i (with h maximal) in the MST T on (V, E, c). In the right-hand side example of Figure 17 the test now succeeds on vertex i : $R_i = \{i, j\}$ and $T_i = M_i$ has edge weight 37, which is lower than 39, the b'-MST weight of L $(= K)$ under deletion of nodes i and j.

When changing in the implied LBD test the object to an edge, say edge $e = (i, j)$, then set R_e may be defined as the union of $\{e\}$ and the set of edges incident to $\{v \mid e \in M_v\}$. Such a test dominates test NSV, as the latter considers one choice for 'tree' T_i : (for high enough h) the path $[k_i..i] \cup \{(i, j)\} \cup [j, k_j]$.

Likewise, one can enhance other inclusion tests SVA and the sensitivity analysis of a linear programming lower bound, e.g. z_{LP} of section 5:

- To fix vertex i and M_i in the solution it is sufficient that the LP bound raises to $c(U)$, if arc-costs of all arcs incident to R_i are raised to infinity.

- In the test 'implied SVA' on vertex i, one continues the ascent after deleting all arcs incident to R_i.

6.4 A Synthesis of Local and Global Inspection

As a preliminary first step, we describe the use of vertex implied edges in the test IRA (or similar global sensitivity tests). Remember that before applying IRA, the dual ascent lower bound $z_D(u, w)$ has been computed, and, with respect to the reduced costs $\tilde{c}$, the shortest path distances $\tilde{d}, \hat{d}$ have been computed. The IRA test condition normally requires for each possible 'Y-triple' of arcs (p, i), (i, v), (i, w) incident to i, that the value

$$z_D(u, w) + \tilde{d}(r, p) + \tilde{c}_{p,i} + \tilde{c}_{i,v} + \tilde{c}_{i,w} + \hat{d}(v, k_v^1) + \hat{d}(w, k_w^1)$$

is at least $c(U)$. Suppose first that the i-implied tree M_i has only one or two edges incident to i. Then the above condition can be relaxed by considering

$\text{TREES}_q^f(\text{vertex } i)$

BEGIN

Put the 0-tree of i, marked, on a Queue;

#marks:=1;

repeat

Remove a tree T from Queue, say an h-tree of i with leaf set L;

    if T is marked **then**

      if $|N(T)| > q$ or $L \subset K$ **then** Exit

      else

        Preprocess a 'full' set of $(h+1)$-trees $T' \supset T$, queueing

        the trees T' that have maximal intersection with $\cup_{v \in T} M_v$

    else *(* when T is unmarked *)*

      if $Z_L > c(T)$ and $z_{RCR}(T) < c(U)$ **then**

        if #marks=f or $|V(T)| > q$ **then** Exit

        else Mark T, put T back on Queue, #marks:=#marks+1;

    until Queue is empty;

    Pseudo-eliminate i;

END

Figure 19: A synthesis of tests

only those Y-triples that involve these edges of M_i. Suppose now that M_i has three or more edges. For any node $m \in M_i$ let $\widetilde{c_m}(M_i)$ denote the total reduced cost of M_i when orienting the edges out of 'root' node m in M_i; further let L_m denote the set of non-special leaf nodes of $M_i \backslash \{m\}$. We consider for all vertices m of M_i, the value

$$z_D(u,w) + \tilde{d}(r,m) + \widetilde{c_m}(M_i) + \sum \{\hat{d}(l,k_l^1)| \, l \in L_m\};$$

it is a lower bound on the cost of a Steiner tree that includes i (and thereby M_i) with on the path from r to i vertex m as point of entry in M_i. Let $z_{RCR}(M_i)$ denote the minimum of these values over all $m \in M_i$. Then one can pseudo-eliminate i if $z_{RCR}(M_i) \geq c(U)$.

Next, with the definitions given above, it is relatively simple, to go a step further, integrating reduced-cost-reachability in test LSD^p. Analogously as $z_{RCR}(M_i)$, for any tree T the value $z_{RCR}(T)$ can be computed; it is a lower bound for having candidate tree T as a part of the Steiner tree. This leads in Figure 6.4 to a synthesis of previously given tests. Remember that $N(T)$ denotes the set of neighboring vertices of T. As test $\text{LSD}^p(i)$, also test $\text{TREES}_q^f(i)$ processes a first-in-first-out queue of local trees. Now, a tree is refuted as candidate subtree of the Steiner tree, when it satisfies the LSD-condition, $Z_L \leq c(T)$, or, when it satisfies the RCR-condition, $z_{RCR}(T) \geq c(U)$. The test is exited, reporting failure, if either a marked tree grows too large, over the limiting parameter q, or, if the number of subfailures, #marks, exceeds the other limiting parameter f; the (redundant) parameter p of test LSD^p is here left out.

In the context of available data $z_D, \tilde{c}, \tilde{d}$ and $\hat{d}$ one can compute for any tree T the value $z_{RCR}(T)$ in $O(|T|^2)$ time. But in $\text{TREES}_q^f(i)$, we consider nested sets of trees; then the value $z_{RCR}(T)$ can be found by update in $O(|T|)$ time. When applied with the same limiting parameters f and q as for $\text{LSD}^p(i)$, the new test has again, per tested vertex, time complexity $O(fq2^q)$.

7 Results and Conclusion

Briefly, we discuss several reduction programs and their design. To give an idea of the potential of the different reduction tests, we tabulate the numerical results of an earlier program. On the same test problems we report on the further reduction by some of the newly presented tests. As yet the dominating tests of section 6 are not yet implemented.

7.1 Reduction Programs

An earlier presented preprocessing program given in [7] reduced with a running time more or less cubic in the number of vertices. Besides the three bottleneck tests this program employed the sensitivity tests of [3], a test called 'change of edge costs', and reachability tests like R_3. On the test problems of randomly drawn graphs, with uniformly random and Euclidean weights, the different tests contributed roughly in order of mentioning. With random weights the reduction performance was considerable, especially on sparse graphs with a high enough fraction of special vertices. This was not so on the other graph types; especially for dense Euclidean instances, the reduction was limited.

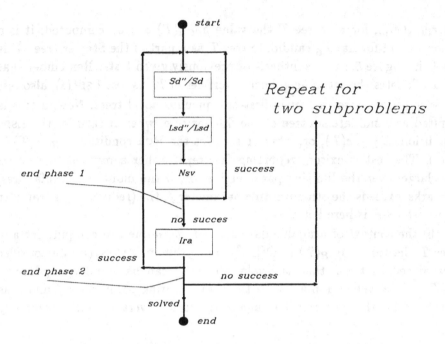

Figure 20: Flowchart of the reduction program of [9]

By the reduction program of [8] such test problems and others were solved up to exact solution, experiencing a running time of square order in the number of vertices. The main improvements here are:

- A lower time complexity order for each of the tests: SD, LSD, and NSV;

- The application of reduced-cost-reachability;

- Better upper bounds by the so-called 'pilot' method.

The latter heuristic method, provides (near) optimal upper bounds $c(U)$, eagerly exploited by the implementation IRA of reduced-cost-reachability. In [9] the implementation of IRA is improved at the lower bound side: by the enumeration of Y-triples and by running the dual ascent for different choices of the root node. We can best show the potential of reduced-cost-reachability by a reprint of the results of the reduction program of [9]. Table 2 provides results after each of three phases in this program; see Figure 20.

In the first phase tests SD'', LSD'' and NSV are applied, using in the first two tests approximated special distances s''. In phase 2 the tests are re-applied with exact s distances in conjunction with test IRA. Typically the tests in the chain of tests SD-LSD-NSV cooperate with each other, preferably in the given order, also with respect to computation time. After test SD the degree of vertices decreases, so they become more amenable for local inspection by LSD; after the actions of (pseudo)elimination of LSD the bottleneck distances increase, which may support test NSV; after NSV some special distances decrease, which may revive test SD, and so on, possibly over more than three iterations.

With regard to the speed of reduction it is always prudent to invoke (dominating) tests of higher time complexity, after a first use of lower complexity tests, to profit also in the reduction process itself from the reduced graph size. Thus, as they are both of higher time complexity order, the heuristic computation and the application of dual ascent in phase 2, benefit from the graph reduction in phase 1. If necessary, when the first as well as the second phase is not able to solve ($|V'| = 1, |K'| = 1, |E'| = 0$), the branch and bound phase is entered; it basically repeats the same reduction process on subproblems, created by branching on vertices selected by test SVA. For a detailed description we refer to [9].

The reduction algorithms in [17] contain significant improvements, leading to a speedier and possibly more effective reduction. On the one hand (approximate) tests are applied in yet a lower time complexity order. More importantly, the time requirement for good heuristic computation is reduced, while their program incorporates also, several methods to produce a lower bound better than the dual ascent lower bound $z_D(u, w)$; thus more effective reduced-cost-reachability tests can be applied.

7.2 Test Problems

Here we describe our four different types of test problems. Three of them, called 'uniform random', 'Euclidean random' and 'incidence random' differ only in the weights on the edges. A connected graph is produced by a random graph generator taking as input desired numbers for $v = |V|$, $k = |K|$ and $e = |E|$. Edge weights c_{ij} for $(i, j) \in E$ are generated according to the desired problem type. For the problems referred to as 'uniform random', the weights are uniform random integer in the interval $[1, ..., 100]$. In the so-called 'incidence random' problems the weight on each edge (i, j) is defined with a sample r from a normal distribution, rounded to integer value with a minimum outcome of 1 and maximum outcome of 100, i.e., $c_{ij} = \min\{\max\{round(r), 1\}, 100\}$. To obtain a graph that is much harder to reduce by preprocessing techniques, three distributions with a different mean value are used. The mean of r is 20 for edges (i, j) with $i, j \notin K$ (no incidence with K), 40 on edges (i, j) with one special end vertex and 60 on edges (i, j) with both ends $i, j \in K$. The standard deviation for each of the three normal distributions is 10. For the 'Euclidean random' problems each vertex is a point in the Euclidean plane with coordinates uniformly random in the interval $(0, 100)$; here the edge weights are given as Euclidean distances rounded up to integer value. As sizes for the above three problem types we choose $v = 80, 160$ and 240. For each size v we choose twenty variants combining five densities of special vertices (with truncation [.] to integer value): $k = [^2logv], [\sqrt{v}], [2\sqrt{v}], [v/4]$ and $[v/2]$ with four edge densities: $e = [3v/2], 2v, [vlogv]$, and $[v(v-1)/4]$. Each variant is drawn five times, giving 100 instances per considered size v. In the 'Euclidean grid' instances the vertices are the intersection points of a rectilinear grid consisting of an equal number of horizontal and vertical lines. Given this number of lines l, the number of vertices is l^2 and the number of edges is $e = 2(l^2 - l)$: the edges are line segments between adjacent intersection points, the weights are Euclidean. We have chosen $l = 9, 13$ and 18. In forming the grid the distance between adjacent horizontal lines and adjacent vertical lines is chosen as a uniform random integer in the interval $[1, ..., 100]$. Finally for a given number k, special vertices were randomly chosen to form the vertex set K, but in this process the new special vertex was chosen (as long as possible) on a 'new' horizontal and on a new vertical line (lines that are as yet without special vertices). Thus for a choice $k \geq l$ the special point coordinates define the grid. For each size v we considered nine special densities, $k \in \{x - 3, x, x + 3 | x = l, 2l, 3l\}$. Drawing five times, this gives 45 instances per size l^2.

7.3   Test results

For the reduction program of [9], summarized in Figure 20, we record in Table 2 the reduction performance at conclusion of phase 1 and 2, by giving the remaining number of (special) vertices and edges as a percentage of the original numbers. The last three columns give for the total program, the number of subproblems solved, the average CPU time and the maximum CPU time in seconds on an Intel Pentium 90 Mhz. Except for the latter value all other data are averages, in each line over 100 instances differing in density of special vertices and edges as discussed above. A column for the CPU times for phase 2 is left out; they differ very little from the average

CPU times of the total program.

The results show an important reduction, especially on uniform random problems. Though the performance of IRA is impressive, the most effective implementation of reduced-cost-reachability is probably not IRA but an implementation with the reduced costs of the optimal linear programming solution. On the other hand one should realize that the success of reduced-cost-reachability (and other such global reduction tests) relies on a small gap between the lower and the upper bound. A relative gap of 5% will not block Ira in smaller sized graphs, but a relative gap of 1% in a very large problem with the upper bound solution U containing thousands of edges will be too great. Unfortunately, the absolute difference between lower an upper bound is bound to grow when difficult problems become larger.

Now we tabulate results on the same test problem beds for the additional use of the new tests, this time without test IRA, to investigate the extent by which tests LSD^p, LSC and NSC are able to reduce further than their forerunners SD, LSD and NSV.

After a first reduction by SD'', LSD'' and NSV as in phase 1 above with results as given in Table 2, we continued by cycling exhaustively SD-LSD-NSV; where, additionally, test SD, when unsuccessful, was replaced by the 'implied' test version as well as by LSD^1(edge) and likewise test NSV, when unsuccessful, was replaced by NSC. The remaining percentages of vertices and edges at the end of this new phase are given in the left hand side columns of Table 3. In the right-hand side columns we give the same data after cycling additionally tests LSD^3(vertex) and LSC^2, the latter being applied heuristically on all vertices and edges. Again all data are averages over 100 instances per line, but CPU times are here measured on an Intel Pentium 166 Mhz. We did not fully optimize the implementation, the given average CPU times are (much) higher than needed, because of the fact that we recalculated from scratch all exact special distances s before starting each new iteration of re-application of tests, whereas one could update these distances in $O(v^2)$ time per edge contraction.

Test NSC dominates NSV, the implied version of SD and test LSD^1(edge) both dominate SD. But the extent by which they reduce further is rather disappointing. A (not tabulated) exhaustive cycling of only exact tests SD, LSD and NSV would have given only little worse results.

The tests LSD^p and LSC dominating LSD are worth while, especially in graphs, that are sparse or have become sparse after SD. Our implementation of test LSD^p coped without difficulty with exact extensions for $p = 3$, with at most six subfailures, $f = 6$, and 12 new tree leaves, $q = 12$. Let us clarify the

size	phase 1				phase 2			total program		
v	%v	%k	%e	cpu	%v	%k	%e	#sp	cpu	max
uniform random										
80	9	10	3	.1	.2	.6	.0	1.0	.1	.3
160	26	18	5	.1	.3	.8	.0	1.2	.2	.7
240	31	18	4	.2	.3	.7	.0	1.0	.3	1.2
320	34	17	4	.3	.1	.3	.0	1.0	.5	1.3
incidence random										
80	66	64	39	.1	3	6	.8	1.8	.3	3
160	66	61	25	.2	6	12	1.1	3.4	1.2	16
240	68	60	16	.3	9	17	1.0	3.9	2.9	34
320	69	59	12	.4	14	24	1.5	6.6	8.3	98
euclidean random										
80	37	29	19	.1	.3	.8	.1	1.0	.1	.3
160	48	33	16	.1	1.3	3.2	.2	1.5	.3	2
240	49	32	13	.2	1.1	2.7	.1	1.6	.5	2
320	51	31	12	.4	1.1	2.7	.1	1.6	.8	6
euclidean grid										
81	25	26	26	.1	1	1	1	2.0	.1	1
169	53	54	55	.1	2	3	2	3.7	.6	3
256	67	69	70	.2	3	6	3	8.5	1.8	21
324	69	69	72	.3	1	2	1	2.3	2.4	7

Table 2: Remaining percentages of vertices\edges in the reduction program of [9]

size	..Sd-Lsd[1]-Nsc					..Lsd[3]-Lsc[2]				
v	%v	%k	%e	its	cpu	%v	%k	%e	its	cpu
uniform random										
80	6	7	1.5	2.2	.1	1	3	.3	3.6	.1
160	18	13	3.2	2.5	.4	4	7	.7	5.1	.4
240	24	14	3.3	2.5	1.0	11	9	1.6	5.4	1.7
320	28	12	3.1	2.6	2.2	13	8	1.6	6.4	3.5
incidence random										
80	52	51	20.2	2.5	.3	34	42	13.7	4.2	.6
160	55	54	12.8	2.4	1.2	43	48	10.4	4.5	2.3
240	61	54	10.4	2.5	3.0	54	51	9.5	3.8	5.3
320	64	55	9.0	2.4	5.7	59	53	8.5	3.6	9.2
euclidean random										
80	32	24	16.1	2.2	.2	7	11	3.2	4.1	.2
160	41	28	13.8	2.4	1.0	18	18	7.2	5.2	1.2
240	43	26	10.9	2.4	2.3	23	17	6.8	5.7	3.0
320	47	27	9.9	2.5	4.6	29	19	7.4	6.2	6.6
euclidean grid										
81	9	13	8.6	3.3	.2	2	6	2.1	3.2	.1
169	30	39	28.1	4.7	1.0	4	14	3.9	5.0	.7
248	41	55	39.5	5.8	3.6	5	16	3.9	8.6	2.4
324	47	57	45.1	6.1	6.8	5	15	4.0	11.6	5.0

Table 3: Further reduction after phase 1 by newly implemented tests

moderate running times. First, there is the effectiveness of the preprocessing operations, which more than decimates the number of tree extensions. Also, we have implemented the lean variant that fixes vertex implied edges. Next, one must realize that the bound $O(fq2^q)$ on the computation per vertex is loose in practice; in case of a test failure, the limit of f subfailures is most likely attained in a situation with the other marked trees still unprocessed on queue. And finally, we did not invoke LSD^3 before LSD^2 had failed on all vertices.

7.4 Summary and Conclusion

We reviewed reduction tests for the SPG, and we formulated other stronger tests. In our presentation, we exemplified the method of relaxing a test condition. Thus not another or a better, but a dominating result is produced. We discerned between local and global techniques. The three bottleneck tests SD, LSD and NSV are in fact hybrid as local adjacency properties are judged by means of bottleneck measures, carrying global information on the entire graph. For the three bottleneck tests we discussed more efficient implementations.

Global tests process information of the entire graph and require more computation time. For a successful application of RCR one needs good lower and upper bounds. Lower and upper bounds are faster and\or better computed in reduced graphs. The results given in Table 2 show that reduced-cost-reachability can be very effective. But RCR is bound to fail in large sized difficult graphs, as then we are not likely to produce a small absolute gap between upper and lower bound. All polynomial time methods are bound to have deficiencies that will repeat locally and accumulate.

This was an incentive to develop further the technique of local inspection of test LSD. Local tests do focus on small portions of the graph, processing information of the local graph structure, executing many tiny computations for obtaining lower and upper bounds in small graphs induced by neighboring vertices. With LSD^p we increase the scope of inspection. We deflected the threat of exponential explosion by preprocessing rules and prudent implementation with limiting parameters.

With the introduction of conditionally optimal edges, the vertex implied edges of section 6, we showed how partial results can be made useful, or to put it differently, that tests should and can be integrated in earlier stage. As an older example with regard to edge exclusion: the local test 'least cost' and the global test MST, can better be synthesized in test SD. Now in section 6.4 we propose to integrate local inspection into global inspection.

With the present improvements in reduction techniques we may have come to a point that a further reduction effort becomes dubious in two ways. Will the amount of time spent be regained? And secondly, even worse, are we sure that the reduced graph is a simplified instance?

When we pseudo-eliminate vertices, it may happen that $|V'|$ decreases, while $|E'|$ increases! Whether this is a simplification depends on the abilities of the main processing algorithm. On the other hand, we might also step out of the system presented in figure 1, by letting preprocessing algorithm π return something more: not only a reduced instance, but a reduced instance with side information, a graph with 'colored' edges.

Initially all edges are uncolored, but with each pseudo-elimination of a vertex v, one could *add* a new color, say μ_v, to each edge of the set of edges that replaces v. By this added color, we save a side result of the LSD analysis that will otherwise get lost: just one edge of this set can be used in an optimal solution. The coloring of edges can first be of use in LSD^p itself. One can further restrict the number of possible tree extensions: in a valid extension one may not encounter the same color twice.

Finally, suppose that after preprocessing, a main algorithm takes over, one that exploits a linear programming model in arc variables y_{ij}. Then, for each different color mark μ_v (i.e. for each pseudo-eliminated vertex) one could add the valid inequality:

$$\sum_{(i,j)\in\mu_v(E)} (y_{ij} + y_{ji}) \leq 1,$$

where $\mu_v(E)$ stands for the set of edges that have (among their set of colors) the color μ_v. During the reduction process the set $\mu_v(E)$ can grow much larger than the set of edges that has replaced vertex v.

References

[1] A.Balakrishnan and N.R. Patel, Problem Reduction Methods and a Tree Generation Algorithm for the Steiner Network Problem, *Networks* 17 (1987) pp. 65-85.

[2] J.E. Beasley, An Algorithm for the Steiner problem in Graphs, *Networks* 14 (1984) pp. 147-159.

[3] J.E. Beasley, An SST based algorithm for the Steiner problem in graphs, *Networks* 19 (1989) pp. 1-16.

[4] S. Chopra, M.R. Rao and E.R. Gorres, Solving the Steiner Tree Problem on a Graph Using Branch and Cut, *ORSA Journal on Computing* 4 (1992), pp.320-335.

[5] E.W. Dijkstra, A Note on Two Problems in Connexion with Graphs, *Numerische Mathematik* 1 (1959), pp. 269-271.

[6] C.W. Duin and A. Volgenant, An Edge Elimination Test for the Steiner Problem in Graphs, *Operations Research Letters* 8 (1989), pp. 79-83.

[7] C.W. Duin and A. Volgenant, Reduction Tests for the Steiner Problem in Graphs, *Networks* 19 (1989), pp. 549-567.

[8] C.W. Duin, Steiner's Problem in Graphs, *PhD Thesis* University of Amsterdam (1993).

[9] C.W. Duin, Reducing the Graphical Steiner Problem with a Sensitivity Test, to appear in *Proceedings of DIMACS workshop on Network Design* (1997)

[10] Duin, C.W. and S. Voss, Efficient Path and Vertex Exchange in Steiner Tree Algorithms, *Networks* 29 (1997), pp. 89-105.

[11] Dreyfus, S.E. and R.A. Wagner, The Steiner Problem in Graphs, *Networks* 1 (1972) pp. 195-207.

[12] M.L. Fredman and R.E. Tarjan, Fibonacci Heaps and Their Uses in improved Network Optimization Algorithms, *Journal of the ACM* 6 (1987), pp. 596-615

[13] Gomory, R.E. and T.C. Hu (1961), Multi-Terminal Network Flows, *Journal of SIAM* 9, pp. 551-556.

[14] Hwang, F.K., D.S. Richards and P. Winter, The Steiner Tree Problem, *Annals of Discrete Mathematics* 53 (1992).

[15] T. Koch and A. Martin, Solving Steiner Tree Problems in Graphs to Optimality, to appear in *Networks*

[16] K. Mehlhorn, A Faster Approximation Algorithm for the Steiner Problem in Graphs, *Information Processing Letters* 27 (1988), pp.125-128.

[17] Polzin, T. and Daneshmand S.V., Improved algorithms for the Steiner Problem in Networks, *Technical Report 06/1998*, Theoretische Informatik Universität Mannheim (1998).

[18] P.M. Spira, On Finding and Updating Spanning Trees and Shortest Paths *SIAM Journal on Computing* 4 (1975), pp. 375-380.

[19] P Winter, The Steiner Problem in Networks: A Survey, *Networks* 17 (1987), pp. 185-212.

[20] P Winter, Reductions for the Rectilinear Steiner tree Problem, *Networks* 26 (1987), pp. 187-198.

[21] Wong, R.T., A dual ascent based approach for the Steiner tree problem in a directed graph, *Mathematical Programming* 28 (1984), pp. 271-287.

A Fully Polynomial Approximation Scheme for the Euclidean Steiner Augmentation Problem

J. Scott Provan
Department of Operations Research
University of North Carolina, Chapel Hill, NC 27599-3180
E-mail: Scott_Provan@UNC.edu

Abstract

The *Euclidean Steiner Augmentation Problem* has as input a set of straight line segments drawn in the Euclidean plane, and has as output the smallest set of straight line segments, in terms of total Euclidean length, whose addition will make the resulting set 2-edge connected. A fully polynomial approximation scheme is given for this problem in the case where the input set is connected. Several extensions and variants are also discussed.

Contents

D.-Z. Du et al. (eds.), Advances in Steiner Trees, 235-253.

1   Introduction

Let $\mathcal{L}$ be the union of a set of straight-line segments drawn in the Euclidean plane. The *length* of $\mathcal{L}$, denoted by $l(\mathcal{L})$, is the total of the Euclidean lengths of its segments. The *graph induced by* $\mathcal{L}$, denoted by $G_{\mathcal{L}} = (V_{\mathcal{L}}, E_{\mathcal{L}})$ is the graph whose vertices are the intersections of noncollinear straight lines, and whose edges are the union of subsegments of the lines in $\mathcal{L}$ between intersection points. It follows that $G_{\mathcal{L}}$ is always planar. For simplicity we will henceforth use $\mathcal{L}$ to refer to $G_{\mathcal{L}}$.

A *bridge* of $\mathcal{L}$ is any edge of $\mathcal{L}$ whose deletion will disconnect $\mathcal{L}$. $\mathcal{L}$ is said to be *2-edge connected* if it has no bridges. The *Euclidean Steiner Augmentation Problem (ESAP)* has as input a set S of m straight line segments, and the goal is to find a set F^* of straight line segments of minimum length so that $S \cup F^*$ is 2-edge connected. Figure 1 shows an instance of ESAP and its solution.

Although there has been no work done on Euclidean augmentation problems explicitly, there has been considerable work done on 2-connectivity and augmentation problems on graphs, and on Euclidean connectivity problems in a more general context. The survey [7] gives a good account of the various versions of the graphical problem, and its importance in telecommunications network design. The graphical analog of ESAP, called the *Graphical Steiner Augmentation Problem (GSAP)*, has as its input an edge-weighted graph G along with a subgraph S of G, and it is required to find a set of edges of G of minimum weight which, when added to S, forms a 2-edge connected graph. Connectivity augmentation problems have been the topic of several papers [1, 2, 3, 5], although these papers deal primarily with minimum cardinality augmentation with no restrictions on the edges that can be added. The general GSAP problem is NP-hard, even when S is a tree [1, 5]. In a recent paper [13] a polynomial-time algorithm is given for GSAP in the case when G is planar and S is connected. That result will be used extensively in this paper.

The 1-connected analog of ESAP is the *Euclidean Steiner tree problem*, and involves connecting a given set $\mathcal{K}$ of points in the plane by a minimum Euclidean-length network. This problem has been studied extensively (see [10], Part I, for an excellent survey), and it will turn out that many of the properties of optimal solutions for the Euclidean Steiner tree problem apply to ESAP as well. The structure of minimum Euclidean length 2- and 3-connected sets spanning $\mathcal{K}$ has also been studied [8, 9].

In this paper, we will be concentrating on *approximation* algorithms for

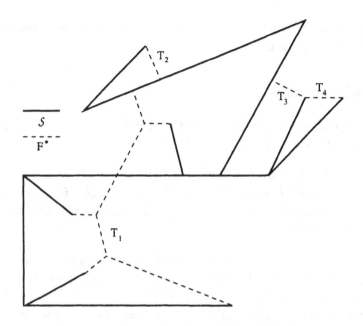

Figure 1: Example of an ESAP Solution

ESAP. Specifically, let S be an instance of ESAP, and let $\epsilon > 0$. A given set $\hat{F}$ is called an *ϵ-approximation* for this instance if

(i) $S \cup \hat{F}$ is 2-edge connected,

(ii) $\frac{l(\hat{F}) - l^*}{l^*} < \epsilon$, where l^* is the optimal length of an ESAP solution for this instance.

A *fully polynomial approximation scheme (FPAS)* for ESAP is an algorithm which, for any instance S and any $0 < \epsilon < 1$, gives an ϵ-approximation for this instance in time polynomial in $1/\epsilon$ and the size of the input describing S. In [12] a FPAS for the Euclidean Steiner tree problem is given when the points lie on the boundary of a convex region, and the technique developed in that paper will also be important here.

The main result of this paper, Corollary 4.4, states that a FPAS for ESAP exists whenever S is a connected set. Section 2 establishes some preliminary results on the structure of ESAP solutions. Section 3 constructs the instance of GSAP whose solution will provide the appropriate ϵ-approximation for ESAP, and Section 4 gives the polynomial-time solution for this instance, using the technique given in [13]. Section 5 gives several important variants of the problem that can also be solved using the technuiques in this paper.

2   Preliminaries

We start with two lemmas describing the structure of a solution for ESAP. The first was proved in [13].

Lemma 2.1 *Let S be an instance of ESAP, and let F be a set of straight line segments. Then F is an optimal solution to ESAP if and only if F is a minimum length forest (union of trees) such that no edge of S is a bridge with respect to $F \cup S$.*

Now let F^* be an optimal solution to ESAP for instance S. F^* is called *proper* if, among all optimal solutions, F^* has the smallest number of (edges) + (vertices that lie on the interiors of edges of S).[1] We can partition F^* into maximal subtrees $T_1, \ldots, T_r$ with the property that the only vertices of S that appear on any T_i are *end* (degree 1) vertices of T_i. We call these T_i *full Steiner subtrees (FSSs)* of F^*, and call any non-end vertex of T_i a *Steiner vertex* of T_i. The solution in Figure 1 has four FSSs, as marked.

Lemma 2.2 *Let F^* be a proper optimal solution to ESAP. Then any FSS T of F^* has the following properties:*

  (i) *Each end vertex of T is either a vertex of S or lies on the interior of an edge e of S with the corresponding line in T perpendicular to e.*

  (ii) *Each Steiner vertex of T has degree exactly 3, does not lie on S, and has each of its adjacent edges meeting at $120°$ angles.*

  (iii) *There is at most one end vertex of T that is not at a vertex of S.*

Proof (i): Let v be an end vertex of T that is not a vertex of S. By the maximality of T, v must be of degree 1 with respect to F^*. Now v must lie on S or at least part of its edge could be deleted without affecting the connectivity of any of the points on S. Thus v must lie on the interior of

---

[1]For clarity we will remark here that the vertices and edges of F^*, S, and $F^* \cup S$ are defined independently; in particular, there may be vertices and edges of $F^* \cup S$ associated with intersecting edges which may not be vertices or edges of F^* or S.

an edge e of S. Further, v must be perpendicular to e, since otherwise v could be moved a small distance in the direction of the acute angle with e to decrease the length of F^* without affecting the 2-connectivity of $F^* \cup S$.

(ii): Let v be a Steiner vertex of T, so that by definition v does not occur at a vertex of S. If v does not lie on any point of S then standard arguments (see, for example, [6]) imply that v must have degree exactly three with each of its adjacent edges meeting at 120° angles.

If v lies on S, then suppose by contradiction that v lies on the interior of some edge e of S, bisecting e into two edges in $F^* \cup S$. First note that v cannot have degree exactly 2 in T, since the two incident edges would not be collinear, and so v could be moved a small distance in the direction of its proper angle, decreasing the length of F^* and contradicting the fact that it is an optimal solution.

Thus there are at least five edges adjacent to v, including the two edges formed by the bisection of e. It follows that there must exist three edges from this set — f_1 and f_2 from T and e' from the bisection of e — satisfying:

(a) f_1 and f_2 are clockwise adjacent, with the angle between them less than 180°;

(b) f_2 and e' are clockwise adjacent, with the angle between them less than 90°.

(see Figure 2). Now perturb F^* into two different forests F_1^* and F_2^* as follows:

(F_1^*) Perturb f_1 and f_2 slightly to meet at a common point v_1 inside the sector spanned by f_1 and f_2.

(F_2^*) Perturb f_2 slightly in the direction of e' to meet e' at a point $v_2 \neq v$.

It is clear from the construction that F_1^* and F_2^* have lengths strictly less than that of F^*, and so neither can maintain 2-connectivity of $F_i^* \cup S$, as this would violate the optimality of F^*. Thus both perturbations produce bridges; in particular, for $i = 1, 2$ the perturbation F_i^* produces partition $(X_i, \bar{X}_i)$ of the vertex set of $F_i^* \cup S$ with $v \in X_i$ and $v_i \in \bar{X}_i$ and admitting at most one edge between the vertices in each set — the bridge for $i = 2$ being (v, v_2). Now if we remove v_i from $\bar{X}_i$ and consider these partitions with respect to $F^* \cup S$, we see that the only edges that can be in $(X_1, \bar{X}_1)$ are f_1, f_2 and the bridge g of $F_1^* \cup S$, and the only edges that can be in

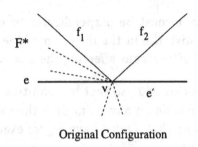

Original Configuration

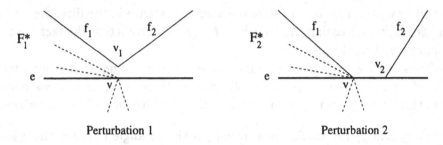

Perturbation 1   Perturbation 2

Figure 2: The two perturbations of F^*

$(X_2, \bar{X}_2)$ are f_2 and e'. Now consider the cuts $(Y_i, \bar{Y}_i)$, $i = 1, 2, 3$, where $Y_1 = \bar{X}_1 \cap X_2$, $Y_2 = \bar{X}_1 \cap \bar{X}_2$, and $Y_3 = X_1 \cap \bar{X}_2$. Since the only four edges that can be in these three cuts are f_1, f_2, g, and e', and the edges f_1, f_2, and e' all start in $X_1 \cap X_2$ it follows that one of the cuts $(Y_i, \bar{Y}_i)$ can contain at most one edge, contradicting the feasibility of F^*. It follows that no Steiner vertex of T can lie on S.

(iii): Suppose that T has two end vertices v_1 and v_2 lying on the interior of edges e_1 and e_2 of S, respectively, so that again v_1 and v_2 are degree one vertices of F^*. Let $P : v_1 = u_0, f_1, u_1, f_2 \ldots, u_{k-1}, f_k, u_k = v_2$ be the path of edges in T between v_1 and v_2. By (i), f_1 and f_k are perpendicular to e_1 and e_2, respectively. Moreover, for $0 < i < k$ each u_i is a Steiner point and so by (ii) u_i has one adjacent edge g_i in addition to the two on P forming $120°$ angles with f_{i-1} and f_i. Consider transforming P by translating each of the edges of P slightly to the left the same distance δ, so that u_0 and u_k remain on e_1 and e_2, respectively, and the points of intersection of each pair f_{i-1} and f_i remain collinear with g_i. The g_i's are then shortened or lengthened appropriately. A simple geometrical argument shows that the total length of the tree T does not change. Further, the 2-connectivity of $F^* \cup S$ will be maintained during this translation up to the first point at which one of the following occurs:

Case 1: v_1 or v_2 intersects a vertex of S;

Case 2: v_1 or v_2 intersects a vertex of F^*;

Case 3: one of the f_i's or g_i's contracts to a point.

Consider the resulting optimal forest F'. If Case 1 occurs then F' has one fewer vertex lying on the interior of an edge of S than does F^*, and if Case 2 or 3 occurs then F' will have one fewer edge than F^*. Both of these contradict the choice of F^* as being proper. □

Corollary 2.3 *Let F^* be a proper optimal solution to ESAP. Then*

(i) *F^* lies entirely inside the convex hull of the vertices of S.*

(ii) *If x is any point on F^* that lies on an FSS having more than one edge, then $l(F^*)$ is at least as great as the maximum of the distances from x to the two nearest vertices of S.*

(iii) *There are at most $4m$ vertices in F^*.*

Proof (i): This follows from the same arguments used to prove it for the Euclidean Steiner tree problem (see [6], Section 3.5).

(ii): Let T an FSS of F^* containing x and having more than one edge. Then T contains at least three end vertices, and so by Lemma 2.2 (iii) T must contain at least two ends that are vertices of S. Since there is a path from x to each of these points in T, then the length of each path must be at least the distance from x to that endpoint. It follows that $l(F^*)$ is at least the maximum of the distances from x to the two nearest vertices of S.

(iii): Consider an FSS T_i of F^* containing n_i vertices of S. By Lemma 2.2 (i) and (ii) we know that the only vertices of T_i that are not vertices of S are either (a) end vertices of T_i perpendicular to the interior of an edge of S, of which there is at most one, or (b) Steiner vertices, which have degree exactly 3. A simple counting argument shows that T_i has at most $2n_i$ vertices. Since the union of the T_i's form a forest then $\sum_i n_i$ is at most twice the number of vertices of S, and since S is connected this is in turn at most m. Thus the number of vertices of F^* is at most $4m$. □

3   An ϵ-approximation for ESAP

In this section we show how a given instance of ESAP can be transformed into a GSAP instance, in such a way that the solution to GSAP on that instance provides ϵ-approximation for ESAP. For instance S of ESAP consisting of m edges, and $\epsilon > 0$, construct the graph $G_{S,\epsilon}$ as follows. The vertex set of $G_{S,\epsilon}$ is made up of the sets V_S, V_P, and V_ϵ. V_S is simply the set of vertices of S. V_P consists of those points v on an edge e of S for which there is some vertex w of S such that the straight line from w to v is perpendicular to e at v. The set V_ϵ consists of variable-density *gridpoints* that will approximate the position of the Steiner points and remaining perpendicular edge incidences. To construct V_ϵ, let D and d be the maximum and minimum distances between any two vertices of S, respectively, and set $\Delta = \frac{d\epsilon}{64m}$. The gridpoints will always have at least one coordinate of the form $a\Delta 2^i$, where a and i are integers with i representing the "coarseness" of the grid at that location. A *free* gridpoint is any point v inside the convex hull of S with coordinates $(a_1\Delta 2^i, a_2\Delta 2^i)$, and is an element of S if and only if the distance from v to the *second-nearest* vertex of S is less than or equal to $(d/2)2^i = d2^{i-1}$. An *edge* gridpoint is any point v on an edge of S with at least one coordinate of the form $a\Delta 2^i$, and likewise is in S if and only if the distance from v to the second-nearest vertex of S is less than or equal to $d2^{i-1}$. By the choice of D and d it follows that i will always be in the range from 0 to $\log_2(D/d)$.

The edges of $G_{S,\epsilon}$ are the straight lines between any pair of vertices in $V_S \cup V_P \cup V_\epsilon$, with the weight of edge (u, v) being its Euclidean length $\|u-v\|$. It follows from the construction of $G_{S,\epsilon}$ that S itself can be represented by a subgraph $\hat{S}$ of $G_{S,\epsilon}$.

Lemma 3.1 *The number of vertices in the graph $G_{S,\epsilon}$ is $O\left(\frac{m^3}{\epsilon^2}\log_2(D/d)\right)$. The number of vertices in the subgraph $\hat{S}$ is $O\left(\frac{m^3}{\epsilon}\log_2(D/d)\right)$.*

Proof The number of vertices in V_S is at most m. The number of vertices in V_P is equal to the number of perpendicular vertex-line pairs, which is at most m^2. The number of free gridpoints can be bounded above by counting, for each $i = 0,\ldots,\log_2(D/d)$ and each $v \in S$, the number of points with coordinates $(a_1\Delta 2^i, a_2\Delta 2^i)$ that are within distance $d2^{i-1}$ of v, since this will clearly include all of the free gridpoints. The points within $d2^{i-1}$ of v are contained in a box $B_v(d2^{i-1})$ of those points each of whose coordinates is within $d2^{i-1}$ of v. The total number of these points is at most

$$\left(\frac{d2^i}{\Delta 2^i}\right)^2 = \left(\frac{64m}{\epsilon}\right)^2 = O\left(\frac{m^2}{\epsilon^2}\right)$$

Summing over the $n \leq m$ vertices of S gives a total of $O(m^3/\epsilon^2)$ free grid-points of the form $(a_1\Delta 2^i, a_2\Delta 2^i)$, and summing over each $i = 0, \ldots, \log_2(D/d)$ gives a total of $O\left(\frac{m^3}{\epsilon^2}\log_2(D/d)\right)$ free gridpoints.

By a similar argument it can be shown that for each edge e of S, each $i = 0, \ldots \log_2(D/d)$, and each vertex v of S, the number of points on e with at least one coordinate of the form $a\Delta 2^i$ that are within distance $d2^{i-1}$ of v is $O(m/\epsilon)$, for a total of $O\left(\frac{m^3}{\epsilon}\log_2(D/d)\right)$ edge gridpoints. Summing over the appropriate types of vertices gives $O\left(\frac{m^3}{\epsilon}\log_2(D/d)\right)$ vertices on S and $O\left(\frac{m^3}{\epsilon^2}\log_2(D/d)\right)$ vertices in all. □

Theorem 3.2 *An optimal solution to GSAP on instance $(G_{S,\epsilon}, \hat{S})$ corresponds to an ϵ-approximation for ESAP on S.*

Proof Let F^* be an optimal solution to ESAP on S. We show that there is a feasible solution $\hat{F}$ to GSAP on $(G_{S,\epsilon}, \hat{S})$ such that

$$\frac{l(\hat{F}) - l(F^*)}{l(F^*)} < \epsilon \qquad\qquad (*)$$

First perturb each vertex v of F^* which is not a vertex of S by moving v to a vertex of $G_{S,\epsilon}$. If v is not on S then v is moved to the nearest free gridpoint of $G_{S,\epsilon}$, and if v is on an edge e of S then v is moved to the nearest edge gridpoint lying on e. Perturb all edges adjacent to v accordingly. It follows that the resulting set $\hat{F}$ is a subgraph of $G_{S,\epsilon}$, and is in fact a feasible solution to GSAP. Now let δ be the greatest distance any vertex of F^* was moved in this process. If $\delta > 0$, then by construction of $G_{S,\epsilon}$ the associated vertex v must be on an FSS of F^* having at least three endpoints, so that by Lemma 2.3, $l(F^*)$ is at least as great as the maximum of the distance from v to the two nearest vertices of S. Call this distance d_2, and set $\mu = \frac{\epsilon}{8m}d_2$. Consider the box $B_v(\mu/2)$ of points each of whose coordinates is within $\mu/2$ of v. From the choice of μ, no point in $B_v(\mu/2)$ can be more than $2d_2$ distance from its second nearest vertex of S. By the construction of V_ϵ this means that all points in $B_v(\mu/2)$ of the form $(a_1\Delta 2^i, a_2\Delta 2^i)$ appear in V_ϵ whenever $2d_2 \leq d2^{i-1}$, or equivalently, $2^i \geq 4d_2/d$. The smallest of these

i's will satisfy $2^i \leq 8d_2/d$, which means that $\Delta 2^i \leq \frac{\epsilon}{8m}d_2 = \mu$. It follows that there must be at least one such point inside $B_v(\mu/2)$, and further, this point can be chosen to be on any edge containing v, in the case where v lies on an edge of S. Since any point in $B_v(\mu/2)$ is within distance μ of v, then v needs to be perturbed no more than distance μ to reach the appropriate vertex of $G_{S,\epsilon}$, that is, $\delta \leq \mu$.

Now by Lemma 2.3(ii), $d_2 \leq l(F^*)$, and so $\mu \leq \frac{\epsilon}{8m}l(F^*)$. By the triangle inequality, each perturbed vertex of $\hat{F}$ can increase the length of any incident edges by at most μ, and so using Lemma 2.3(iii), the maximum total increase in the length of all edges is at most

$$4m(2\mu) \leq \epsilon l(F^*).$$

and the inequality ($*$) follows. The optimal solution to GSAP will therefore clearly also satisfy ($*$), and hence will be an ϵ-approximation to ESAP. □

4 Solving GSAP for $(G_{\mathcal{S},\epsilon}, \hat{S})$

In [13] a polynomial algorithm for GSAP is given for instance (G, w, S) with G a planar graph, w a set of nonnegative edge weights on G, and S a connected subgraph of G. In this section, we modify the algorithm in [13] so that it solves the instance $(G_{\mathcal{S},\epsilon}, \hat{S})$ constructed in Section 3. This section relies heavily on the material in [13], and the reader is referred to that paper for a complete treatment of the techniques and proofs.

We start by making two assumptions about the instance $(G_{\mathcal{S},\epsilon}, \hat{S})$ given in Section 3:

A1: $\hat{S}$ *is a tree*, since we can always choose a spanning tree $\hat{T}$ of $\hat{S}$ and set the weights of the edges in $\hat{S} \setminus \hat{T}$ to zero. The optimal solution to the resulting instance of GSAP will also be a solution for $(G_{\mathcal{S},\epsilon}, \hat{S})$.

A2: *No edge of the optimal solution crosses an edge of S exactly at right angles*, since we can perturb the appropriate vertices of S slightly to prevent this without affecting the complexity arguments of the paper.

Now the tree $\hat{S}$ can be traversed clockwise by a closed walk $W = \sigma_0, \sigma_1, \ldots, \sigma_r = \sigma_0$ of vertices of $\hat{S}$, so that no point of $\hat{S}$ lies immediately to the left of W in this walk. Figure 3 gives a clockwise traversal of $\hat{S}$ (numbering only the vertices of S and the end vertices of F^*). Each vertex of $\hat{S}$ is repeated in this sequence a number of times equal to its degree; we will consider each of the σ_i's as a distinct element. For any edge e of $G_{\mathcal{S},\epsilon} \setminus \hat{S}$ that has an endpoint on $\hat{S}$, we will refer to that endpoint as σ_j if σ_j is the vertex of W passed when e comes into $\hat{S}$ from the left.

For two vertices σ_i, σ_j on W, the *interval* $[\sigma_i, \sigma_j]$ is defined to be that portion $\sigma_i, \sigma_{i+1}, \ldots, \sigma_j$ of W appearing clockwise from σ_i to σ_j (the indices for σ will always be taken modulo r). For technical convenience we take the interval $[\sigma_i, \sigma_{i+r}]$ to represent the entire walk W; $[\sigma_i, \sigma_i]$ will represent the one point path containing σ_i. Also, for $i \neq j$ we set $(\sigma_i, \sigma_j] = [\sigma_i, \sigma_j] \setminus \{\sigma_i\}$ and $[\sigma_i, \sigma_j) = [\sigma_i, \sigma_j] \setminus \{\sigma_j\}$. The edges traversed by $[\sigma_i, \sigma_j]$ can be partitioned into two sets: one, $P[\sigma_i, \sigma_j]$, being the set of edges in the unique path in $\hat{S}$ from σ_i to σ_j, and the other, $A[\sigma_i, \sigma_j]$, being the set of distinct edges of $[\sigma_i, \sigma_j]$ that are not in $P[\sigma_i, \sigma_j]$. In other words, $A[\sigma_i, \sigma_j]$ is that portion of $\hat{S}$ attached to $P[\sigma_i, \sigma_j]$ from the left when traversing $P[\sigma_i, \sigma_j]$ from σ_i to σ_j. For the interval $[\sigma_{11}, \sigma_{32}]$, in Figure 3, $P[\sigma_{11}, \sigma_{32}]$ consists of the edges $(\sigma_{11}, \sigma_{12})$, $(\sigma_{12}, \sigma_{19})$, $(\sigma_{19}, \sigma_{22})$, $(\sigma_{22}, \sigma_{23})$, $(\sigma_{23}, \sigma_{24})$, and $A[\sigma_{11}, \sigma_{32}]$

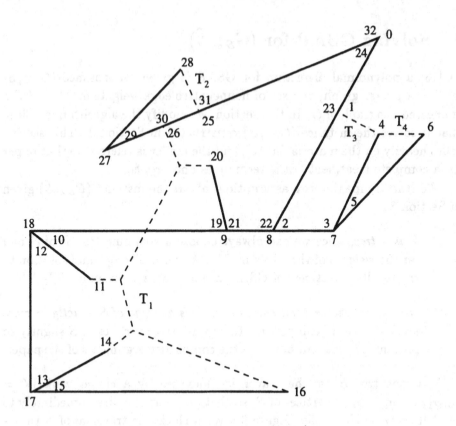

Figure 3: Numbering of $\hat{S}$

consists of the edges $(\sigma_{12}, \sigma_{13})$, $(\sigma_{13}, \sigma_{14})$, $(\sigma_{13}, \sigma_{16})$, $(\sigma_{19}, \sigma_{20})$, $(\sigma_{24}, \sigma_{25})$, $(\sigma_{25}, \sigma_{26})$, $(\sigma_{26}, \sigma_{27})$, $(\sigma_{27}, \sigma_{28})$.

We next prove an important lemma pertaining to the "essential planarity" of the optimal solution to ESAP.

Lemma 4.1 *Let F^* be an optimal solution to ESAP. If T and T' are FSSs of F^*, then the two sets of end vertices of T and T' (which are necessarily on $\hat{S}$) lie in disjoint intervals of $\hat{S}$ in a clockwise traversal of $\hat{S}$.*

Proof Suppose not, that is, suppose that there exist end vertices u, v on T and u', v' on T' that are distinct and lie in the order u, u', v, v' in a clockwise traversal of $\hat{S}$. Let Γ be the path in T from u to v, and let Γ' be the path in T' from u' to v'. By the planarity of F^*, it must be that either Γ crosses $P[u', v']$ or Γ' crosses $P[u, v]$. By symmetry suppose the first occurs, and let y be the point of crossing. By Lemma 2.2 and the definition of FSS we know that y cannot be a vertex of either F^* or S. Let e and e' be the edges of Γ

and $\hat{S}$, respectively, that cross at y, so that by Assumption A2, e and e' are not perpendicular. Suppose that we split e at y and move each side slightly in the direction of the acute angle, forming new set F'. Since e' is part of a cycle in $T \cup \hat{S}$ then we cannot create a bridge edge by this process, and so $\hat{S} \cup F'$ remains 2-connected. Further, the total length of F' is less than that of F', contradicting the fact that F^* is optimal. The lemma follows.   □

Now let $[\sigma_i, \sigma_j]$ be an interval of $\hat{S}$ and let F be a set of edges of G. We call F *one-sided* with respect to $[\sigma_i, \sigma_j]$ if it is incident to $\hat{S}$ only at elements of $[\sigma_i, \sigma_j]$. Denote by F^a the set $F \cup P[\sigma_i, \sigma_j] \cup A[\sigma_i, \sigma_j] \cup \{(\sigma_i, \sigma_j)\}$. Finally, F is called a $[\sigma_i, \sigma_j]$-*feasible* forest if F is one-sided and $A[\sigma_i, \sigma_j]$ contains no bridges with respect to F^a. In Figure 3, for example, the union of the trees T_1 and T_2 forms a $[\sigma_{11}, \sigma_{32}]$-feasible forest. It follows that the optimal solution to GSAP will then be a minimum weight $[\sigma_0, \sigma_r]$-feasible forest.

We next give functions that will produce minimum weight $[\sigma_i, \sigma_j]$-feasible forests. As a technical matter, these may actually be computing only an ϵ-approximation for the ESAP version of the problem, but for simplicity of presentation we will speak as though we were actually solving the GSAP problem. For $[\sigma_i, \sigma_j]$-feasible forest F define an $\hat{S}$-*component* to be a maximal subtree of F having no non-end vertices in common with $\hat{S}$. (A $\hat{S}$-component corresponds to an FSS for the associated ESAP problem.) For $[\sigma_i, \sigma_j]$ an interval, and v a vertex of $G_{\mathcal{S}, \epsilon}$, we define the following four auxiliary functions associated with special types of $[\sigma_i, \sigma_j]$-feasible forests:

$\mathbf{S}([\sigma_i, \sigma_j]) = $ minimum weight of a $[\sigma_i, \sigma_j]$-feasible forest;

$\mathbf{B}([\sigma_i, \sigma_j], v) = $ minimum weight of a $[\sigma_i, \sigma_j]$-feasible forest having σ_i, σ_j, and v in the same $\hat{S}$-component and such that v is of degree at least 2 or equal to σ_i or σ_j;

$\mathbf{C}([\sigma_i, \sigma_j], v) = $ minimum weight of a $[\sigma_i, \sigma_j]$-feasible forest having σ_i, σ_j, and v in the same $\hat{S}$-component;

$\mathbf{D}([\sigma_i, \sigma_j], v) = $ minimum weight of a $[\sigma_i, \sigma_j]$-feasible forest having σ_j and v in the same $\hat{S}$-component;

We also use the notation

$$[\sigma_i, \sigma_j]_+ = \{[\sigma_r, \sigma_s]: \sigma_r, \sigma_s \in [\sigma_i, \sigma_j], [\sigma_r, \sigma_s] \neq [\sigma_i, \sigma_j], \sigma_r \neq \sigma_s, \text{ and}$$
$$P[\sigma_r, \sigma_s] \text{ and } P[\sigma_i, \sigma_j] \text{ have at least one vertex in common}\}$$

Proposition 4.2 *For any interval $[\sigma_i, \sigma_j]$, $\mathbf{S}([\sigma_i, \sigma_j]))$ can be computed using the following set of equations:*

$$\mathbf{S}([\sigma_i, \sigma_j]) = \min_{[\sigma_r, \sigma_s] \in [\sigma_i, \sigma_j]+} \{\mathbf{S}([\sigma_i, \sigma_r]) + \mathbf{B}([\sigma_r, \sigma_s], \sigma_r) + \mathbf{S}([\sigma_s, \sigma_j])\} \quad (1)$$

$$\mathbf{B}([\sigma_r, \sigma_s], v) = \min_{\sigma_p \in [\sigma_r, \sigma_s)} \{\mathbf{C}([\sigma_r, \sigma_p], v) + \mathbf{D}([\sigma_p, \sigma_s], v)\} \qquad (2)$$

$$\mathbf{C}([\sigma_r, \sigma_p], v) = \min_{u \in V} \{\|u - v\| + \mathbf{B}([\sigma_r, \sigma_p], u)\} \qquad (3)$$

$$\mathbf{D}([\sigma_p, \sigma_s], v) = \min_{\sigma_q \in (\sigma_p, \sigma_s]} \{\mathbf{S}([\sigma_p, \sigma_q]) + \mathbf{C}([\sigma_q, \sigma_s], v)\} \qquad (4)$$

with these values being 0 whenever $A[\sigma., \sigma.] = \emptyset$.

The proof of this is essentially the same as that given in [13] (Proposition 1). Although the proof is too lengthy to give in full here, we do give a brief motivation for the proof by considering the recursive decomposition of a given optimal solution (or more generally a $[\sigma_i, \sigma_j]$-feasible forest) F for 2EC. Refer to Figure 4 for clarification. Roughly speaking, F is decomposed in four stages. In the first stage we find the extreme ends of an $\hat{S}$-component T of F, in terms of the maximal interval it spans. The forest F_B spanned by T may contain other components, which must be discovered recursively. This is the role of Equation (1). In the second stage we pick a vertex v on T that is either a vertex of $\hat{S}$ or is of degree 2. F_B can now be split at v into two nonempty forests: the forest F_C spanned by the "leftmost arm" of T with respect to v, and the remaining forest F_D to the "right" of F_C. This is the role of Equation (2). The portion F_C is either made up of v alone (which is dispensed with trivially) or has a component connected to v by a single arc. In the third stage we split the subforest into this arc together with the remaining subforest. This is the role of Equation (3). In the final stage, we separate F_D into the portion spanned by the component of F_D containing v, and the remaining portion of F_D to the "left" of this component — that is, between this portion of T and F_C. This is the role of Equation (4). When the partition is done correctly, F is recursively broken into smaller forests, each of which has the properties of one of the four types given above. The minimum weight forests of each type can then be determined for each interval and each vertex v and put together to form the optimal solution to 2EC.

The reader should refer to [13] for details of the proof. Although that paper assumes that $F^* \cup \hat{S}$ is planar, Lemma 4.1 and the structure of the perturbation of F^* given in Section 3 insure that the ϵ-approximate solution can be built using the same set of decompositions. The following algorithm

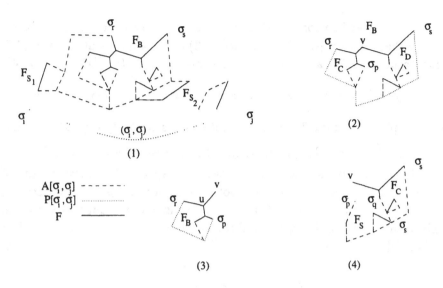

Figure 4: Example of recursive forest decomposition

for finding the length of an optimal solution to GSAP on instance $(G_{S,\epsilon}, \hat{S})$ is also taken from [13].

GSAP Algorithm

for $i = 1, \ldots, r - 1$ **do**

    for $j = 0, \ldots, r - 1$ **do**

        compute $\mathbf{S}([\sigma_j, \sigma_{j+i}])$ using (1)
        for each $v \in V$ compute $\mathbf{D}([\sigma_j, \sigma_{j+i}], v)$ using (4)
        for each $v \in V$ compute $\mathbf{B}([\sigma_j, \sigma_{j+i}], v)$ using (2)
        for each $v \in V$ compute $\mathbf{C}([\sigma_j, \sigma_{j+i}], v)$ using (3)

compute and return $\mathbf{S}([\sigma_0, \sigma_r])$ using (1).

Theorem 4.3 *The GSAP Algorithm finds the solution for instance $(G_{S,\epsilon}, \hat{S})$ in $O(n^2 k^2)$ time, where n is the number of vertices of $G_{S,\epsilon}$ and k is the number of vertices of $\hat{S}$.*

We add here that irrational distances pose no problem, for we can always compute the numbers $||u-v||$ in Equation (3) to the appropriate polynomial-length accuracy to obtain the correct ϵ-approximation (see [12], Section 4).

Corollary 4.4 *ESAP has a FPAS whenever the input set S is connected.*

Proof Applying Lemma 3.1 we have that n is of size $O\left(\frac{m^3}{\epsilon^2}\log_2(D/d)\right)$, and k is of size $O\left(\frac{m^3}{\epsilon}\log_2(D/d)\right)$, where m is the number of edges of S and d and D are the minimum and maximum distances, respectively, between pairs of vertices S. Since these numbers are polynomial in the input size describing S — and $1/\epsilon$ — then the GSAP algorithm produces an ϵ-approximation to ESAP in time polynomial in $1/\epsilon$ and the size of the input describing S. The theorem follows.   □

5   Summary and Extensions

This paper gives a FPAS for ESAP in the case where the input set S is connected. It is based on the algorithm of [13] for the GSAP. The requirement of connectivity is essential. A recent result [11] shows that finding a minimum Euclidean length 2-connected set spanning a given set of points in the plane is NP-hard; in fact, even finding a FPAS for this problem is not possible unless P=NP. It follows immediately that finding a FPAS for the general ESAP problem is also not possible unless P=NP.

We should remark that the FPAS given here is not particularly satisfactory for practical purposes, being roughly of the order m^{12}/ϵ^6. The bottleneck seems to be the number of gridpoints required in the construction of $G_{S,\epsilon}$. A decrease in the complexity exponent for the number of these points could result in as much as a *four-fold* decrease for the exponents in the final complexity, and so any efficiencies gained in gridpoint density can improve the FPAS dramatically.

We end by outlining some fairly straightforward variants to ESAP for which this technique applies.

Finding ϵ-approximations with respect to $F^* \cup S$: If we measure our approximation accuracy with respect to the length of the entire resulting set $F^* \cup S$, rather that just the length of the added edges F^*, the ϵ-approximation becomes much simpler. The FPAS given in [12] for the Euclidean Steiner tree problem applies directly (substituting the GSAP algorithm of Section 4) giving an $O(m^6/\epsilon^4)$ algorithm to find an ϵ-approximation for $F^* \cup S$. The main difficulty with approximating F^* is that the length of F^* can be significantly smaller than that of the original set S, and so more sophisticated techniques are needed for this type of approximation.

The Euclidean Steiner Vertex-Connectivity Augmentation Problem: In this variant of the problem the set $F^* \cup S$ is required to be *2-vertex connected*, that is, has no point whose removal will disconnect $F^* \cup S$. It turns out that ESAP essentially solves the vertex-connectivity problem as well. This is because any solution to ESAP can be made 2-vertex connected by the addition of edges of any arbitrarily small length, and so can be used to construct an ϵ-approximation for the vertex-connectivity variant.

The Euclidean Spanning Augmentation Problem: This is the variant where Steiner vertices are not allowed. There are two versions of this problem, the first requiring that edges of F^* connect points of S, and the second requiring that these edges actually connect *vertices* of S. For the first variant it can be shown, using similar techniques as for ESAP, that the resulting solution F^* is planar and satisfies the "essential planarity" conditions of Lemma 4.1. Thus the GSAP Algorithm of Section 4 applies — in fact, Equation 1 is the only necessary equation — the resulting complexity becomes $O(m^4)$. The second version is slightly more complex, since Lemma 4.1 does not necessarily apply, but by adding the points of crossing of every vertex-to-vertex line segment with S we can still obtain an $O(m^8)$ algorithm for this problem. We note that these are *not* ϵ-approximations, but can be made to be any number of decimal places close to the length of the optimal solution — although not necessarily producing *the* optimal solution, due to the problem of efficiently comparing irrational numbers.

We emphasize that the complexities given here for ESAP and all of its variants are fairly crude, and can in all likelihood be improved substantially. We leave this as a topic of further research.

References

[1] K.P. Eswaran and R.E. Tarjan, Augmentation problems, *SIAM J. Comput.* **5** (1976), pp. 653–665.

[2] A. Frank, Augmenting graphs to meet edge-connectivity requirements, *SIAM J. Disc. Math.* **5** (1992), pp. 25–53.

[3] A. Frank, Connectivity augmentation problems in network design, in J.R. Birge and K.G. Murty (eds.) *Mathematical Programming: State of the Art 1994*, (Ann Arbor, The University of Michigan Press, 1994) pp. 34–63.

[4] H. Frank and W. Chou, Connectivity considerations in the design of survivable networks, *IEEE Trans. Circuit Theory* **CT-17** (1970), pp. 486–490.

[5] Fredrickson, G.N. and J. JáJá, Approximation algorithms for several graph augmentation problems, *SIAM Jour. Comp.* **10** (1982), pp. 189–201.

[6] E.L. Gilbert and H.O. Pollak, Steiner minimal trees, *SIAM J. Appl.Math.* **18** (1984), pp. 1–55.

[7] M. Grötschel, C.L. Monma, and M. Stoer, Design of survivable networks, in M.O. Ball, T.L. Magnanti, C.L. Monma, and G.L. Nemhauser (eds.) *Network Models*, (Handbooks in Operations Research and Managements Science, Volume 7, New York, North-Holland, 1995) pp. 617–672.

[8] D.F. Hsu and X.-D. Hu, *On shortest two-edge connected Steiner networks with Euclidean distance*, (Technical Report, Graduate School of Information Science, Japan Advanced Institute of Science and Technology, Ishikawa, Japan, 1994).

[9] D.F. Hsu and X.-D. Hu, *On shortest three-edge connected Steiner networks with Euclidean distance*, (Technical Report, Graduate School of Information Science, Japan Advanced Institute of Science and Technology, Ishikawa, Japan, 1994).

[10] F.K. Hwang, D.S. Richards, and P. Winter, *The Steiner Tree Problem*, (North-Holland, New York 1992).

[11] E.L. Luebke and J.S. Provan, *On the Structure and Complexity of the 2-Connected Steiner Network Problem*, Technical Report UNC/OR TR99-2, Department of Operations Research, University of North Carolina, Chapel Hill, 1999.

[12] J.S. Provan, *Convexity and the Steiner tree problem*, Networks 18 (1988), pp. 55–72.

[13] J.S. Provan and R.C. Burk, *Two-connected Augmentation Problems in Planar Graphs*, J. Algorithms 32 (1999), pp. 87–107.

Effective Local Search Techniques for the Steiner Tree Problem

A.S.C. Wade
School of Information Systems,
University of East Anglia,
Norwich NR4 7TJ, UK
E-mail: asw@sys.uea.ac.uk

V.J. Rayward-Smith
School of Information Systems,
University of East Anglia,
Norwich NR4 7TJ, UK
E-mail: vjrs@sys.uea.ac.uk

Contents

D.-Z. Du et al. (eds.), *Advances in Steiner Trees*, 255-281.
© 2000 *Kluwer Academic Publishers.*

1 Introduction

Steiner's Problem in Graphs (SPG) involves connecting a given subset of a graph's vertices as cheaply as possible. More precisely, given a graph $G = (V, E)$ with vertices V, edges E, a cost function $c : E \to \mathbf{Z}^+$, and a set of special vertices, $K \subseteq V$, a Steiner tree is a connected subgraph, $T = (V_T, E_T)$, such that $K \subseteq V_T \subseteq V$, and $|E_T| = |V_T| - 1$. The problem is to find a Steiner tree T which minimises the cost function, $c(T) = \sum_{e \in E_T} c(e)$. Such a tree is referred to as a minimal Steiner tree.

A minimal Steiner tree is not normally a minimum spanning tree on just the special vertices, it also spans some non-special vertices; the vertices $V_T \setminus K$ are referred to as Steiner vertices.

All Steiner vertices must have degree ≥ 2; it is clear that any Steiner vertex with degree one can be removed from T, resulting in a Steiner tree T' with $c(T') < c(T)$.

The vertices V_T can be partitioned into two sets, the key and non-key vertices. We define the set of key vertices $V_{key}(T)$ in Steiner tree T to be

$$V_{key}(T) = \{v \in V_T \mid v \in K \ \vee \ d_T(v) \geq 3\}$$

where $d_T(v)$ is the number of edges incident to vertex v in subgraph T. Key vertices are either special vertices or Steiner vertices acting as junctions where two or more paths meet. A key path in Steiner tree T is a simple path

$$\begin{aligned} p = \ & < v_1, v_2, \ldots, v_n > \text{ such that} \\ & (v_i, v_{i+1}) \ \in E_T, 1 \leq i < n, \\ & v_1, v_n \ \in V_{key}(T) \text{ and} \\ & v_j \ \notin V_{key}(T), 1 < j < n. \end{aligned}$$

In other words, a key path connects two key vertices via zero or more intermediate non-key vertices. The set of key paths $Q = kp(T)$ consists of all key paths in T.

In Figure 1, special vertices are highlighted. The minimum Steiner tree consists of the four key paths

$$Q = \{< 1, 0, 2 >, < 2, 3 >, < 2, 4 >, < 3, 6 >\},$$

vertices 0 and 2 are Steiner vertices, and vertex 2 is the only non-special key vertex. All feasible Steiner trees can be uniquely represented by their set Q of key paths; the Steiner tree is trivially constructed from Q by taking the union of the paths, let $T(Q)$ denote such a tree. A minimal Steiner tree will have key paths that are the shortest paths between its key vertices. It is straightforward to show that a Steiner tree contains at most $2|K| - 2$ key vertices, and therefore contains between $|K| - 1$ and $2|K| - 3$ key paths.

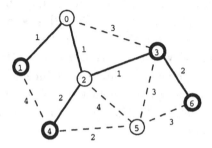

Figure 1: Steiner's problem in graphs

The Steiner problem in graphs is NP-hard [10]. Despite this result there exist special cases solvable in polynomial time. Of particular relevance in a local search context are those distinguished by the number of special vertices, $|K|$.

- $|K| = 2$ With only two special vertices the SPG reduces to finding the shortest path between the two vertices, a problem solvable in $O(|V|^2)$ time using Dijkstra's algorithm [4], or $O(|E| \lg |V|)$ time using a modified version of the algorithm.

- $|K| = 3$ The tree will consist of two paths linking the special vertices, or three paths linking the special vertices and a further intersection vertex. Given the intersection vertex, the solution is simply the union of the three shortest paths from this vertex to the special vertices. As there are a maximum of $|V|$ choices of intersection vertex, and the shortest paths can be found in $O(|E| \lg |V|)$ time, an SPG instance with $|K| = 3$ can be solved in $O(|V|(|E| \lg |V|))$ time using this technique.

- $|K| = |V|$ Where all vertices are special the SPG coincides with the Minimum Spanning Tree (MST) problem, solved in $O(|E| \lg |E|)$ or better time by variants on Prim's or Kruskal's algorithms [14].

These special cases form the basis of the neighbourhood search described later.

1.1   Heuristics for SPG

Since first formulated [9], the SPG has been the subject of a great deal of research effort. A number of exact techniques have been applied to SPG, e.g. Branch and Bound [1], Branch and Cut [3], spanning tree enumeration [9], and Dynamic Programming [7]. Despite their success these techniques all suffer from exponential worst case running time. Hence there is interest in heuristic approaches, generally aiming to reduce computational effort at the expense of guaranteed optimal solutions. A number of heuristics fall into three main categories, reduction techniques, constructive algorithms, and local search variants.

Reduction methods are applied to instances of combinatorial optimisation problems in an effort to reduce problem size. SPG provides ample scope for devising methods for reducing the values $|V|$ and $|E|$. The general approach is to identify edges or vertices guaranteed not to belong to a minimal Steiner tree; eliminating these from G results in an equivalent but smaller instance. Alternatively, identifying edges or vertices guaranteed to form part of a minimal Steiner tree allows a partial solution to be generated. The reduced instance will have fewer candidate solutions than the original, and hence will be easier to solve. Indeed, for some small instances based on sparse graphs or with high special vertex density, applying a combination of reduction techniques may be sufficient to find optimal solutions [6].

Constructive algorithms for SPG are based on connecting vertices in K by inserting paths between subtrees. An algorithm starts with U consisting of a forest of disjoint single element trees covering K, and at each iteration selecting two trees in U and joins them via a path. After $|K| - 1$ iterations the algorithm terminates with U consisting of a single Steiner tree. One of the first such algorithms proposed was the Shortest Path heuristic (SPH) [18]. In SPH all components of U remain singleton, apart from the tree

$T_{v_{start}}$ containing the arbitrarily chosen starting vertex v_{start}. For each constructive step the shortest paths from $T_{v_{start}}$ to all vertices in $K \setminus V_{T_{v_{start}}}$ are calculated using Dijkstra's algorithm, and the vertex having the shortest such path is connected to $T_{v_{start}}$ by that path.

The Average Distance Heuristic, [16], improves on SPH's straightforward greedy algorithm by including a form of heuristic lookahead. At each step, a function identifies a vertex, $v \in V \setminus V_U$, such that the average distance to at least two trees in U is minimised; v is considered to be the vertex heuristically closest to the forest U. With v identified, the two vertices v_1, v_2 in V_U with the shortest paths to v are found, and the trees T_{v_1}, T_{v_2}, and vertex v connected via these shortest paths.

The local search approach of Dowsland [5] is based on the key path (KP) Steiner tree representation. The initial solution is generated using the SPH algorithm, or a randomised SPH variant. Local search is applied to this solution using a neighbourhood function based on the exchange of key paths. If the solution is represented by its key paths, $Q = kp(T)$, and an arbitrary key path, p, is removed from Q, then the resulting graph consists of two disconnected trees. The trees can be optimally reconnected by considering each component as a single vertex and finding the shortest path between them. Dowsland uses an $O(V^2)$ shortest path algorithm.

The removal of p and the reconnection of the two trees by the shortest path between them is described by Dowsland as a 1-opt move. Dowsland uses this 1-opt neighbourhood in a steepest descent local search algorithm. An extended KP neighbourhood is also presented, referred to as 2-opt. Two paths are removed from Q, disconnecting the tree into three sub-trees. These can be optimally reconnected, in a similar manner to the 1-opt case, by treating each sub-tree as a single special vertex forming a new SPG instance with $|K| = 3$. As we showed previously this is solvable in $O(|V|(|E|\lg|V|))$ time, although Dowsland used a $O(|V|^3)$ implementation. Testing Steepest Descent with 1-opt and 2-opt neighbourhoods on a set of 100 vertex problems showed 2-opt consistently found better solutions, but with significantly longer execution times.

Verhoeven, in [19], uses the same representation and 1-opt neighbourhood for his SPG neighbourhood search algorithm. A Random Descent algorithm is used, with initial solutions generated by the SPH algorithm. Verhoeven's implementation includes an efficient $O(|E|\lg|V|)$ shortest path algorithm which is better than Dowsland's $O(|V|^2)$ for relatively sparse graphs. This makes it feasible to apply the algorithm to much larger, in terms of $|V|$, SPG instances. Detailed results for this algorithm are presented later.

Another SPG heuristic is the Genetic Algorithm (GA) presented in [12]. In this paper, the representation and neighbourhood function are based on the identification of Steiner vertices. A solution is represented by a bitstring of length $|V|$, with each bit i corresponding to a vertex $v_i \in V$. A 1 in position i means vertex v_i is present in the Steiner tree represented, and conversely a 0 indicates v_i is excluded. Note that to represent a valid Steiner tree it is necessary for bit i to be 1 for all $v_i \in K$, hence the bitstring essentially identifies a set, W, of Steiner vertices. A new SPG instance $(G' = (V', E'), K')$ can be generated from $G, W,$ and K; thus:

$$G' = sg(G, K \cup W),$$
$$K' = K \cup W,$$

where the function $sg(G, V)$ defines the subgraph induced in graph G by the vertices V. i.e. $V' = V$, $E' = \{(v_1, v_2) \in E \mid v_1, v_2 \in V\}$.

The generated SPG instance has the property that all the vertices are special vertices and hence can be solved to optimality in $O(E \lg E)$ time using Kruskal's MST algorithm. Let $T(W)$ denote the solution of the generated instance. This Steiner vertex representation does not ensure $T(W)$ is a Steiner tree for the original SPG instance. There are two possibilities for representing invalid Steiner trees. First, the subgraph G' may consist of more than one component, and if K is not spanned by a single such component then no Steiner tree exists in G'. Second, and less crucially, $T(W)$ may contain Steiner vertices of degree one.

A pool of Steiner vertex representation solutions is maintained. GA search is applied to the population; child solutions are generated by standard crossover and mutation operators. Solutions representing infeasible Steiner trees due to disconnection are priced out of the search; a large penalty is applied to the fitness function of any such solution. Solutions with Steiner vertices of degree one are tolerated on the assumption that the GA search will discourage such solutions. Good quality results are presented for the Beasley B set problems, but the algorithm is rather less successful on the larger instances (C and D sets).

Another genetic algorithm with a vertex based representation is presented in [8]. Although the bitstring used is similar to [12], the relationship between a bitstring and resulting tree is different. The vertices specified by the bitstring are used as parameters for the heuristic algorithm of [11]. Good quality results are presented for the Beasley C, D, and E sets. Execution time is several hours for the larger instances, despite all instances being pre-processed with a SPG reduction algorithm. Furthermore, the technique requires calculation and storage of the shortest paths between all pairs of vertices, resulting in a $O(|V|^2)$ memory requirement which limits scalability.

A simulated annealing algorithm for a related problem, the directed Steiner problem on networks, is presented in [15]. A similar representation and cost function to [12] is used. The neighbourhood function is based on adding or removing single vertices from the set of Steiner vertices. Results are presented for small instances with at most 80 vertices.

2 Effective SPG Local Search algorithms

We present four local search algorithms for SPG. The first, Steiner Vertex Search (SVS), is a Simulated Annealing algorithm based on a Steiner vertex representation. The second, Key Path Search (KPS), is a Random Descent algorithm based on the key path representation. The third, Dual Topology Search (DTS), is a straightforward combination of the SVS and KPS. The fourth algorithm, Adaptive Topology Search (ATS), more closely integrates the two representations into a single, adaptive, local search algorithm.

For each algorithm, the necessary neighbourhood functions are explained first, followed by description of the pivoting rule used, and discussion of performance. The algorithms are tested on the 78 SPG instances from the Beasley SPG problem sets [2]. The instances are randomly generated, ranging in size from the relatively small B set (100 vertices, 200 edges), to the larger E set instances (2500 vertices, 62500 edges). The optimal solutions for all instances, except E-18, are known due to the exact algorithms of [1, 3]. Details of each instance are given in Tables 1 and 2, columns $|V|, |E|, |K|$, and Opt give the number of vertices, edges, special vertices, and cost of best known solution respectively.

2.1 Steiner vertex neighbourhood

The first neighbourhood function is based on a similar Steiner vertex (SV) representation to that of [12]. A solution is represented by its set, W, of Steiner vertices selected from $\overline{K} = V \setminus K$. The set of solutions neighbouring the solution W is defined by the function $N_s : \mathcal{P}(\overline{K}) \to \mathcal{P}(\mathcal{P}(\overline{K}))$ where

$$N_s(W) = \{W \setminus \{v\} \mid v \in W\} \cup \{W \cup \{v\} \mid v \in \overline{K} \setminus W\}.$$

This is a straightforward neighbourhood with $N_s(W)$ containing solutions generated by adding or removing a vertex from W, the set of Steiner vertices. Evaluation of a solution follows the approach in [12], i.e. taking the MST of the subgraph induced in V by the vertices $W \cup K$. Generation of a single element of $N_s(W)$, as required by a single iteration of a random descent algorithm, is a polynomial operation. Selecting a vertex for addition or removal from W takes $O(1)$ time, and the subgraph and MST operations are combined into a single $O(|E| \lg |E|)$ operation, an extension of Kruskal's algorithm. Hence a single random descent step takes $O(|E| \lg |E|)$ time.

Given W representing a Steiner tree, $N_s(W)$ may still contain solutions representing invalid Steiner trees. There are three possible cases. First, removing a Steiner vertex may result in a disconnected induced subgraph. Second, an added Steiner vertex may form a single element disconnected component. Third, an added Steiner vertex may become a Steiner vertex of degree one in the resulting tree. The first two cases are handled by the same mechanism; the cost function is modified to

$$c(T) = \sum_{e \in E_T} c(e) + (C - 1) \sum_{e \in E} c(e)$$

where C is the number of components in the graph $T = sg(G, K \cup W)$. The penalty term ensures any disconnected graph will always have a cost greater than a connected graph. Invalid solutions of the third type are tolerated because an invalid Steiner tree containing Steiner vertices of degree one can always be transformed to a valid Steiner tree by random descent with this neighbourhood. Alternatively, all degree one Steiner vertices can be removed in $O(|V| \lg |E|)$ time using a dedicated algorithm, such as the *prune* algorithm described in [16].

2.2 Improving $N_s(W)$

The number of solutions in the neighbourhood of a typical solution is one of the factors determining the execution time of a local search algorithm; proving a solution to be locally optimal requires enumeration of its entire neighbourhood. The neighbourhood $N_s(W)$ of a solution in SV representation contains at most $|\overline{K}|$ solutions, hence proving local optimality takes

$O(|\overline{K}|(|E| \lg |E|))$ time. Although it is difficult to improve on this upper bound, the average case can be improved by modifying the neighbourhood function so that it includes less invalid Steiner trees. By modifying $N_s(W)$ to

$$\{W \setminus \{v\} \mid v \in W\} \cup$$
$$\{W \cup \{v\} \mid v \in \overline{K} \setminus W,$$
$$(v, v_i), (v, v_j) \in E, \ v_i, v_j \in W \cup K, \ i \neq j\}$$

type two invalid solutions are excluded, and the frequency of type three invalid solutions is reduced. The new vertex v must share at least two edges with vertices in the existing tree. Selecting such a vertex, or determining no such vertex exists, is easily done in $O(|E|)$ time. For SPG instances with sparse graphs or a low ratio of special vertices the resulting neighbourhood size is often reduced by an order of magnitude, with an obvious increase in performance.

Using a simple pivoting rule, such as random descent, solution quality is unaffected by the modified neighbourhood; all solutions excluded from the neighbourhood represent cost increases, and would not be moved to. However, solutions representing cost increases are vital when using higher level pivoting rules, such as Simulated Annealing or Tabu Search. Theoretically a neighbourhood need only include a single higher cost solution for SA or TS to function. However, in practice, the modified neighbourhood is too restrictive for effective Simulated Annealing; we found Steiner vertices with degree one played an important role in the search progress, by enabling new paths to be formed vertex by vertex. The difficulty is resolved by modifying N_s to

$$\{W \setminus \{v\} \mid v \in W\} \cup$$
$$\{W \cup \{v\} \mid v \in \overline{K} \setminus W, \ (v, v_i) \in E, \ v_i \in W \cup K\}$$

relaxing the constraint on v's connection to $W \cup K$ to at least one edge, rather than two. The new neighbourhood includes solutions with degree one Steiner vertices, but still precludes an added vertex forming a disconnected component. Further references to $N_s(W)$ refer exclusively to this version of the SV neighbourhood.

There is potential for improving the time complexity of evaluating solutions by applying delta evaluation. Given solutions W and $W' \in N_s(W)$, delta evaluation can be used if knowledge concerning the solution of W can accelerate evaluation of W'. Given graphs G and G' differing only in a single vertex and incident edges, Spira's algorithm [17] can generate the MST of G' in $O(|V|)$ time given the MST of G. An $O(|V|)$ algorithm obviously represents an improvement over $O(|E|\lg|E|)$. However our evaluation algorithm possesses small leading coefficients (the edge list is sorted once only) and has proved sufficiently fast working with the Beasley problem sets. For larger instances, especially with denser graphs, delta evaluation with Spira's algorithm could significantly improve running times.

2.3   Steiner vertex search

Initial experiments with N_s used the random descent pivoting rule. Starting solutions were randomly generated; for each vertex in $\overline{K}$, a fair coin is flipped to determine whether or not the vertex is to be included in the initial solution W. The algorithm was found to be capable of generating optimal or near optimal solutions to the majority of the B and C problem sets. However, the expectation that a single run, from initial solution to local optimum, will find a near optimal solution is small. Producing consistently good results requires a number of independent searches, and discouragingly, as the size $(|V|)$ of instances increases the number of runs required increases rapidly.

To improve the expected solution quality for a single run, a more advanced pivoting rule is used. Steiner Vertex Search (SVS), presented in Figure 2, is a Simulated Annealing algorithm working with the N_s neighbourhood function. SVS's main parameters are W, the initial solution, *temp*, the initial temperature, and *reduce*, the factor to reduce *temp* by at every iteration.

The *terminate* condition is based on a measure of search diversity since the last change in solution cost. A set of vertices *dset*, and integer variable *diversity* are updated after each accepted move thus:

> **if** $\Delta \neq 0$
> > *diversity* := 0;
> > *dset* := $\emptyset$;
> > **else if** $v \notin dset$ where $\{v\} = (W \setminus W') \cup (W' \setminus W)$
> > *diversity* := *diversity* + 1;
> > *dset* := *dset* $\cup \{v\}$;

The *terminate* condition returns TRUE if after $|\overline{K}|$ proposed moves no cost improvement is found, and the value *diversity* has not increased. This stopping condition will terminate the search quickly if local optimum consisting of a single solution is entered and the temperature is too low to escape. However a plateau, consisting of a large number of equal cost solutions, will be searched more thoroughly.

```
algorithm SVS (W, temp, reduce);
    X := ∅;
    while N_s(W) \ X ≠ ∅ and not terminate do
        W' :∈ N_s(W) \ X;
        Δ := c(T(W)) − c(T(W'));
        if rand(0, 1) < exp(Δ/temp) then
            W := W';
            X := ∅;
            update terminate variables;
        else
            X := X ∪ {W'};
        temp := temp * reduce;
```

Figure 2: The SVS algorithm

The algorithm was tested on all the Beasley SPG problem sets with the following parameters: Randomly generated initial solutions, $temp = 3.0$, $reduce = 0.999975$. For each instance, ten independent searches were completed. For each instance, the deviation from the known optimum of the best Steiner tree found is given in the column SVS in Tables 1 and 2. For clarity, a deviation of zero is denoted by a missing value. Optimal solutions are found for the whole B set, 12 of the 20 C set instances, but for both the D and E sets optimal solutions are found for only 5 of the 20 instances. Total deviation from the best known solutions over all instances is 207. For the only open problem in the Beasley sets, E-18, the current best known solution of 572 is reduced to 567.

Simulated Annealing is often associated with a requirement to carefully hand tune parameters to obtain good quality results. However, we found SVS's temperature and cooling schedule parameters quite robust; the single set of parameters were successfully used for all 78 instances. For many instances it is possible to reduce SVS's execution time by hand tuning the cooling schedule, but significantly improving solution quality by tuning the cooling schedule was found to be difficult.

2.4   Key path neighbourhood

The KP representation, as described in section 1, represents a Steiner tree by its set of constituent key paths $Q \subset \mathbf{Q}$, where $\mathbf{Q}$ is the set of all possible key paths in G. The KP based neighbourhood function $N_k : \mathcal{P}(\mathbf{Q}) \rightarrow \mathcal{P}(\mathcal{P}(\mathbf{Q}))$ is

$$N_k(Q) = \{Q_i \cup Q_i' \cup sp(V_{Q_i}, V_{Q_i'}) \mid Q_i \cup \{p_i\} \cup Q_i' = Q\}$$

where Q_i, Q_i' are the two components formed by removal of path $p_i \in Q$, $V_{Q_i}, V_{Q_i'}$ are the vertices in these components, and the function sp returns the shortest path in G between two (disjoint) sets of vertices. Each neighbour of Q is generated by removing a key path from Q and optimally reconnecting the resulting components. The function sp is a straightforward variant of Dijkstra's single source shortest path algorithm running in $O(|V| \lg |E|)$ time.

The set Q contains at most $2|K| - 3$ key paths, each one a candidate for removal, so Q has at most $2|K| - 3$ neighbours. Removal of a path which is already the shortest path between components leads to reconnection by the same path, hence $N_K(Q)$ only contains solutions of cost less than or equal to $T(Q)$. This makes N_K unsuitable for Simulated Annealing or Tabu Search algorithms.

2.5   Key Path Search

The Random Descent algorithm using the N_K neighbourhood is shown in Figure 3, this follows closely the algorithm applied by Verhoeven in [19]. Key Path Search (KPS) takes a single parameter, the initial solution. As with SVS, KPS terminates on finding a local optimum. The KP representation is only defined for valid Steiner trees, hence a simple random solution generator of the type used with SVS is unsuitable. Initial solutions are generated using the SPH constructive heuristic described earlier.

KPS was tested on the same Beasley SPG instances as SVS. For each instance ten independent searches were completed. For each instance, the deviation from the known optimum of the best Steiner tree found is given in column KPS in Tables 1 and 2. In terms of number of optimal solutions found, KPS performs comparably with SVS: 15 B set, 9 C set, 7 D

```
algorithm KPS (Q);
   X := ∅;
   while (N_s(Q) \ X ≠ ∅) do
      Q' :∈ N_k(Q) \ X;
      if c(T(Q')) < c(T(Q)) then
         Q := Q';
         X := ∅;
      else
         X := X ∪ {Q'};
```

Figure 3: The KP procedure

set, and 6 E set, with a total deviation over all instances of 411. KPS uses heuristically generated initial solutions, and their quality must be considered when analysing the results. The SPH heuristic performs relatively well on instances with smaller numbers of special vertices. Of the 37 instances apparently solved by KPS, 27 of these were solved by SPH alone without recourse to local search.

SVS and KPS did not solve the same instances; only 24 of the 54 instances solved by either technique were solved by both. It is possible to characterise instances for which one technique is likely to outperform the other. For sparse graphs, and small numbers of special vertices, KPS is likely to outperform SVS. For dense graphs, and higher percentages of special vertices, SVS typically outperforms KPS.

A possible explanation for KPS's relatively poor performance on some instances can be found in the length of key paths forming the Steiner trees. An SPG instance with a small number of special vertices will contain a small number (bounded above by $2|K| - 3$) of key paths, and (assuming uniformly distributed special vertices) these will be relatively long. Conversely, instances with high special vertex density will have a higher number of, typically short, key paths. Analysis of solutions for the Beasley instances confirmed this; the longest key paths belonged to instances with 5 or 10 special vertices, whereas for instances with 50% special vertices the majority of key paths consist of a single edge and only a small percentage are longer than 2 edges. Short key paths reduce the effectiveness of N_k; removing a key path of length one and reconnecting the components cannot reduce the number of Steiner vertices.

Longer key paths may also explain why SVS performs less well on instances with a small number of special vertices. It is easy to imagine a situation where the current solution differs from the optimum by a single key path of length n. Moving to the optimal solution requires a single step with N_k, but N_s may require at least $2n - 2$ steps, in the case of the paths sharing no non-key vertices. Therefore, as the length of the key path increases, it becomes less probable that SVS will be able to find an acceptable ordering of moves to achieve optimality.

2.6   When is a local optimum not a local optimum?

The neighbourhoods N_s and N_k are not exact, as can be shown with simple examples; and based on results for the Traveling Salesman Problem [20] it seems unlikely that an exact neighbourhood of polynomial size could be found for SPG. Given an inexact neighbourhood, the role of a local search algorithm is to find local optima (which may also be global optima), but then to escape the local optima to continue the search process. There are two main techniques for escaping from local optima. First, allow uphill moves to be accepted in some controlled manner. This is the approach used in Simulated Annealing, where uphill moves are accepted probabilistically, and in Tabu Search where non-tabu moves may be accepted irrespective of cost.

The second technique is to change the topology of the search space, so the current solution is no longer a local optimum. Search space topology is defined by the neighbourhood function used; therefore, if we change the neighbourhood function, we change the topology. The difficulty is determining how to modify the neighbourhood so as to include a lower cost solution. Although such a modification can always be made (unless the local optimum is also a global optimum), given that a problem is NP-Hard, there is no generally applicable method so to do.

Modifying search space topology is the approach used in Variable Depth Search; if a solution is found to be a local optimum with respect to the current neighbourhood, the neighbourhood is extended to include the neighbours of the neighbours of the current solution. The neighbourhood is repeatedly extended until a lower cost solution is found, or until the neighbourhood becomes too large to enumerate. This growth in neighbourhood size is the limiting factor for Variable Depth Search; if the basic neighbourhood contains $O(n)$ solutions then the neighbourhood extended k times will contain $O(n^k)$ solutions. The Lin-Kernighan heuristic [13] shows that for the TSP, using sophisticated neighbourhood reductions can make a restricted form of Variable Depth Search a practical technique.

Ignoring for now the SA aspect of SVS, our local search algorithms rely on changing the neighbourhood to help avoid becoming stuck in local optima. Unfortunately, N_s and N_k are defined on fundamentally different SPG representations, so it is not possible to apply both neighbourhoods directly. The answer is to define additional functions, N_{sk} and N_{ks}, that transform solutions between the two representations. Figure 4 illustrates the relationship between the two representations and neighbourhood functions.

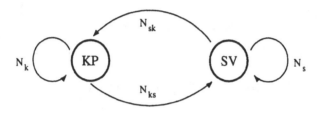

Figure 4: Moving between N_s and N_k

There exist Steiner trees which are locally optimal with respect to N_s but not N_k and vice-versa, as can be shown with straightforward examples.

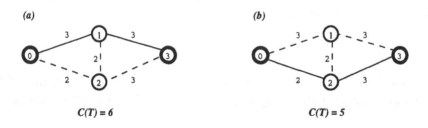

Figure 5: Local optima w.r.t. N_s only

Figure 5 is an example of a local optimum with respect to N_s but not N_k. The current solution (a) in SV representation is $W = \{1\}$ and $c(T(W)) = 6$, and the SV neighbourhood is, $N_s(W) = \{\{\}, \{1,2\}\}$, and contains only solutions of cost greater than 6. In KP representation the current solution is $Q = \{< 0,1,3 >\}$, and $N_k(Q)$ contains the lower cost solution $\{< 0,2,3 >\}$ shown in Figure 5 (b). The lower cost solution is formed by removing the key path in Q and reconnecting vertices 0 and 3 by the shortest path between them.

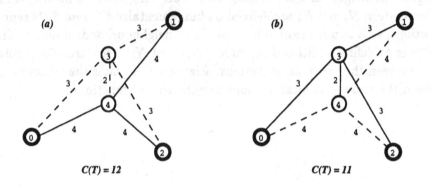

$C(T) = 12$ $C(T) = 11$

Figure 6: Local optima w.r.t. N_k only

An example of the reverse case, a local optimum w.r.t. N_k but not N_s is shown in Figure 6. In KP representation the current solution (a) is $Q = \{< 0,4 >, < 1,4 >, < 2,4 >\}$, and the neighbourhood $N_k(Q)$ is empty; if any path, p, is removed from Q the shortest path reconnecting the components is simply p. In SV representation the current solution $W = \{4\}$ is not a local optima because $N_k(W)$ contains the solution $\{3,4\}$ representing the lower cost tree in Figure 6 (b).

2.6.1   The N_{ks} function

Translating a solution in KP representation to a related solution in SV representation is carried out by the function $N_{ks} : \mathcal{P}(\mathbf{Q}) \rightarrow \mathcal{P}(\overline{K})$ where

$$N_{ks}(Q) = \{v \in \overline{K} \mid v \in p \in Q\}$$

The solution generated by N_{ks} is in SV representation so its cost is determined as described in section 1.1. We can treat N_{ks} as another neighbourhood function, albeit mapping to a single solution. The reasoning here is that the Steiner trees $T(Q)$ and $T(N_{ks}(Q))$ typically differ; the two trees will have identical Steiner vertices, but potentially different edge sets. The

SV representation ensures $T(N_{ks}(Q))$ is an MST whereas the edges in $T(Q)$ may represent an arbitrary tree. Hence we have the following relationship

$$c(T(N_{ks}(Q))) \leq c(T(Q)) \quad \forall Q \in \mathbf{Q}.$$

Applying N_{ks} is an $O(|E| \lg |E|)$ operation; the vertices in W can obviously be determined in $O(|V|)$ time, and the subgraph and MST operations in $O(|E| \lg |E|)$.

2.6.2 The N_{sk} function

Translating a solution in SV representation to a related solution in KP representation is carried out by the function $N_{sk} : \mathcal{P}(\overline{K}) \to \mathcal{P}(\mathbf{Q})$.

$$N_{sk}(W) = \begin{cases} kp(prune(T(W))) & \text{if } T(W) \text{ is connected,} \\ \emptyset & \text{otherwise} \end{cases}$$

where *prune* denotes the algorithm in [16] for removing all degree one Steiner vertices from a tree. A SV solution, W, with $T(W)$ consisting of more than one component, has no corresponding KP solution. In practice, with the local search algorithms we apply, this is not a problem because the current solution very rarely represents a graph with more than one component. The *prune* algorithm is required because Steiner vertices of degree one cannot be represented in KP representation, but as mentioned earlier these vertices are desirable in local search using the SV representation. For all W representing connected trees, we have the relationship

$$c(T(N_{sk}(W))) \leq c(T(W)).$$

Applying N_{sk} is an $O(|E| \lg |E|)$ operation; the *prune* algorithm's complexity is dominated by a MST operation, and the kp function simply walks the resulting tree returning each key path found.

The solution spaces defined by the SV and KP representations are not homogeneous. There exist trees in both solution spaces that have no exact counterpart in the other. Figure 7 shows the relationship between the SV and KP solution spaces. Solutions generated by N_s and N_k are arbitrary members of SV and KP respectably, however solutions generated by N_{sk} and N_{ks} are constrained to the intersection between SV and KP. Both SV and KP can represent the minimal Steiner tree for any SPG instance, therefore the minimal Steiner tree must lie in the intersection between the representations. This relationship does not imply a local search algorithm constrained to searching the KP, SV intersection will necessarily outperform an unconstrained search. However, constraining the solution returned by a local search algorithm to this intersection is clearly always beneficial.

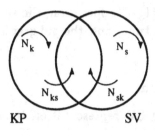

Figure 7: Relationship between the N_k and N_s solution spaces

2.7   Dual Topology Search

The first algorithm, Dual Topology Search (DTS), is a simple iterative combination of SVS and KPS. It works on the assumption that the the local optimum converged on by SVS (respectively KPS) will not represent a local optimum for KPS (respectively SVS). The main parameters for DTS are a starting state, SV or KP, and an initial feasible solution in the corresponding representation. Assuming the initial state is SV, DTS will apply SVS to the current solution until a local optimum is found. The N_{sk} neighbourhood function is then used to convert the solution to KP representation, and KPS is applied. If any improvement is made by KPS, N_{ks} is used to convert back to SV representation, and SVS is applied again. These steps are repeated until the current solution is a local optimum with respect to both neighbourhood functions. The *temp* parameter of the SVS algorithm is not reset for each application of the algorithm.

DTS was tested on the same Beasley SPG instances as SVS and KPS. For each instance ten independent searches were completed, five starting in each state. The SPH heuristic was used to generate all initial solutions. The *temp* and *reduce* parameter values of 3.0 and 0.999975 respectively are the same used for SVS in isolation. The results for DTS are presented in Tables 1 and 2. As DTS is essentially a high level composition of SVS and KPS, we would expect it to performs at least as well as either algorithm in isolation. The results confirm this; in all cases where either SVS or KPS found optimal solutions, DTS also converges to optimally.

Of the 24 instances unsolved by either SVS or KPS, the DTS algorithm improves on 12 of the solutions found but, of these, only 4 are solved to optimality. Although the dual topology approach does improve solution quality, the gain is slightly disappointing. The experiments showed only a small number of successful topology switches were made during each search. For the majority of the instances where an improvement was found only one or two successful topology switch were made. The maximum number of successful switches, six, was observed for instance E-9.

2.8 Adaptive Topology Search

Experience with DTS showed that changing the search space topology was beneficial to the search process, however the number of switches DTS could make was limited. To improve on the results of DTS we propose an algorithm which enforces regular representation changes during the course of the search.

The obvious strategy would be to allow topology switches at every iteration, however there are two factors limiting this approach. First, each topology switch incurs some overhead and is at least as computationally expensive as a conventional move. Therefore, too many switches will results in extended running times. Second, the Simulated Annealing, vital for consistently good results with N_s, is effectively disabled by the pruning of degree one Steiner vertices inherent in applying N_{sk}.

In ATS (Figure 8) the frequency of topology switches and the amount of time spent searching in each topology is controlled by the two variables *SVmoves* and *KPmoves*. The values of *SVmoves* and *KPmoves* are initialised to be proportional to the upper bounds on their respective neighbourhood sizes. For *SVmoves* this is $|\overline{K}|$ multiplied by a parameter α, and for *KPmoves* $2|K| - 3$ multiplied by parameter β. The actual search is carried out by calls to slightly modified versions of algorithms SVS and KPS. The algorithms are both modified to accept a parameter limiting the maximum number of proposed moves to be made.

As the search progresses the ratio between *SVmoves* and *KPmoves* is adaptively modified according to the success of KPS. Success is defined as finding a new best solution within *KPmoves* proposed moves. If KPS is successful, *KPmoves* is increased by a factor *rate* and *SVmoves* decreased, otherwise *KPmoves* is decreased by $(1/rate)$ and *SVmoves* increased. ATS

algorithm ATS (*temp, reduce, α, β, rate*)
> *temp*   : initial temperature for SA
> *reduce*   : cooling factor, < 1
> α : factor proportional to SVmoves
> β : factor proportional to KPmoves
> *rate*   : rate to alter representation bias, ≥ 1

> *initialise*(W);
> $SV\,moves = \alpha\,|\overline{K}|;$
> $KP\,moves = \beta\,(2|K| - 3);$
> **while not** *terminate* **do**
>> SVS($W, temp, reduce, SV\,moves$);
>> $Q := N_{sk}(W);$
>> $Y := \{v \in W \mid v \notin p \in Q\};$
>> KPS($Q, KP\,moves$);
>> **if** $(c(T(Q)) < bc)$ **then**   {*bc* is cost of best solution found so far}
>>> $KP\,moves := KP\,moves \times rate;$
>>> $SV\,moves := SV\,moves \times (1/rate);$
>>
>> **else**
>>> $KP\,moves := KP\,moves \times (1/rate);$
>>> $SV\,moves := SV\,moves \times rate;$
>>
>> $W := N_{ks}(Q);$
>>
>> $$W := W \cup \left\{\; v_1 \in Y \;\middle|\; \begin{array}{l} \exists \text{ path } < v_1, v_2, \ldots, v_n >, \\ (v_i, v_{i+1}) \in E, \\ v_i \in Y,\; 1 \leq i < n, \\ v_n \in W \cup K \end{array} \right\};$$

Figure 8: Adaptive Topology Search

terminates according to the same stopping conditions described for SVS in Section 2.3. The *initialialise* procedure generates a solution using the SPH heuristic, and to this additional Steiner vertices are added by flipping a fair coin for each vertex in V not already included.

2.8.1 SA Warm Start

To achieve good results, SVS requires a large number of proposed moves to build a tree containing many degree one Steiner vertices. The problem is that in changing to KP representation the degree one Steiner vertices are removed. To allow the highest possible number of topology switches, without seriously impairing the effectiveness of the SVS algorithm, the concept of warm starting is introduced. Warm starting allows high quality solution attributes generated by SVS to be retained while applying KPS. After *SVmoves* iterations of SVS the warm start set, Y, is defined by

$$Y = \{v \in W \mid v \notin p \in N_{sk}(W)\}.$$

Thus Y contains all degree one Steiner vertices represented in W, but not in $N_{sk}(W)$. After *KPmoves* iterations of KPS, and the tree is converted back to SV representation, the warm start vertices in

$$\left\{ v_1 \in Y \;\middle|\; \begin{array}{l} \exists \text{ path } < v_1, v_2, \ldots, v_n >, \\ (v_i, v_{i+1}) \in E, \\ v_i \in Y,\; 1 \le i < n, \\ v_n \in W \cup K \end{array} \right\}$$

are added to the solution tree. Restricting the warm start vertices in this way ensures the tree generated will not be disconnected. With reference to Figure 7, the SA warm start allows key path moves to be applied to a tree in the SV solution space without restricting the resulting tree to the KP, SV intersection.

ATS was tested on the same 78 Beasley SPG instances as the other algorithms. For each instance ten independent searches were completed. The parameters used were as follows $temp = 3.0$, $reduce = 0.999975$, $\alpha = 4$, $\beta = 2$, and $rate = 1.1$. Results are presented in Tables 1, and 2. For each instance, column ATS gives the deviation from the known optimum of the lowest cost Steiner tree found. Column ATS_{ave} gives the average deviation over all ten runs. Column ATS_{Time} gives the average execution time of a single search, from initial solution to best solution found, in CPU seconds (DEC Alpha 600/255).

3   Conclusions

This paper has examined the application of local search methods to generate high quality heuristic solutions for the Steiner problem in graphs. Local search algorithms were developed based on two fundamentally different solution representations, one path based and the other vertex based. Theoretical results were presented showing the representations defined different search space topologies. This feature was exploited by combining the two representations within a Simulated Annealing framework. An adaptive technique was used to control the proportion of moves applied in each representation.

The results for ATS represent a substantial improvement over the other three algorithms. All instances in the B,C, and D problem sets are solved to optimality. For the open problem, E-18, the best known solution generated by SVS is further improved. Furthermore, all remaining E set instances are solved to optimality except for E-4, E-8, and E-13. Computational requirements are relatively small. The B set instances typically take under 1 second, C set under 100 seconds, D set under 4 minutes, and the E set under 14 minutes. ATS is a stochastic algorithm, multiple runs often converge on different solutions, the column ATS_{ave} provides an indication of this variation in solution quality over a number of runs. The values are expressed in absolute cost units; expressed as a percentage, the average deviation for the majority of instances is much less than 1%. The largest percentage deviation is 3.6% for instance E-6.

The parameters for ATS appear fairly robust, they were used successfully for all 78 instances. As with SVS in isolation, hand tuning the parameters allowed execution time to be reduced for a number of instances. By using much slower cooling schedules instances E-4, E-8, and E-13 can all be solved to optimality in around one hour of computation. Although we could not better the result for problem E-18 using a hand tuned cooling schedule, it remains possible that a solution with a cost less than 565 does exist.

Solution quality and running times of ATS compares favourably with other techniques from the literature. The Genetic Algorithm approach in [12] appears best suited to smaller SPG instances, successfully solving all of the B set, but only 3 C set, and 3 D set instances. Results are not presented for the E set. Although not directly comparable due to use of different machines, the GA's computational requirements appear much larger than ATS's. For example the average GA search time for the D set instances is approximately one hour, compared to less than 60 seconds for ATS.

The GA of Esbensen [8] performs better, solving all C set, and 17 D set instances, but has only limited success on the E set, failing to solve 9 instances. Again the GA approach appears to require greater computational resources, average execution time reported for the E set instances is around 4 hours compared with ATS's 4 minutes. Esbensen only presents results for instances preprocessed using problem reduction methods. It is probable that ATS's performance could be further improved by working with similar reduced instances.

Comparing the performance of the ATS heuristic with exact solution methods is difficult, as they involve very different techniques. The exact methods are capable of generating provably optimal solutions for the majority of the Beasley instances, however their worst case running time means they cannot generate solutions for some larger instances. The Branch and Bound algorithm of [1] did not terminate within 6 hours of Cray X-MP/48 CPU time for six E set instances, and for instance E-18 the Branch and Cut algorithm of [3] failed to terminate within 10 days of Vax8700 CPU time. The ATS algorithm is capable of finding optimal solutions to all Beasley instances, without excessive runtime requirements. It would be interesting to compare directly ATS and Chopra's exact techniques on similar hardware. Indeed, it may be profitable to combine the approaches to exploit the speed of ATS and the quality guarantees of Branch and Cut.

The authors would like to thank Marco Verhoeven for making available a code for the Steiner tree key path representation. The work of the first author has been supported by a grant from the EPSRC.

References

[1] J.E. Beasley. An SST-based algorithm for the Steiner problem in graphs. *Networks*, 19:1–16, 1989.

[2] J.E. Beasley. OR-Library: Distributing test problems by electronic mail. *Journal of the Operations Research Society*, 41:1069–1072, 1990.

[3] S. Chopra, E.R. Gorres, and M.R. Rao. Solving the Steiner tree problem on a graph using branch and cut. *ORSA Journal on Computing*, 4(3):320–335, 1992.

Inst	$|V|$	$|E|$	$|K|$	Opt	SVS	KPS	DTS	ATS	ATS_{ave}	ATS_{Time}
B-1	50	63	9	82						1
B-2			13	83						1
B-3			25	138						1
B-4	50	100	9	59						1
B-5			13	61						1
B-6			25	122	2					1
B-7	75	94	13	111						1
B-8			19	104						1
B-9			38	220						1
B-10	75	150	13	86						1
B-11			19	88	2					1
B-12		-	38	174						1
B-13	100	125	17	165						1
B-14			25	235						1
B-15			50	318						1
B-16	100	200	17	127						1
B-17			25	131						1
B-18			50	218	1					1
C-1	500	625	5	85	1					1
C-2			10	144	6					1
C-3			83	754	4					4
C-4			125	1079		1	1			4
C-5			250	1579		1				5
C-6	500	1000	5	55	2					2
C-7			10	102						1
C-8			83	509	2	1				13
C-9			125	707	3	3	1		0.3	25
C-10			250	1093		1			0.8	36
C-11	500	2500	5	32	2				0.2	8
C-12			10	46						4
C-13			83	258					0.1	27
C-14			125	323		5				38
C-15			250	556		1				68
C-16	500	12500	5	11	1				0.2	26
C-17			10	18		1				18
C-18			83	113		7				63
C-19			125	146		9				62
C-20			250	267		2				94

Table 1: Results for Beasley problem sets B and C

Inst	$|V|$	$|E|$	$|K|$	Opt	SVS	KPS	DTS	ATS	ATS$_{ave}$	ATS$_{Time}$
D-1	1000	1250	5	106	1					1
D-2			10	220	16					1
D-3			167	1565	8	2	2			17
D-4			250	1935	6					19
D-5			500	3250	2	5	2			31
D-6	1000	2000	5	67	3	3				5
D-7			10	103						3
D-8			167	1072	8	12	6		1.5	29
D-9			250	1448	4	7	4			35
D-10			500	2110	1	8			0.1	56
D-11	1000	5000	5	29	3	1	1		0.6	18
D-12			10	42						4
D-13			167	500	1	6	2		0.3	83
D-14			250	667	1	4			0.3	120
D-15			500	1116		10				147
D-16	1000	25000	5	13	2					46
D-17			10	23					0.2	49
D-18			167	223	1	7	1		1.5	108
D-19			250	310	1	20			0.6	134
D-20			500	537		7				204
E-1	2500	3125	5	111						9
E-2			10	214	13	13	13		2.5	18
E-3			417	4013	24	17	3		3.8	98
E-4			625	5101	9	10	7	1	4.1	119
E-5			1250	8128	2	2	1		0.5	214
E-6	2500	5000	5	73	17				2.6	24
E-7			10	145	3				1.3	21
E-8			417	2640	21	18	13	4	5.4	115
E-9			625	3604	10	12	2		2.4	163
E-10			1250	5600	3	21	2		1.0	336
E-11	2500	12500	5	34						76
E-12			10	67	6				0.8	91
E-13			417	1280	8	23	10	2	5.0	345
E-14			625	1732	5	22	5		1.6	370
E-15			1250	2784		19			0.4	845
E-16	2500	62500	5	15	2				0.5	209
E-17			10	25	5	1	1		0.2	305
E-18			417	572	-5	46	-5	-7	-5.2	784
E-19			625	758		45			1.0	621
E-20			1250	1342		21				406

Table 2: Results for Beasley problem sets D and E

[4] E.W. Dijkstra. A note on two problems in connexion with graphs. *Numerische Math.*, 1:269–271, 1959.

[5] Kathryn A. Dowsland. Hill-climbing, Simulated Annealing and the Steiner problem in graphs. *Engineering Optimization*, 17:91–107, 1991.

[6] C.W. Duin and A. Volgenant. Reduction tests for the Steiner problem in graphs. *Networks*, 19:549–567, 1989.

[7] S.E. Dreyfus and R.A. Wagner. The Steiner problem in graphs. *Networks*, 1:195–207, 1971.

[8] H. Esbensen. Computing near-optimal solutions to the Steiner problem in a graph using a genetic algorithm. *Networks*, 26:173–185, 1995.

[9] S.L. Hakimi. Steiner's problem in graphs and its applications. *Networks*, 1:113–133, 1971.

[10] R. M. Karp. Reducibility among combinatorial problems. In R.E. Miller and J.W. Thatcher, editors, *Complexity of Computer Computations*, pages 85–103. Plenum Press, 1972.

[11] L. Kou, G. Markowsky, and L. Berman. A fast algorithm for Steiner trees. *Acta Informatica*, 15:141–145, 1981.

[12] A. Kapsalis, V.J. Rayward-Smith, and G.D. Smith. Solving the graphical Steiner tree problem using genetic algorithms. *J. Oper. Res. Soc.*, 44:397–406, 1993.

[13] S. Lin and B.W. Kernighan. An effective heuristic algorithm for the Traveling Salesman Problem. *Operations Research*, 21:498–516, 1973.

[14] B.M.E. Moret and H.D. Shapiro. *Algorithms from P to NP. Volume I: Design and Efficiency.* Benjamin-Cummings, Menlo Park, CA, 1991.

[15] L. Osborne and B. Gillett. A comparison of two Simulated Annealing algorithms applied to the directed Steiner problem on networks. *ORSA Journal on Computing*, 3(3):213–225, 1991.

[16] V.J. Rayward-Smith and A. Clare. On finding Steiner vertices. *Networks*, 16:283–294, 1986.

[17] P.M. Spira. On finding and updating spanning trees and shortest paths. *SIAM Journal on Computing*, 4:375–380, 1975.

[18] H. Takahashi and A. Matsuyama. An approximate solution for the Steiner problem in graphs. *Mathematica Japonica*, 24:573–577, 1980.

[19] M.G.A. Verhoeven, E.H.L. Aarts, and M.E.M. Severens. Local search for the Steiner problem in graphs. In V.J. Rayward-Smith, editor, *Modern Heuristic Search Methods*. J. Wiley, 1996.

[20] M Yannakakis. The analysis of local search problems and their heuristics. In *Proceedings of the seventh Symposium on Theoretical Aspects of Computer Science*, volume LNCS 415, pages 298–311, Berlin, 1990. Springer-Verlag.

Modern Heuristic Search Methods for the Steiner Tree Problem in Graphs

Stefan Voß
Technische Universität Braunschweig,
Institut für Wirtschaftswissenschaften, Abt-Jerusalem-Straße 7,
D - 38106 Braunschweig, Germany
E-mail: stefan.voss@tu-bs.de

Contents

D.-Z. Du et al. (eds.), Advances in Steiner Trees, 283-323.
© 2000 *Kluwer Academic Publishers.*

1   Introduction

Given an edge-weighted graph, the Steiner tree problem in graphs is to determine a minimum cost subgraph spanning a set of specified vertices. More specifically, consider an undirected connected graph $G = (V, E)$ with vertex set V, edge set E, and nonnegative weights associated with the edges. Given a set $Q \subseteq V$ of basic vertices *Steiner's problem in graphs* (SP) is to find a minimum cost subgraph of G such that there exists a path in the subgraph between every pair of basic (or required) vertices. In order to achieve this minimum cost subgraph additional vertices from the set $S := V - Q$, so-called Steiner vertices, may be included. Since all edge weights are assumed to be nonnegative, there is an optimal solution which is a tree, a so-called *Steiner tree*.

Correspondingly, *Steiner's problem in directed graphs* (SPD) is to find a minimum cost directed subgraph of a given graph that contains a directed path between a root node and every basic vertex. Note, that any instance of the SP may easily be transformed into an instance of the SPD by replacing each undirected edge (i, j) with two arcs $[i, j]$ and $[j, i]$ directed opposite each other both having the same weight as the original edge.

Applications of the SP and the SPD are frequently found in the layout of connection structures in networks as, e.g., in topological network design, location science, and VLSI design. Even the uncapacitated facility location problem turns out to be a special case of the SPD. Only some well-known special cases of the SP are polynomially solvable. For instance, if $|Q| = 2$ the problem reduces to a shortest path problem and in the case where $Q = V$ the SP reduces to the *minimum spanning tree* (MST) problem. Restricting the number of Steiner vertices in a solution to a constant number (say 1) leads to polynomial solvability, too. On the other hand a large number of special cases turn out to be NP-hard (see e.g. [8]), and therefore, in the general case the SP is an NP-hard problem.

Due to the problem's complexity, one is interested in developing efficient heuristic algorithms to find good approximate solutions. Furthermore, as the SP reveals great importance with respect to its fundamental structural properties, algorithms are also investigated and developed for this already NP-hard 'special case' of more general network design problems. Judging from the recent effort for improved ideas in Steiner tree heuristics, an ever increasing interest with respect to heuristics for both problems, the SP as well as the SPD can be observed. Especially the use of modern heuristic search concepts including metaheuristics such as tabu search, genetic algorithms, among others, has led to considerable results that deserve special mention.

The purpose of this paper is to provide a survey on recent heuristic search methods for the SP.[1] First approaches for determining initial feasible solutions are considered and then local search based methods are described.

In Section 2, we present a survey on construction methods including greedy approaches and worst case analysis results. Furthermore, a look ahead implementation designed to overcome the greedy trap is investigated. The main drawback of algorithms such as deterministic exchange procedures is their inability to continue the search upon becoming trapped in a local optimum. This suggests consideration of modern heuristic search techniques for guiding known strategies to overcome local optimality. Following this theme, we investigate the application of local search techniques such as the tabu search metastrategy for solving Steiner's problem in graphs (Section 3). Furthermore, we provide some ideas which have not yet been fully exploited with respect to the SP (e.g. chunking within genetic algorithms and the ant system). Finally, some summary conclusions are drawn (Section 4).

2 Construction Methods

In this section, we survey existing approaches for determining initial feasible solutions for the SP. First, a general framework based on common building blocks of those heuristics is presented. It is shown that different well-known heuristics are using respective building blocks and may be compared by means of a unifying heuristic measure. Different aspects of comparability and some enhancements are considered. Before going into detail, however, we briefly mention the importance of reduction techniques.

2.1 Reduction Techniques

In a large number of investigations, it has been shown that a very important step towards solving the SP is to perform some preprocessing routines [4, 18, 19, 20, 79, 91]. Among these routines are the following, just to mention some simple ones: Any vertex $v \in S$ of degree 1 can be removed and any vertex $v \in S$ of degree 2 can be replaced by a single edge. Any edge with its

---

[1]For earlier surveys including extensive numerical results the reader is referred to [22, 80]. Comprehensive surveys on the SP up to the begin of the nineties are [38, 79, 89]. Here additional references to earlier sources may be found. As we do not intend to describe exact algorithms the reader may consider [49, 62] for excellent recent results.

weight larger than a shortest path between its end vertices can be removed. If an edge between two basic vertices appears in an MST on the entire graph, then it belongs to an optimal solution and the graph can be contracted along this edge.

Recent results in applying reduction techniques are reported in [18, 62].

2.2 Extending and Connecting by Shortest Path Insertion

A central theme in Steiner tree heuristics is the use of some principles known from MST algorithms together with the computation of shortest paths. The general idea is to build up a feasible solution by insertion of shortest paths. Based on this observation, we consider a classification scheme that helps to establish some common principles as well as some dissimilarities within different algorithms (cf. [22, 80]). Two main ideas may be distinguished:

- **Single Component Extension**
  Start with a *partial solution* $T = (\{w\}, \emptyset)$ consisting of a single vertex w, the so-called root arbitrarily chosen from Q, if not stated otherwise. Extend T to a feasible solution by successive insertion of at most $|Q|$ shortest paths to basic vertices.

- **Component Connecting**
  Start with a *partial solution* $T = (Q, \emptyset)$ consisting of $|Q|$ singleton components. Expand T to a feasible solution by repeatedly selecting components which are connected by shortest paths.

By distinguishing these two main concepts, a classification of the algorithms is given. The first may be considered as a special case of the second one where all components except one must remain singleton. To clarify the concepts, they may be related to the steps of two famous MST algorithms. That is, the single component extension concept may be related to the steps of Prim's MST algorithm [65]. When in each step of the component connecting approach two nearest components are connected, this method corresponds to the steps of Kruskal's MST algorithm [47].

The above ideas may also be considered as follows. Besides a specific partial solution to start with, different heuristics arise according to the number of built-up steps used to obtain a feasible solution T. Most frequently one of the following rules is applied: In 1BASIC, we build by single component extension (e.g. cheapest insertion). In kBASIC, k given components (a basic vertex not yet part of a partial solution may be viewed as a component) are connected by use of shortest paths.

Furthermore, heuristics differ in the amount of knowledge used in the computation for calculating a feasible solution. Together with the specific implementation for extending or connecting components and including one of the above building blocks, we get different algorithms.

Before presenting the two most well-known heuristics for the SP, a unified approach based on the component connecting idea is revisited. Let $T_0, ..., T_\sigma$ denote the components of a partial solution T and $d(v, T_i)$ be the minimum of the shortest distances between v and any node of T_i in G. That is, for d_{ij} being the weight of a shortest path $P(i, j)$ between nodes i and j in G $d(i, T) = \min\{d_{ij} | j \in V_T\}$ where V_T denotes the vertex set of T. A vertex of degree 1 in T is called leaf. Whenever ties occur, they are broken arbitrarily if not stated otherwise.

If the vertex set V_T of an optimal solution is known, the task is to compute an MST of the subgraph of G induced by V_T. The set $V_T - Q$ of Steiner vertices may be heuristically determined by means of a heuristic function f (see e.g. [67, 68]):

HEUM: *Heuristic Measure*
(1) Start with $T = (Q, \emptyset)$ comprising $|Q|$ basic vertices (subtrees of the final subgraph).
(2) **While** T is not connected,
    do choose a vertex v using a function f and unite the two components of T which are nearest to v by combining them with v via shortest paths (the vertices and edges of these paths are added to T).

The general idea of the heuristic measure relates to the original ideas of artificial intelligence (especially pre-knowledge based systems) and the corresponding view on heuristics [57]. Usually f is minimized over all vertices where $f(v)$ is used to numerically express the attractivity of v to be part of a solution. Intuitively $f(v)$ gives a measure of proximity of v to elements of a subset of vertices in T, or the least average distance of v to a set of vertices in T, respectively. In that sense we may easily modify Step 2 of HEUM by uniting more than two components, say three [80], at a time by means of kBASIC.

By choosing a suitable function f, e.g.

$$f_1(v) := \min_{1 \leq i \leq \sigma} \{d(v, T_0 \cap Q) + d(v, T_i \cap Q)\},$$

the distance network or shortest distance graph heuristic (SDISTG, see e.g. [12, 46, 58]) relates to HEUM. Likewise the cheapest insertion method (CHINS, [76]) is obtained by

$$f_2(v) := \min_{1 \leq i \leq \sigma} \{d(v, T_0) + d(v, T_i)\}$$

where, opposite to f_1, the non-basic vertices in T_0 are active. That is, CHINS gives special emphasis to the basic vertices and the partial solution under development whereas SDISTG uses only the basic vertices as its knowledge base. As a synthesis this clearly explains the restrictedness of the SDISTG approach when compared to CHINS and is clearly underlined in a large variety of numerical experiments (see e.g. [68, 80]). A more general function incorporating a component connecting approach is

$$f_3(v) := \min_{1 \leq t \leq \sigma} \{\frac{1}{t} \sum_{i=0}^{t} d(v, T_i) | d(v, T_i) \leq d(v, T_j), 0 \leq i < j \leq \sigma\}$$

(cf. [22, 68, 80]).

One of the most interesting aspects of CHINS is the influence of how the root w is chosen. A repetitive approach results in the following modified algorithm called CHINS-V: *Cheapest insertion (all roots)*. CHINS is applied for every vertex (even possible Steiner vertices) defined as root w and the cheapest of all evaluated solutions gives the overall solution. Correspondingly, in CHINS-Q, CHINS is applied for every basic vertex as root w and in CHINS-2Q, CHINS is applied for every pair p, q of basic vertices where the root component is initialized as $P(p, q)$ and again the overall solution is the cheapest of all solutions obtained. Further modifications include repetitive applications of CHINS (see Section 2.4) as well as the incorporation of 3BASIC into the framework of CHINS [80, 91].

SDISTG has received considerable attention in the literature with respect to an efficient implementation even including a posterior improvement step (cf. [29, 45, 53]). That is, in the original version of SDISTG a subsequent improvement procedure MST+P is applied which may be used in general after having performed any Steiner tree heuristic (see e.g. [68]):

MST+P: *Minimum spanning tree and pruning*

(1) Given a solution with vertex set V_T, construct an MST $T' = (V'_T, E'_T)$ of the subgraph of G induced by V_T.

(2) **While** there exists a leaf of T' that is a Steiner vertex, **do** delete that leaf and its incident edge.

When constructing the complete graph for vertex set $Q' \cup \{k\}$ for all non-basic vertices k where $|Q'| = 3$ we obtain a polynomial algorithm 3BASIC, which is useful as a subroutine in a component connecting approach: Determine with 3BASIC for every triple of components in the partial solution the cost to connect these three components. Execute the connection of lowest cost by corresponding path insertion (decreasing the number of components by two) and repeat these two steps until a feasible solution is obtained. (A somewhat different description of 3BASIC is given in [10]).

It should be noted that a common distinction within the area of heuristics is that of *single pass* versus *multiple pass* approaches depending on whether the proceeding underlying a specific heuristic calculates a feasible solution through a single pass through the given data or through multiple runs. Whereas CHINS itself belongs to the first category its modifications presented above like CHINS-V belong to the second. With respect to HEUM in general and SDISTG there is no clear difference between these classes as the function evaluation is consecutively performed on the whole vertex set in each iteration. Considering the improvement part of SDISTG, we may interpret the method as a multiple pass approach.

A more versatile multiple pass approach may be described as follows [17]. Consider as input a solution T given by a heuristic, say H, and let the *crucial set* of T be defined as the set $V_T^c := Q \cup \{i \in V_T - Q | \gamma_T(i) \geq 3\}$, where $\gamma_T(i)$ is the degree of vertex i in T. Then for a number of iterations (e.g. as long as improved solutions are obtained) V_T^c could be taken as the new set of basic vertices for another pass of the heuristic retaining a new solution T' which repetitively replaces T. This multiple pass approach may also be implemented while allowing V_T^c to change each iteration in a minimal fashion. While each Steiner tree heuristic may serve as a candidate for heuristic H, SDISTG and CHINS have been investigated more thoroughly within this framework than others (see e.g. [1, 35, 54]).

Another idea is to perturb the data such that the heuristic H is definitely finding a different solution and to project the results onto the original data (see e.g. [79]). In the following sections some of these ideas will be reconsidered.

2.3   Enhancements

In this section, we briefly give some ideas on additional topics that have been treated previously or are part of additional studies still under way.

Combining 3BASIC with SDISTG may lead to the following algorithm. It successively applies 3BASIC to every three-element subset of Q (or later a modified set of basic vertices) and then it effectively contracts the most promising of such subsets, to eventually obtain a new Steiner vertex after each contraction (the basic idea of this algorithm is due to Zelikovsky [96]):

SDISTG-3: *Shortest distance graph with 3BASIC*

(1) Start with $T = (Q, \emptyset)$ and denote by $T_0, ..., T_\sigma$ the components of T; $S' := \emptyset, r := 1$.

While ($\sigma \geq 2$ and $r > 0$) do

(2) **For all three element subsets $\{i, j, k\}$ of indices from $\{0, \ldots, \sigma\}$ do**

Apply 3BASIC to obtain a vertex v_{ijk} minimizing the heuristic function $f_4(v_{ijk}) := \sum_{h \in \{i,j,k\}} d(v_{ijk}, T_h \cap Q)$ and obtain T_{ijk} from T by uniting T_i, T_j and T_k (by combining them with each other and v_{ijk} through 3BASIC). Let $Z_{MST}(T)$ be the length of an MST of the complete graph with the vertex set corresponding to the components of T and edge weights $d(T_p, T_q) := \min\{d_{ij} | i \in T_p \cap Q, j \in T_q \cap Q\}$. $Z_{MST}(T_{ijk})$ is given, correspondingly.

(3) Define $r := \max\{Z_{MST}(T) - Z_{MST}(T_{ijk}) - f_4(v_{ijk}) | \forall \{i, j, k\} \subseteq \{0, \ldots, \sigma\}\}$ and $(ijk)*$ as *argmax* of subsets $\{i, j, k\}$ leading to r. If $r > 0$ then $T := T_{(ijk)*}, S' := S' \cup \{v_{(ijk)*}\}$.

endwhile

(4) Apply SDISTG to the problem with modified set $Q \cup S'$ of basic vertices.

The main idea of SDISTG-3 is to compute, based on a specific heuristic measure function, a sequence of promising Steiner vertices that have to be included into the solution and then to apply SDISTG on this modified set of basic vertices. (Combining 3BASIC with different algorithms for SP, especially HEUM and CHINS, has already been proved to be a reasonable approach. The heuristic function f_4 used in SDISTG-3 is a restricted version of a function used in [10, 80].)

Considering naive implementations of the above mentioned algorithms we obtain the following complexities: $O(|Q| \cdot |V|^2)$ for CHINS, $O(|Q|^2 \cdot |V|^2)$

for CHINS-Q and $O(|Q|^3 \cdot |V|^2)$ for CHINS-2Q as well as $O(|Q| \cdot |V|^2)$ for SDISTG and $O(|Q|^3 \cdot |V| + |V|^3)$ for SDISTG-3.

With respect to clever implementations, these running times can be reduced (see e.g. [53, 97] for SDISTG and SDISTG-3). An overall idea in this respect is to use efficient shortest path and MST computations as e.g. presented by Gabow et al. [30]. With respect to the repetitive applications of CHINS, however, we can reduce the complexity in a different setting by simply rearranging some calculations. We explain this for CHINS-Q (the same principle applies to other multiple pass approaches as well).

First, observe that only shortest paths between vertices of V and Q are necessary in CHINS. For CHINS-Q we can calculate all such shortest paths once at forehand. With all such paths available CHINS consumes only $O(|Q| \cdot |V|)$ time (cf. [24]). Therefore, for CHINS-Q (and also for the better variant CHINS-V) a time complexity equal to that of a single pass of CHINS follows.

2.4 Pilot Heuristic

In the previous section we compared SDISTG and CHINS. While SDISTG in each iteration only uses paths between basic vertices, CHINS takes into consideration all nodes inserted in previous iterations. However, this is still myopic without any look ahead feature. That is, in our opinion, one of the weaker aspects of CHINS and similar heuristics is the fact that new paths are selected only based on shortest distances without evaluating the corresponding paths themselves. The algorithms fail to judge all intermediary nodes on these paths as junction nodes for future iterations. Only after inclusion, one hopes to exploit these connective qualities.

In this respect, HEUM with function f_3 is a first improvement. Another idea is to incorporate some sort of tie breaking rule whenever more than one shortest path of the same length exists (cf. [24, 43]). However, it seems to be desirable that an evaluation function accounts for overlaps between paths and calculates in advance some connective consequences of including the paths associated with the evaluated node. This can be done by a look ahead method or a so-called pilot heuristic (see Duin and Voß [23]).

In the pilot heuristic the impact of the vertices will be evaluated by additional Steiner tree calculations before inserting them. At any time within a component connecting approach, the partial solution $T = (V_T, E_T)$ induces an alternative smaller SP: assume the remaining components as basic vertices, while the non-basic vertex set is $V - V_T$. In this problem, we can tentatively run, as a heuristic measure, CHINS for various root nodes. When this is done for all possible root nodes while applying 2BASIC the complexity order is $O(|Q|^2 \cdot |V|^2)$.

To formulate in general terms, the basic idea of the pilot method is described as follows [23, 22]: By looking ahead with a heuristic H as a 'pilot heuristic', one cautiously builds up a partial solution M, the master solution. Separately for each $e \notin M$, the pilot heuristic extends a copy of M by tentatively including e. Let $p(e)$ denote the objective value of the solution obtained by pilot H for $e \in E - M$, and let e_0 be the most promising element according to the pilot heuristic, i.e., $p(e_0) \leq p(e)$, $\forall e \in E - M$. Element e_0 is included into the master solution M changing it in a minimal fashion. On the basis of the changed master solution M new pilot calculations are started for each $e \notin M$, providing a new solution element e_0', and so on. This process could proceed, e.g., until further pilot calculations do not lead to an improvement.

In various combinatorial contexts, a multiple pass method may be helpful, e.g., if providing the information on elements of a high quality solution can let a heuristic obtain better results. CHINS-V reviewed above also enhances the solution of a (single pass) heuristic by trial and error. The pilot method on the SP, applied with CHINS as pilot heuristic, starts similarly, but continues. At any time the so-called master solution, a partial solution T, is a forest of trees: $T_1, T_2, ..., T_s$ covering the vertices of Q. In copies of this graph, with the components conceived of as basic vertices (as if contracted into a single vertex), one runs, separately for each vertex v CHINS with root v. Let vertex $v*$ produce the best objective function value. Then two components of the master solution T are permanently joined in a rerun of CHINS from $v*$, this time in the original graph for two iterations (or one if $v*$ is a basic node). Repetition of this process constitutes a pilot method for the SP denoted as PH_{CHINS}. Alternatively, one can interpret PH_{CHINS} as a synthesis of HEUM with f_3 and CHINS: the formula $f(v)$ is replaced by $p(v)$, the objective function value returned by subroutine CHINS with start v. Heuristic PH_{CHINS} can produce better solutions than CHINS-V, because it can find more than one (local) improvement.

As a more accurate pilot heuristic we may use, e.g., RECONSTRUCT from Duin and Voß [24]. Given a feasible solution represented as a shortest distance tree (i.e., the edges represent shortest paths in the original graph) spanning Q, the aim is to improve it by explicit shortest distance edge exchanges while preserving feasibility (see the edge-oriented transformation presented in Section 3). This process will redefine the crucial set of the solutions.

PH_{RECON}: *Pilot method with RECONSTRUCT*

(1) Denote by (Q, d) the distance graph for Q, which is the complete graph with vertex set Q and edge weights d_{ij} for all $(i, j) \in Q$, and start with the crucial set M being the MST of (Q, d); *iter* := 0.

Repeat

(2) *iter* := *iter* + 1.

(3) **For all** vertices v not in the present crucial set M **do**
   (a) initialize v's pilot solution by S_v := MST of $(V_c(M) \cup \{v\}, d)$;
   (b) apply RECONSTRUCT(S_v) to transform S_v to a pilot solution;
   (c) set $p(v)$ to $d(S_v)$, which is the weight of the subtree S_v.

(4) Select a vertex v_0 with minimum objective function value, i.e. $v_0 := \arg\min p(v)$.

(5) **If** $p(v_0) < d(M)$ **then** modify the master solution M to include v_0:
   (a) define M_0 as the MST of $(V_c(M) \cup \{v_0\}, d)$;
   (b) **if** $d(M_0) \geq d(M)$ **then**
        RECONSTRUCT(M_0), but exit this subroutine as soon as: $d(M_0) < d(M)$;
   (c) $M := M_0$.

until $p(v_0) \geq d(M)$.

Within RECONSTRUCT vertex exchanges are performed by either rejecting or accepting a vertex v as new *crucial vertex* (basic vertex or Steiner vertex of degree 3 or more) in a candidate solution. In the latter case, v enters the candidate solution together with incident paths at the expense of other paths, and possibly also former crucial vertices are then expelled by v. When using RECONSTRUCT as a pilot, the aspect of vertex exchange invites us to loosen the system of a pilot method. Previously PH_{CHINS} extended, on the basis of pilot results, a master solution being a partial solution respectively a partial crucial set. Now we just seek to modify a master solution on the basis of pilot results to a solution of lower cost (with eventually additional Steiner vertices).

With respect to numerical results, different faster variants of the pilot heuristic proved to be successful. Considering a single iteration of a pilot method, the effort may be enormous: $O(|V|)$ times the operations for the underlying heuristic summing up to a significant time complexity. Therefore, ideas to reduce the running times (eventually allowing for slightly worse solution quality) are investigated in [23]. Examples are to limit the number of iterations *iter* by modifying the master solution each time to a greater extent, to obtain pilot values in iteration *iter* for a limited number of vertices, say *pilots(iter)* or to apply short cuts in the calculation of pilot values, i.e. to approximate them.

As an example, we suggest a drop-policy where the pilot is executed in iteration $iter = 1$ for all non-basic vertices, but in each subsequent iteration the number of pilots is halfed (rounded up to integer). One half of the remaining vertices is to drop off in each iteration, e.g. those vertices with a pilot result being among the 50% worst results in that iteration. That is, a drop policy lets former pilot results rule out unattractive vertices. This approach shall be denoted as PH_{DROP}.

Like PH_{DROP} also SDISTG-3 modifies a solution on the basis of a pilot where the pilot value is calculated as a MST such that the respective vertex under consideration has degree 3 and based on the respective decision some vertices will not be considered furtheron.

A different policy may be described as follows. At the start of each new pilot iteration, none of the vertices is a priori ruled out. But each time, in a prepilot phase, one is to filter out a number of candidate vertices among which a vertex, say e_0, is to be determined. For all vertices a much quicker prepilot procedure is run where the vertices with the lowest values are candidates for a full pilot run. Duin and Voß [23] use as a prepilot procedure a shortened version of RECONSTRUCT, processing by the above mentioned subroutine only the vertices of the shortest paths from v to adjacent vertices in S_v. The method is called PH_{FILTER}.

2.5 Dual Ascent

Obviously, finding Steiner arborescences, i.e. feasible solutions for the SPD, is a more general problem than the SP having additional real world applications (e.g., in drainage systems when unidirectional flows are necessary). Nevertheless, it has not yet received the same amount of interest as its undirected version. Because, according to e.g. Chopra et al. [11] as well as [79], the directed version of a mathematical formulation has some advantages over its undirected counterpart, it is believed that the SPD will achieve more interest in the future.

Wong [93], as the SPD's most influential source describes a dual ascent algorithm based on a multicommodity network flow formulation of the SPD.

Besides the implementations in [80, 93] more recently very successful implementations based on improved data structures have come up by [62, 63]. The dual ascent procedure produces as a side result reduced arc costs. The arcs with reduced costs zero define a *support graph*, a subgraph G' of G of promising arcs. Good upper bounds are obtained by running heuristics previously mentioned in this support graph. (Even simple algorithms like MST+P are then efficient; cf. [80]).

Subsequently different ideas including reduction techniques and generalizations of algorithmic ideas from the SP have been developed (see [21, 50, 79]). In [79, 82] the worst case performance of several heuristics for the SPD including directed versions of CHINS, SDISTG and SPATH is investigated. With respect to this criterion it is shown that none of the described algorithms for the SPD dominates any other. Contrary to the SP even one of the simplest algorithms has the same worst case error ratio as the more sophisticated ones. Indeed, the best bound which can be given for any of the heuristics considered is $Z_T/Z_{opt} \leq |Q|$, which is quite disencouraging. The proof idea is simply that the length of a shortest path from the root w to any basic vertex is at most Z_{opt}, i.e., $Z_T \leq |Q| \cdot Z_{opt}$. Furthermore, examples can be found to show that this bound is tight. Finally, it should be noted that all of these algorithms are strongly incomparable (see Section 2.6.1).

Specialized approaches have been developed and investigated for special cases of the SPD, e.g., when graphs are assumed to be acyclic [26, 66, 84, 98, 99]. Furthermore, for the general case most ideas may be directly transfered from the SP to the SPD (see e.g. [55, 79]).

2.6 Results

Comparing different Steiner tree heuristics has been part of experimental as well as theoretical research for a long time. Results from the literature are reported distinguishing theoretical as well as numerical results.

2.6.1 Theoretical Results

For a given instance, let Z_i and Z_j denote the objective function values derived by algorithms A_i and A_j, respectively. A_i and A_j are called *incomparable* if instances with $Z_i < Z_j$ as well as instances with $Z_i > Z_j$ exist. If A_i and A_j are incomparable independently of how ties are broken we call them *strongly incomparable*. The algorithms above, namely SDISTG and

CHINS are strongly incomparable. However, this definition of incomparability does not give any idea on which algorithms could be preferred against others because even the simplest ones SPATH (i.e., computing the shortest path from a root w to any basic vertex) and MST+P on V (i.e., computing a minimum spanning tree on G and then successively pruning any Steiner vertex with degree one) have to be added to this list of strongly incomparable algorithms (cf. [79, 88]). For further results the reader is referred to [60, 61].

Alternatively, algorithms can be compared with respect to their worst case performance. Unfortunately, nearly all of the heuristics given in the literature (including CHINS and SDISTG) have a worst case bound of 2, i.e., $Z_T/Z_{opt} \leq 2(1 - 1/|Q|)$, where Z_T and Z_{opt} denote the objective function values of a feasible solution and an optimal solution, respectively (see e.g. [12, 46, 58, 59, 60, 80, 87]).

Zelikovsky [96] proved that SDISTG-3 has a worst case bound of 11/6 instead of 2 which turned out to be a significant breakthrough in Steiner tree heuristics. At the same time, a slightly different approach had been considered by Berman and Ramaiyer [7]. Even earlier Minoux [54] discussed a similar idea in a different context.

Subsequently, based on Zelikovsky's heuristic [96] a number of papers considered even better heuristics with respect to the respective worst case bounds. Among them, Karpinski and Zelikovsky [42] add a certain type of preprocessing which allows a reduction to approximately $1.644 < 1 + \ln 2$ instead of 11/6. For further references on worst case results see [15, 97].

2.6.2 Numerical Results

In the literature, a specific set of test problems available through e-mail (see [5]) has been considered by several authors. Although these problem instances tend to be very sparse graphs, we will adopt them here for reason of comparability. There are currently four sets of problem instances (B, C, D, E) available with different numbers of vertices and edges ranging from 50 to 2500 nodes and 63 up to 65000 edges with integer edge weights as well as various numbers of basic vertices. For all instances the optimal solution values are provably known (see e.g. [11, 48, 62, 63]).

Even large problem instances of the traveling salesman problem have been modified and used within the scientific community to obtain suitable data for the SP (see [78]). For an algorithm computing problem instances with known optimal solutions see [44].

		B	C	D
CHINS-Q	avg(%)	0.78	2.13	2.12
	worst(%)	2.50	13.30	8.50
	#opt	10	6	7
CHINS-Q-I	avg(%)	0.24	1.58	1.52
	worst(%)	2.20	10.6	6.30
	#opt	15	6	7
HEUM-I	avg(%)	1.17	2.76	1.47
	worst(%)	4.70	9.30	6.90
	#opt	9	4	3
Pilot method [17, 23]				
PH_{DROP}	avg(%)	0.00	0.22	0.27
	worst(%)	0.00	2.65	2.24
	#opt	18	14	12
PH_{FILTER}	avg(%)	0.00	0.11	0.17
	worst(%)	0.00	0.88	0.13
	#opt	18	14	14

Table 1: Numerical results for construction methods

More complicated instances have been devised and investigated by Duin and Voß [17, 24, 23]. In these so-called incidence problems, the weight on each edge (i, j) is defined with a sample r from a normal distribution, rounded to integer value with a minimum value of 1 and a maximum value of 500, i.e., $c_{ij} = \min\{\max\{1, \text{round}(r)\}, 500\}$.

However, to obtain a graph that is much harder to reduce by preprocessing techniques such as given in [17, 21, 79], three distributions with a different mean value are used. Any edge (i, j) is incident to none, to one, or to two basic vertices. The mean of r is 100 for edges (i, j) with $i, j \notin Q$ (no incidence with Q), 200 on edges (i, j) with one special end vertex and 300 on edges (i, j) with both end vertices $i, j \notin Q$. The standard deviation for each of the three normal distributions is 5. For given $|Q|$, various variants may be chosen combining different densities of basic vertices and different edge densities.

There is a large number of studies describing numerical results on different Steiner tree heuristics. For a comprehensive comparison of heuristics see [22, 62, 63, 68, 80, 91]. Summarizing the numerical results in these sources with respect to the heuristics described above SDISTG is clearly outperformed by CHINS and HEUM with the function f_3, for instance. In Table 1, a summary of some of these results is presented. For the included heuristics are given the average deviation (expressed as percentage) from the optimum value (avg(%)), the maximum deviation (expressed as percentage) from the optimum value (worst(%)) and the number of optimum solutions achieved (#opt). Note that the suffix -I denotes the application of the posterior improvement procedure MST+P. For CHINS-Q-I it has been applied to all $|Q|$ solutions under consideration and the best values are referred to.

Furthermore, various modifications and repetitive applications of CHINS and HEUM as well as the dual ascent procedure of Wong [93] (see [62, 63, 80]) turn out to produce even more competitive results with respect to solution quality. Note that the time complexity of HEUM with f_3, $O(|V|^3)$, is not competitive with that of CHINS-Q if $|Q| \ll |V|$.

With respect to the results of Table 1 it may be deduced that repetitive versions of CHINS as well as HEUM with the posterior improvement phase provide reasonable results in short computation time. (Detailed CPU times are not reported as the calculations for the algorithms surveyed in this paper have been performed in different programming languages on various platforms usually taking fractions of a second on todays PCs.) Moreover, the additional effort necessary for the pilot methods gain a considerable improvement in solution quality.

With respect to the pilot method additional results are given in Table 2. Each entry gives the percentage of average or worst cases over a sample of 100 problem instances that have been randomly generated. In column Incidence the graphs are built following the above description with the number of vertices up to 320. As we do not provably know the optimal solutions in ten cases the deviation values are calculated with respect to the best known lower bounds calculated by Duin [17]. Column Random considers the values for 100 random graphs that are built according to the same basic concept as the B-E data, however, with the number of vertices and edges chosen similar to the incidence instances. For Euclidean instances a certain number of points (according to the values for the incidence problems, too) is randomly chosen and connected by means of Euclidean distances. Even the small values given in Table 2 indicate that the incidence problems are harder to solve than other instances and that the different pilot approaches are able to find almost optimal solutions in most cases.

Though quite simple, the above mentioned reduction techniques (see Section 2.1) may already be effective. Applying them together with some more sophisticated rules described in [19, 20] we obtain, e.g., the following results: All 18 problem instances of data set B can be solved to optimality solely by applying reduction techniques. With more versatile reduction

	weights	Random	Euclidean	Incidence
Pilot method [23]				
PH_{RBCON}	avg(%)	0.00	0.02	0.46
	worst(%)	0.23	0.87	5.67
PH_{DROP}	avg(%)	0.03	0.03	0.60
	worst(%)	1.17	0.93	5.77
PH_{FILTBR}	avg(%)	0.05	0.06	0.48
	worst(%)	1.63	0.97	4.57

Table 2: Additional results for the pilot method

techniques even larger instances from data sets C, D, and E may be solved to optimality [62].

3 Improvement Procedures

Various improvement procedures have been investigated for the SP. The first idea is the pruning routine MST+P where, for a given feasible solution $T = (V_T, E_T)$ of SP, an MST T' of the subgraph of G induced by V_T is constructed and then successively all Steiner vertices of degree 1 are deleted. Numerical results show that with this simple approach considerable improvements are possible [22, 68, 80].

Following ideas from Section 2.4, we may incorporate deterministic path exchanges into insertion methods. In the sequel, we follow a slightly different concept in the sense that posterior exchanges will be performed with respect to a given feasible solution T in a local search environment. We first describe the main ideas of local search for the SP. These may be applied within any of the modern heuristic search concepts which are considered subsequently.

3.1 Local Search

For the application of local search for the SP two different neighbourhood definitions are distinguished. Both neighbourhoods may be used as building blocks in various metastrategies.

In an *edge-oriented transformation* an edge $(i,j) \in E_T$ is deleted from T together with all resulting degree 1 Steiner vertices. The resulting two disconnected components of T are reconnected by a shortest path between them. (To prevent from immediately reconsidering the same solution it may be useful to disallow the immediate reinsertion of (i, j) into T.)

Let $(x_1, \ldots, x_{|S|})$ be a binary vector with x_i denoting that node i is included $(x_i = 1)$ or excluded $(x_i = 0)$ from the solution. The principle of a *node-oriented transformation* is to choose a vertex $i \in S$ and define $x_i := 1 - x_i$. Then, an MST of the graph induced by $Q \cup \{i \in S | x_i = 1\}$ is recomputed. The node-oriented transformation may be extended to a so-called λ-interchange for a positive integer λ where for a transformation at most λ nodes may be exchanged at a time. However, even for $\lambda = 2$ this neighbourhood is considerably enlarged in size.

Within the literature, often an edge-oriented transformation is described (see e.g. [39, 72]). This neighbourhood may be extended to exchanging so-called key paths (cf. [14, 77, 79, 86]). Given a Steiner tree T, a *key path* is defined as a path in T of which all intermediate vertices are Steiner vertices of degree two with respect to T and whose end vertices are basic vertices or Steiner vertices of degree three or more (crucial vertices). Obviously an optimal Steiner tree consists of at most $2|Q| - 3$ key paths and at most $|Q| - 2$ crucial vertices. A somewhat different neighbourhood may be advised by eliminating crucial vertices of a given solution and reconnecting the resulting components by means of shortest paths.

Variable neighbourhoods as well as combinations of node-oriented and edge-oriented neighbourhoods are possible, too. Recently, Wade and Rayward-Smith [86] have proposed very interesting and effective neighbourhood definitions which allow changes of the means of transformation within local search algorithms for the SP. Furthermore, the idea of variable neighbourhoods is investigated by [28].

Duin and Voß [24] perform an edge-oriented transformation during the built-up steps of a cheapest insertion method by incorporating deterministic path exchanges. An exception is the approach of Minoux [54] who successively adds vertices as long as they give an improvement to the solution under consideration. Also, related to the node-oriented transformation is the approach within SDISTG-3, i.e., choosing additional Steiner vertices as long as shortcuts seem to be possible, as well as the algorithm described in [1].

Comparing the effort of evaluating all possible exchanges within either of the neighbourhoods shows, at first glance, some advantage for the node-oriented approach. It should be noted, however, that the latter approach allows for some infeasibilities whenever the induced graph is not connected (in that case the objective function value should be ∞). Furthermore, various authors have observed that local search based approaches, especially when based on the node-oriented transformation, may be too time consuming. Reasons for this observation may be non-reduced data (e.g. vertices of degree two may be exchanged without any impact on the objective function value) or the necessity of recalculating MSTs. Clever ideas for implementing the node-oriented transformation are given in [31, 70] (see Section 3.5 below). Furthermore, candidate list strategies may be helpful, too. Depending on the relation of nodes and edges, of course, the picture may change. Especially, when implemented in a clever way, the edge-oriented approach may lead to inspiringly short CPU times, too [77].

Given a feasible solution for the SP or the SPD both transformation approaches may be applied in a first-fit or best-fit manner. Especially in a first-fit approach a circular neighbourhood search may be advantageous, i.e., exchanges that have not yet been explored in one iteration are explored first in the next iteration [56]. In addition one may set a flag for those transformations that did not give an improvement when explored and to reconsider them only when no improvement based on the remaining transformations is possible.

The main drawback of local search procedures is their inability to continue the search upon becoming trapped in a local optimum. This suggests consideration of recent techniques for guiding these heuristics to overcome local optimality. Following this theme, in the subsequent sections we sketch some applications of local search techniques within the framework of modern heuristic search concepts such as simulated annealing, genetic algorithms and tabu search for the SP.

It should be noted that recently considerable effort has been made with respect to the development of object-oriented frameworks or class libraries for local search based heuristics. Although not tailored to the SP or the SPD these problems have been among those investigated as a testbed (see e.g. [2, 27]). Of course the basic concept of the pilot method may be applied in a local search environment as well.

3.2 Greedy Randomized Adaptive Search Procedure

A greedy randomized adaptive search procedure (GRASP) is a multiple pass approach where each pass consists of two phases: constructing a feasible solution and then performing a local search. In the construction phase, e.g., feasible solutions may be built iteratively like in CHINS. However, to obtain diversity in each iteration the next element to be added is determined randomly choosing one of the best possible paths instead of the overall best one.

GRASP approaches for the SP are implemented by [51, 52]. Instead of CHINS, the authors use SDISTG for the construction phase. Within the implementation of Mehlhorn [53] for the SDISTG, they modify the MST calculation (following the procedure of Kruskal [47]) in a way not necessarily to choose a shortest path between two basic vertices but one with a larger weight. The local search phase is based on the node-oriented transformation. However, to speed up the calculations at first only adding Steiner vertices is considered. The evaluation of eliminating Steiner vertices is explored only in those cases where no improvements are possible by adding Steiner vertices. In order to further accelerate the local search phase the authors also evaluate possible transformations based upon candidate lists of promising transformations, which are periodically updated.

3.3   Simulated Annealing

Applications of simulated annealing with respect to the SP can be found in Dowsland [14], Schiemangk [72] as well as Wade and Rayward-Smith [86]. For an adaptation to the SPD see [55]. A somewhat simplified version of simulated annealing has been described under the acronym of threshold accepting, where, different from simulated annealing, deteriorations of the objective function are accepted as long as they are not beyond a given threshold value. Dueck et al [16] apply threshold accepting to the SP with rectilinear distances.

In Dowsland [14], the usual Boltzmann distribution and a geometric cooling schedule were used together with an edge-oriented transformation approach.[2] She achieved best results with a starting temperature of 0.3 which was reduced by a factor of $\frac{3}{4}$ after a prescribed number of 500 to 1000 iterations (depending on the temperature decrease and allowing an overall maximum of 5684 iterations). The test data were randomly generated graphs with 100 vertices, real valued edge weights, a density (i.e. ratio of actual edges to the maximum number of possible edges) of $\frac{1}{2}$, $\frac{3}{4}$, and 1 as well as $|Q| \in \{30, 50, 80\}$. The simulated annealing algorithm is tested while initializing either randomly or by applying CHINS. The algorithms are compared with two local search methods exchanging one or two paths at a time (cf. the edge-oriented transformation) as well as HEUM. HEUM

---

[2]Although we summarize numerical results in Section 3.7, we provide the results from [14] here, because she investigated a different testbed compared to other researchers.

shows an average deviation of $1.1 - 1.3\%$ compared to the results obtained by simulated annealing indicating that the simulated annealing approach can be effective with respect to solution quality.

The currently most effective simulated annealing algorithm for the SP is the one developed by Wade and Rayward-Smith [86] which uses different ideas for developing local search techniques. Basically their simulated annealing implementation regularly changes its search space topology by changing the problem representation and the related neighbourhood transformation.

3.4 Genetic Algorithms

Genetic algorithms (GA) are a class of adaptive search procedures based on the principles derived from the dynamics of natural population genetics. For first references on genetic algorithms for the SP see [36, 37]. Further approaches are presented by [25, 34, 41, 85].

Based on our above mentioned idea of a node-oriented transformation, the SP may be represented by a bit string of size $|S|$ where the entry in position i refers to Steiner vertex i either included or excluded from the solution. The fitness of any solution is computed by Prim's algorithm [65] restricted to the graph induced by the basic vertices plus all included Steiner vertices. For a population size of ten, Kapsalis et al. [41] define a breeding cycle as follows. A new generation is obtained by successively choosing parents and replacing them by new solutions determined by applying the genetic operators crossover and mutation with probabilities 0.9 and 0.02, respectively. The best results have been obtained by initializing the gene pool either randomly or MST-seeded and applying the well-known roulette strategy for selecting solutions to join the gene, or parent pool. Numerical results for the B-data show that using one or another parameter setting the genetic algorithm finds optimal solutions in all of the test problems with at most 4000 seconds for each single run on an AppleMacIIfx.

Whereas Kapsalis et al. [41] use an identical order of genes in every chromosome, in Esbensen [25] each gene is represented by a pair of variables, one representing the number of the gene (node) and the second the allele of the gene (whether it is included or not), in order to apply an inversion operator. A critical point in designing any GA is the treatment of chromosomes derived by crossover that do not represent a feasible solution. One way to overcome this deficiency is the use of penalty functions which lead to a worse fitness for an infeasible solution [41]. Instead Esbensen [25] uses a heuristic, i.e. SDISTG, for decoding a feasible solution for any possible chromosome. The SDISTG heuristic is applied for decoding by not only spanning the basic nodes, but also selected Steiner nodes. With this it is guaranteed that every feasible solution may be obtained. (One might argue, based on our exposition in Section 2.2, that it may be advantageous to replace SDISTG by CHINS. On the other hand the solutions might be too similar and this might lead to premature convergence. The real merit in this respect might yet be explored more fully.)

The crossover operator used for the SDISTG decoder is the uniform crossover, first introduced by Syswerda [75], who shows that this operator is superior to both one- and two-point crossover in most cases. Reeves [69, p. 173] remarks, that the inversion operator, which is used in [25], has not often been found significantly useful.

As a new concept, we investigate the combination of genetic algorithms with the concept of chunking [85]. Chunking may be referred to as a basic learning concept used to keep certain properties in mind, but also to structure complex systems.

When faced with a complex structure, one strategy is to form groups of solution attributes, i.e., to find higher level relationships in the search space [92]. This may lead to a smaller search space, however, eventually at the expense that some combinations of attributes are no longer exploited. Optimal solutions can therefore only be found if the chunks that have been formed during the optimization process belong to respective solutions.

Formally, if a solution x for an optimization problem can be represented by a list of s attributes, a chunk C is defined as a nonnegative vector of attribute indexes [92]. The chunk value v depends on the attributes included in the chunk C and on the solution x, because v need not be identical for a chunk that occurs in different solutions. A solution x can contain more than one chunk, but these should be pairwise disjoint.

The problem that arises using any crossover operator is that promising schemata might be destroyed and the search process is slowed down. To cope with this problem, a modified GA starts with relatively short, so called *building blocks* that are combined to form longer strings, which cover all features of the underlying problem.

This approach contrasts to the fixed length representation of a simple GA. Chunks, in the context of genetic algorithms, define schemata. Here, chunks will not be generated randomly. The general idea of the chunking approach is to speed up the process of evolution by combining chunks of attributes that seem to be superior and to define schemata being responsible for the solution quality. This is done by additionally analyzing each solution to form a few "interesting" chunks, which will be separated with only a small probability by the crossover operator.

The question that arises is: how to find interesting chunks for the SP. If a solution T is given for a specific graph, the Steiner nodes included can be separately investigated. First the Steiner nodes of the solution are considered separately. For each Steiner node i the value $vsn(i)$ is calculated. Different approaches for computing vsn are possible, as e.g., by using $\gamma_T(i)$. A different approach is to differentiate between neighbours being either basic or Steiner vertices as Steiner vertices appear to be more interesting the more basic vertices they connect to the tree.

In both ideas presented so far, the costs associated with a Steiner node were not taken into consideration. A simple approach, which only looks at the costs, is to select the average costs of adjacent edges in the solution tree for $vsn(i)$. Again, we can differentiate between basic and Steiner vertices. In the GA investigated in the results below, we do not compute the average costs, but rather a value that is lower for Steiner nodes incident with more basic nodes than other Steiner vertices. Given a parameter α, this can be accomplished, e.g., by minimizing

$$vsn_T(i) = \frac{\alpha \sum_{j \in V_T \cap Q} c_{ij} + \sum_{j \in V_T \cap S} c_{ij}}{\gamma_T(i)}.$$

After the evaluation of the Steiner nodes of a given solution, the values $vsn(i)$ will be used to create chunks. A chunk can only be defined by Steiner nodes that define a subtree within the solution. The Steiner nodes of an individual, which do not exceed a certain threshold value, will be stored in a list of chunk candidates of that specific individual. This threshold is derived from a parameter stating how many Steiner nodes, say 15%, should be included in chunks of a solution at an average.

After a threshold is computed for the first population, it is used throughout the entire algorithm to determine the list of chunk candidates for every individual. The chunk list is split into pairwise disjoint chunks, which are defined as connected subtrees of the solution, such that every chunk C is not adjacent to any other chunk in the solution.

3.5 Tabu Search

Following a steepest descent mildest ascent approach, any transformation may either result in a best possible improvement or a least deterioration of the objective function value. To prevent the search from cycling between the same solutions endlessly, in tabu search some exchanges will be made tabu, i.e. forbidden for a specific number of iterations. In the sequel we refer to a tabu search approach based on the node-oriented neighbourhood definition as it is investigated in a number of papers for the SP (cf. [22, 31, 70, 83]) as well as for some generalizations of the SP [94, 95]. To the best of our knowledge applying a tabu search approach based on the edge-oriented neighbourhood definition still lacks thorough consideration.

The first published reference on tabu search for the SP is [22] where the reverse elimination method (REM) is considered, however, with limited testing. In the REM, a dynamic tabu strategy is performed that assigns any exchange a tabu status for at most one iteration at a time. The aim is to prevent any solution from being revisited in future iterations by elaborating a necessary and sufficient condition. With respect to the SP, any transformation or move is evaluated under the MST+P routine. All node exchanges are recorded in a list called running list.

Intensification and diversification are among the most versatile strategic components of local search. Usually being utilized in longer term strategies diversification aims at creating solutions within regions of the search space not previously investigated. On the other hand intensification tries to aggressively lead the search into promising regions by hopefully incorporating good attributes or elements. Sondergeld and Voß [73] investigate a specific intensification and diversification approach for the SP that strategically seeks a diversified collection of solutions by scattering multiple search trajectories over the solution space. The intensification strategy is based on strategic restarts of search strategies on so-called elite solutions previously encountered throughout the search. Depending on a periodic switching between diversification and intensification, multiple search levels may be distinguished.

The basic star-shaped diversification approach (SSA) using the REM is to consider multiple lists where the disjointness of solutions can be assured in any case. The first trajectory is built by the conventional REM performed on a given starting solution (constructed by, e.g., CHINS). Starting from the same solution, for each of the remaining trajectories the running lists of "previous" trajectories may successively (by means of consecutive so-called runs) be connected to the current running list in reverse order. That is, pseudo-trajectories are constructed representing fictitious search paths from the final (i.e. current) solutions of all other running lists or trajectories over the starting solution up to the current solution of the actual running list. These connected running lists together with the current running list determine which moves have to be set tabu. By this approach, it is guaranteed that the current trajectory is not entering other trajectories; that is, duplication of a solution of any of the previous trajectories is avoided.

The star-shaped approach itself can be extended by consecutively performing multiple star-shaped runs starting from different starting solutions [73]; this may be called a multi-level star-shaped approach (MSSA). In each level one star-shaped search starts from the same starting solution. If within this search process a new better (or, in a weaker formulation: a new promising) solution is found, this solution now becomes the starting solution for the next star-shaped search level. In the case that such a new better solution could not be found MSSA might use any other (maybe randomly generated) starting solution. Within this context we can see the close relationship and mutual dependency between intensification and diversification. The SSA provides diversification outgoing from a unique, i.e. the same, starting solution and thereby an intensified search within the neighbourhood of this starting solution, whilst the multi-level SSA accentuates a more global diversification aspect.

Gendreau et al. [31] present a static tabu search approach with diversification based on long term memory.[3] Starting from a certain number of feasible solutions constructed by CHINS-V-I they perform a tabu search on all these solutions and record the best overall result. Then a diversification mechanism is applied. The general idea of their continuous diversification consists of removing those vertices that have been a long time within considered solutions while favoring the introduction of vertices as Steiner nodes that have not yet been chosen. A second way to diversify, a so-called path diversification, is performed whenever no improvements have been obtained for a certain number of iterations. Here a path between two nodes i and j is eliminated from a given solution T where $d_T(i,j) - d(i,j)$ is largest for all nodes $i, j \in V_T$ ($d_T(i,j)$ denotes the shortest distance between i and j in T).

An important contribution of Gendreau et al. [31] is their careful consideration of efficient data structures for manipulating given trees when adding or dropping nodes to or from a given solution/tree. As the computational effort for exploring a complete neighbourhood may still be enormous, addi-

---

[3] It should be noted that the authors determine the tabu list length value or tabu tenure at random within given intervals, say from the interval 20-40 for graphs with $|V| \geq 351$. According to the classification in [83] this is still a static approach.

tional ideas for reducing it are given by Ribeiro and de Souza [70]. They describe transformation estimations, elimination tests, and neighbourhood reduction techniques that may be helpful to speed up the local search, leading to a faster tabu search algorithm than that of Gendreau et al. [31] without loosening too much the solution quality.

3.6 Miscellaneous

In this section, we consider different other approaches that recently have been investigated by ourselves (ant systems) as well as other researchers (relating the SP to the satisfiablity problem, neural networks) without having obtained a breakthrough in solution quality.

Ant Systems

Intelligent agent systems provide a useful paradigm for search processes designed to solve complex problems. These systems are particularly relevant for parallel processing applications, and also offer useful strategies for serial heuristic search. As a special case, ant colony methods have recently attracted attention for their application to several types of optimization problems, especially those with a "graph related formulation". These methods can be interpreted as applying specific formulas (to monitor "ant traces") that embody components of strategic principles that are fundamental to adaptive memory programming (AMP, [33]) processes, as notably represented by tabu search.

One of the recently proposed methods in the intelligent search category is the so-called ant system, a dynamic optimization process reflecting the natural interaction between ants searching for food (see, e.g., [13]). The ants' operations are influenced by two different kinds of search criteria. The first one is the local visibility of food, i.e., the attractivity of food in each ant's neighborhood. Additionally, each ant's way through its food space is affected by the other ants' trails as indicators for possibly good directions. The intensity of trails itself is time-dependent: As time goes by, parts of the trails "are gone with the wind", meanwhile the intensity may increase by new and fresh trails. With the quantities of these trails changing dynamically an "autocatalytic" optimization process is started with the hope of forcing the ants' search into most promising regions. This process may be conceived as a form of interactive learning, which can easily be modelled for most kinds of optimization problems by using simultaneously and interactively processed search trajectories. For instance, with respect to the SP and the SPD the greedy choice underlying CHINS reflects some local visibility.

Ant systems show an evident analogy and relationship with other approaches commonly applied within the frame of intelligent search and constitute an example of the class of methods that repeatedly execute a particular heuristic on strategically modified instances of an underlying main problem.

There are basically two principles for implementing the ant system: In the case of a constructive (path-oriented) approach trails are laid during a process that corresponds to constructing a feasible solution. In the case of a transition (neighbourhood based) approach trails are laid which indicate that specific moves (neighbourhood transitions) are performed more or less often during the search.

Satisfiability Related Approach

The weighted max-sat problem asks for a propositional variable assignment that maximizes the weighted sum of the satisfied clauses (i.e. conjunction of variables) of a given formula. This is a related version to the max-sat problem (all clauses weighted equally) which is a generalization of the well-known satisfiability problem. Recently there has been active research on heuristic algorithms for the satisfiability problem where most of these algorithms are also applicable for the (weighted) max-sat problem. Jiang et al. [40] use their so-called GSAT algorithm to get an approximate solution for the SP. That is, they develop an encoding of instances of the SP as propositional formulas. The straightforward translation leads to a variable number quadratically in the number of edges. So they use an alternative encoding that is only linearly dependent on the number of edges. Their idea can be described as follows: We may precompute a number of potentially promising subsolutions - here the k shortest paths between pairs of basic nodes (restricted to pairs consecutive in an arbitrarily chosen order). This leads for each pair of basic nodes to a clause that consists of all such path variables (that are set to 1 iff the path is used in the solution). Further, they introduce a variable for each edge of the graph (variable set to 1 iff the edge is used in the solution) and clauses that guarantee that either each edge variable is set to 1 or the path is not included (i.e., the path variable is set to 0) in the solution. All these clauses considered are weighted prohibitively high ('hard constraints'), so all must be fulfilled in an acceptable solution. Besides, there are 'soft constraints': single variable clauses with negated edge variables. Those clauses are weighted with the respective edge lengths.

The results show that the application of such a general framework (weighted max-sat) on specific problems like the SP leads to considerably worse results than those of special algorithms. Jiang et al. [40] use the benchmark instances from data sets C, D, and E (see Table 3 for a summary of the results). The described running times are rather fast (the times to pre-compute the k best paths between all basic nodes are not given). There is the problem to choose a reasonable value for k; dependent on the specific problem instance, k was chosen up to 150 (less for the larger instances).

Neural Networks

Neural networks have not yet been proven to be successful, let alone competitive when dealing with the SP. First attempts to apply, e.g., the concepts of the Hopfield model are reported in [6, 64, 100]. However, results are yet only reported for problem instances in fairly small graphs with up to 17 basic vertices. The only exception is [32] who uses a neural network approach to perform transformations in a node-oriented fashion based on solutions obtained by initial feasible solutions determined by using SDISTG and HEUM. Results are reported for graphs with 25 and 100 nodes without comparing to proven optimal solutions.

3.7 Numerical Results

Local search approaches have been investigated by a number of researchers. The most important results based on the key path concept are given by Verhoeven [77, 78]. He starts with a feasible solution generated by CHINS and compares two different neighbourhoods based on key path exchanges and the elimination of crucial vertices as described in Section 3.1. The local search approach is embedded in a first-fit circular neighbourhood search. In the first rows of Table 3 results are provided as follows. The algorithm is run ten times and the indicated values are refering to the best of these runs.

Additional studies based on the edge-oriented neighbourhood are performed by, e.g., [14, 71]. Some results based on the node-oriented transformation can be found in [22]. Table 3 gives results for the improvement procedures described in the previous sections (see the description of the data sets as well as the notation in Section 2.6.2). Whereas in this table results are based on applying the different approaches on the given data, in Table 4 we report on some results for considerably reduced data. (Again we have to admit that it is doubtful to present comparative numerical results as different authors reduced the data by means of different reduction techniques. In Table 3 we report the quality of two tabu search approaches which may be distinguished in the sense that a full exposition of the approach (F-Tabu) is compared to a curtailed one (P-Tabu).

With respect to some of the REM-based descriptions above, we applied the REM from a solution provided by CHINS-V-I. For $\lambda = 1$, 2500 iterations have been performed. As the neighbourhood considerably enlarges for $\lambda = 2$, the number of iterations has been reduced to 200. As the REM provides a necessary and sufficient condition to prevent from revisiting previously investigated solutions, a simple diversification arises when a tabu duration larger than 1 is defined. That is, any attribute that becomes tabu for one iteration in the original REM now obtains a tabu status for a number of iterations equal to the tabu duration. As the purpose of the star-shaped approaches is to diversify the search, we have even chosen an enlarged tabu duration of 10 and 100 (given in brackets). The results show that the REM gives reasonable results without having to think about too many parameters in the original version (only the number of iterations and the neighbourhood and its respective evaluation). However, additional topics like some intensification as well as diversification strategies have to be observed to get further improvements, as indicated through the additional results of Table 3.

Further ideas may incorporate a modified candidate list strategy as follows. We have applied the following idea applying the heuristic framework of Fink and Voß [27]. Within one complete calculation of the node-oriented neighbourhood we store a certain percentage of all possible transformations in a candidate list CL, say 20%. Depending on the number of elements in CL in the next, say $0.4 \times |CL|$ iterations, only the elements from the candidate list are explored before the next full cycle of all possible transformations is considered. Following this strategy with a reactive tabu search approach we obtain optimal solutions for all but two C- and D-data within a time limit of two hours. Almost the same succes is observed for the REM with this candidate list strategy.

Numerical results for the ant system are not yet competitive compared to the other approaches presented. However, preliminary results on a combination of the ant system paradigm with a dynamic tabu search implementation (based on the REM) are quite promising.

The genetic algorithm of Esbensen [25] has been applied in 10 runs and the results are the best over those 10 runs. Furthermore, different to all other entries in Table 3, Esbensen provides only results for reduced problem instances, i.e., his method has been applied to the reduced graph obtained after having applied those reduction techniques given in Section 2.1 above. This leads to considerably different preconditions when comparing the results. Nevertheless, the results seem to be astonishingly good when compared to previous genetic approaches for the SP. The combination of the GA with chunking has been applied for a certain parameter α and 5 runs where the best result is taken. Compared to the tabu search approaches, we seem to have only slightly worse results.

The satisfiability based approach provides solutions that are not competitive even with the construction methods and their respective results presented in Section 2.6.2.

4   Conclusions

In this paper, we have surveyed recent developments in Steiner tree heuristics considering, for instance, new ideas with respect to look ahead, tabu search, and the investigation of worst case analysis results. Both, heuristics for finding initial feasible solutions as well as improvement procedures usually rely on the calculation of shortest paths. As a synthesis from related research not only devoted to heuristics we may conclude that no matter what algorithmic approach is followed the existing data needs to be preprocessed and reduced to avoid too many calculations. Additionally, the more information is used (in a meaningful way; see e.g. the pilot heuristic as well as the structural advantage of the cheapest insertion heuristic CHINS over the shortest distance graph heuristic SDISTG) the better solutions may be obtained. While most ideas refer to direct calculations of shortest paths the reasoning of the dual ascent heuristic is somewhat different and may be worth further consideration.

With respect to worst case analysis results, we should note that the search for a constant c which separates polynomial approximability from non-approximability (and with that NP-hardness) is a general topic that deserves considerable interest for related problems as well. In a recent paper Arora [3] considers the approximability of the Euclidian traveling salesman problem which is also applicable to the corresponding Euclidean Steiner problem. This approach constitutes (for the first time) a polynomial approximation scheme for these problems (which, however, is not applicable to the more general metric problem versions as well as the problem in graphs).

		C	D	E
Local Search				
Key path exchange [77]	avg(%)	-	1.42	1.30
	worst(%)	-	8.39	7.92
	#opt	-	7	8
Crucial vertex exchange [77]	avg(%)	-	0.50	0.71
	worst(%)	-	3.55	4.40
	#opt	-	11	10
Tabu Search				
REM ($\lambda = 1$)	avg(%)	0.05	0.32	-
	worst(%)	0.31	2.26	-
	#opt	15	8	-
SSA(10) [73]	avg(%)	0.07	0.34	-
	worst(%)	1.16	1.93	-
	#opt	17	9	-
MSSA(10) [73]	avg(%)	0.08	0.29	-
	worst(%)	1.16	2.26	-
	#opt	16	9	-
MSSA(100) [73]	avg(%)	0.18	0.25	-
	worst(%)	1.16	1.79	-
	#opt	13	11	-
Genetic Algorithms				
SDISTG+chunking ($\alpha = 0.75$) [85]	avg(%)	0.18	0.41	-
	worst(%)	1.13	1.21	-
	#opt	12	8	-
Satisfiability [40]	avg(%)	4.77	5.23	5.93
	worst(%)	15.04	17.49	18.47
	#opt	8	8	6

Table 3: Numerical results for improvement procedures

		C	D	E
Tabu Search				
P-Tabu, Gendreau et al. [31]	avg(%)	0.10	0.16	0.50
	worst(%)	0.88	1.35	2.66
	#opt	16	14	9
F-Tabu, Gendreau et al. [31]	avg(%)	0.02	0.11	0.31
	worst(%)	0.31	0.90	1.60
	#opt	18	14	10
Genetic Algorithms				
Esbensen [25]	avg(%)	0.00	0.11	0.42
	worst(%)	0.00	1.34	3.55
	#opt	20	17	11
SDISTG+chunking [85]	avg(%)	0.08	0.29	-
	worst(%)	0.98	1.21	-
	#opt	16	8	-

Table 4: Numerical results for improvement procedures (reduced data)

Arora [3] shows, that for an arbitrarily chosen $\epsilon > 0$ there exists an $(1 + \epsilon)$-approximate solution with a special structure: "very few" edges cross each line in a recursively chosen partition of the plane (extensible to higher dimensions). As yet a straightforward implementation might be too slow to be of practical interest. However, the impact for the SP and the SPD needs to be investigated.

In a different setting, further research could be to perform some *target analysis*, i.e., a careful analysis of differences from feasible solutions and the optimal solution (whenever available). In this sense problems with known optimal solutions are of great help [5, 44].

Furthermore, the development of efficient heuristics for special cases could be useful and trying to project the findings onto the SP and SPD. An example might be the consideration of the rectilinear Steiner arborescence problem where even the problem's complexity is still open (see e.g. [26, 84, 98, 99]), or the general case of undirected rectilinear graphs (see e.g. [74, 83, 90]).

Most of the algorithms described in this paper may be investigated with respect to parallel implementations which has not yet received comprehensive interest (see e.g. [9, 52, 77] for exceptions). Another area worth to be investigated is the recalculation of MSTs when only slight modifications are made (e.g. in- or excluding a vertex and its incident edges, see e.g. [54, 31]

for some ideas). This might help especially in speeding up any of the metastrategies when based on a node-oriented transformation. Finally, especially in any of the metastrategies as well as in multiple pass approaches the clever use of information gained in previous iterations still needs careful investigation.

Acknowledgement

The fruitful collaboration and joint work with Cees Duin and Andreas Fink is appreciated.

References

[1] M.J. Alexander and G. Robins, New performance-driven FPGA routing algorithms, Working Paper, University of Virginia (1995).

[2] A.A. Andreatta, S.E.R. Carvalho and C.C. Ribeiro, An object-oriented framework for local search heuristics, Working Paper, Catholic University of Rio de Janeiro (1998).

[3] S. Arora, Polynomial-time approximation schemes for Euclidean TSP and other geometric problems, Princeton University (1996), to appear in *Journal of the ACM*.

[4] A. Balakrishnan and N.R. Patel, Problem reduction methods and a tree generation algorithm for the Steiner network problem, *Networks* Vol. 17 (1987) pp. 65-85.

[5] J.E. Beasley, OR-Library: distributing test problems by electronic mail, *Journal of the Operational Research Society* Vol. 41 (1990) pp. 1069-1072.

[6] I. Bennour and J. Cloutier, Steiner tree approximation using neural networks, in: *Congress Canadian sur l'integration a tres grande echelle (CCVLSI)*, Halifax (1992) pp. 268-275.

[7] P. Berman and V. Ramaiyer, Improved approximations for the Steiner tree problem. *Journal of Algorithms* Vol. 17 (1994) pp. 381-408.

[8] M. Bern and P. Plassmann, The Steiner problem with edge lengths 1 and 2, *Information Processing Letters* Vol. 32 (1989) pp. 171-176.

[9] G.-H. Chen, M.E. Houle and M.T. Kuo, The Steiner problem in distributed computing systems, *Information Sciences* Vol. 74 (1993) pp. 73-96.

[10] N.P. Chen, New algorithms for Steiner tree on graphs, in: *Proceedings of the IEEE International Symposium on Circuits and Systems* (1983) pp. 1217-1219.

[11] S. Chopra, E.R. Gorres and M.R. Rao, Solving the Steiner tree problem on a graph using branch and cut, *ORSA Journal on Computing* Vol. 4 (1992) pp. 320-335.

[12] E.A. Choukhmane, Une heuristique pour le probleme de l'arbre de Steiner, *R.A.I.R.O. Recherche Operationelle* Vol. 12 (1978) pp. 207-212.

[13] M. Dorigo, V. Maniezzo and A. Colorni, Ant system: Optimization by a colony of cooperating agents, *IEEE Transactions on Systems, Man, and Cybernetics* Vol. B-26 (1996) pp. 29-41.

[14] K.A. Dowsland, Hill-climbing, simulated annealing and the Steiner problem in graphs, *Engineering Optimization* Vol. 17 (1991) pp. 91-107.

[15] D.-Z. Du and Y. Zhang, On better heuristics for Steiner minimum trees, *Mathematical Programming* Vol. 57 (1992) pp. 193-202.

[16] G. Dueck, T. Scheuer and H.-M. Wallmeier, The C.H.I.P.-Algorithm, Working Paper, IBM Germany, Heidelberg (1991).

[17] C. Duin, *Steiner's problem in graphs*, Ph.D. thesis, Amsterdam (1993).

[18] C.W. Duin, Preprocessing the Steiner problem in graphs, this volume (1999).

[19] C.W. Duin and A. Volgenant, An edge elimination test for the Steiner problem in graphs, *Operations Research Letters* Vol. 8 (1989) pp. 79-83.

[20] C.W. Duin and A. Volgenant, Reduction tests for the Steiner problem in graphs, *Networks* Vol. 19 (1989) pp. 549-567.

[21] C.W. Duin and A. Volgenant, Reducing the hierarchical network design problem, *European Journal of Operational Research* Vol. 39 (1989) pp. 332-344.

[22] C.W. Duin and S. Voß, Steiner tree heuristics - a survey, in: H. Dyck-hoff, U. Derigs, M. Salomon and H.C. Tijms (eds.) *Operations Research Proceedings*, (Springer, Berlin, 1994) 485-496.

[23] C.W. Duin and S. Voß, The pilot method - a strategy for heuristic repetition with application to the Steiner problem in graphs, Working Paper, Univ. Amsterdam (1995), to appear in *Networks*.

[24] C.W. Duin and S. Voß, Efficient path and vertex exchange in Steiner tree algorithms, *Networks* Vol. 29 (1997) pp. 89-105.

[25] H. Esbensen, Computing near-optimal solutions to the Steiner problem in a graph using a genetic algorithm, *Networks* Vol. 26 (1995) pp. 173-185.

[26] A. Fink and S. Voß, A note on the rectilinear Steiner arborescence problem, Working Paper, TH Darmstadt (1994).

[27] A. Fink and S. Voß, Generic metaheuristics application to industrial engineering problems. Working Paper, TU Braunschweig (1998), to appear in *Computers & Industrial Engineering*.

[28] A. Fink and S. Voß, Candidate list strategies and variable neighbourhood search. Working Paper, TU Braunschweig (1999).

[29] R. Floren, A note on "A faster approximation algorithm for the Steiner problem in graphs", *Information Processing Letters* Vol. 38 (1991) pp. 177-178.

[30] H.N. Gabow, Z. Galil, T. Spencer and R.E. Tarjan, Efficient algorithms for finding minimum spanning trees in undirected and directed graphs, *Combinatorica* Vol. 6 (1986) pp. 109-122.

[31] M. Gendreau, J.-F. Larochelle and B. Sansó, A tabu search heuristic for the Steiner tree problem, Working paper, CRT, Montreal (1998).

[32] A. Ghanwani, Neural and delay based heuristics for the Steiner problem in networks, *European Journal of Operational Research* Vol. 108 (1998) pp. 241-265.

[33] F. Glover and M. Laguna, *Tabu Search*, (Kluwer, Boston, 1997).

[34] P. Guitart and J.M. Basart, A genetic algorithm approach for the Steiner problem in graphs, in: *EUFIT 98 Proceedings*, (Elite Foundation, Aachen, 1998) pp. 508-512.

[35] P. Guitart and J.M. Basart, A high performance approximate algorithm for the Steiner problem in graphs, in: M. Luby, J. Rolim and M.J. Serna (eds.) *Randomization and Approximation Techniques in Computer Science*, Lecture Notes in Computer Science Vol. 1518, (Springer, Berlin, 1998) pp. 280-293.

[36] J. Hesser, R. Männer and O. Stucky, Optimization of Steiner trees using genetical algorithms, in: J.D. Schaffer (ed.) *Proceedings of the Third International Conference on Genetic Algorithms*, (Morgan Kaufmann, San Mateo, 1989) pp. 231-236.

[37] J. Hesser, R. Männer and O. Stucky, On Steiner trees and genetic algorithms, in: J.D. Becker, I. Eisele and F.W. Mündemann (eds.) *Parallelism, Learning, Evolution*, Lecture Notes in Computer Science Vol. 565, (Springer, Berlin, 1991) pp. 509-525.

[38] F.K. Hwang, D.S. Richards and P. Winter, *The Steiner Tree Problem*, (North-Holland, Amsterdam, 1992).

[39] Y.-B. Ji and M.-L. Liu, A new scheme for the Steiner problem in graphs, in: *Proceedings of the IEEE International Symposium on Circuits and Systems* (1988) pp. 1839-1842.

[40] Y. Jiang, H. Kautz and B. Selman, Solving problems with hard and soft constraints using a stochastic algorithm for MAX-SAT, Working Paper, AT&T Bell Laboratories (1995).

[41] A. Kapsalis, V.J. Rayward-Smith and G.D. Smith, Solving the graphical Steiner tree problem using genetic algorithms, *Journal of the Operational Research Society* Vol. 44 (1993) pp. 397-406.

[42] M. Karpinski and A. Zelikovsky, New approximation algorithms for the Steiner tree problems, *Electronic Colloquium on Computational Complexity* TR95-003 (1995), and *Journal of Combinatorial Optimization* Vol. 1 (1997) pp. 47-65.

[43] B.N. Khoury and P.M. Pardalos, A heuristic for the Steiner problem in graphs, *Computational Optimization and Applications* Vol. 6 (1996) pp. 5-14.

[44] B.N. Khoury, P.M. Pardalos and D.-Z. Du, A test problem generator for the Steiner problem in graphs, *ACM Transactions on Mathematical Software* Vol. 19 (1993) pp. 509-522.

[45] L.T. Kou, On efficient implementation of an approximation algorithm for the Steiner tree problem, *Acta Informatica* Vol. 27 (1990) pp. 369-380.

[46] L. Kou, G. Markowsky and L. Berman, A fast algorithm for Steiner trees, *Acta Informatica* Vol. 15 (1981) pp. 141-145.

[47] J.B. Kruskal, On the shortest spanning subtree of a graph and the travelling salesman problem, *Proc. Amer. Math. Soc.* Vol. 7 (1956) pp. 48-50.

[48] A. Lucena, Steiner problem in graphs: Lagrangean relaxation and cutting-planes, *Bulletin of the Committee on Algorithms* Vol. 21 (1992) pp. 2-7.

[49] A. Lucena and J.E. Beasley, A branch and cut algorithm for the Steiner problem in graphs, *Networks* Vol. 31 (1998) pp. 39-59.

[50] N. Maculan, P. Souza and A. Candia Vejar, An approach for the Steiner problem in directed graphs, *Annals of Operations Research* Vol. 33 (1991) pp. 471-480.

[51] S.L. Martins, P.M. Pardalos, M.G.C. Resende and C.C. Ribeiro, Greedy randomized adaptive search procedures for the Steiner problem in graphs, in: P. Pardalos, S. Rajasekaran and J. Rolim (eds.) *Randomization Methods in Algorithm Design*, DIMACS Series in Discrete Mathematics and Theoretical Computer Science 43, (AMS, Providence, 1998) pp. 133-145.

[52] S.L. Martins, C.C. Ribeiro and M.C. de Souza, A parallel GRASP for the Steiner problem in graphs, in: Lecture Notes in Computer Science Vol. 1457, (Springer, Berlin, 1998) pp. 285-297.

[53] K. Mehlhorn, A faster approximation algorithm for the Steiner problem in graphs, *Information Processing Letters* Vol. 27 (1988) pp. 125-128.

[54] M. Minoux, Efficient greedy heuristics for Steiner tree problems using reoptimization and supermodularity, *INFOR* Vol. 28 (1990) pp. 221-233.

[55] L.J. Osborne and B.E. Gillett, A comparison of two simulated annealing algorithms applied to the directed Steiner problem on networks, *ORSA Journal on Computing* Vol. 3 (1991) pp. 213-225.

[56] C.H. Papadimitriou and K. Steiglitz, *Combinatorial Optimization: Algorithms and Complexity*, (Prentice-Hall, Englewood Cliffs, 1982).

[57] J. Pearl, *Heuristics: Intelligent Search Techniques for Computer Problem Solving*, (Addison-Wesley, Reading, 1984).

[58] J. Plesnik, A bound for the Steiner tree problem in graphs, *Math. Slovaca* Vol. 31 (1981) pp. 155-163.

[59] J. Plesnik, Worst-case relative performance of heuristics for the Steiner problem in graphs, *Acta Math. Univ. Comenianae* Vol. 60 (1991) pp. 269-284.

[60] J. Plesnik, Heuristics for the Steiner problem in graphs, *Discrete Applied Mathematics* Vol. 37/38 (1992) pp. 451-463.

[61] J. Plesnik, On some heuristics for the Steiner problem in graphs, in: J. Nesetril and M. Fiedler (eds.), *Fourth Czechoslowakian Symposium on Combinatorics, Graphs and Complexity*, (Elsevier, Amsterdam, 1992) pp. 255-257.

[62] T. Polzin and S. Vahdati Daneshmand, Algorithmen für das Steiner-Problem, Diploma thesis, University of Dortmund (1997).

[63] T. Polzin and S. Vahdati Daneshmand, Improved algorithms for the Steiner problem in networks, Working paper, University of Mannheim (1998).

[64] C. Pornavalai, N. Shiratori and G. Chakraborty, Neural network for optimal Steiner tree computation, *Neural Processing Letters* Vol. 3 (1996) pp. 139-149.

[65] R.C. Prim, Shortest connection networks and some generalizations, *Bell Syst. Techn. J.* Vol. 36 (1957) pp. 1389-1401.

[66] S.K. Rao, P. Sadayappan, F.K. Hwang and P.W. Shor, The rectilinear Steiner arborescence problem, *Algorithmica* Vol. 7 (1992) pp. 277-288.

[67] V.J. Rayward-Smith, The computation of nearly minimal Steiner trees in graphs, *Int. J. Math. Educ. Sci. Technol.* Vol. 14 (1983) pp. 15-23.

[68] V.J. Rayward-Smith and A. Clare, On finding Steiner vertices, *Networks* Vol. 16 (1986) pp. 283-294.

[69] C.R. Reeves, Genetic Algorithms, in: C.R. Reeves (ed.), *Modern Heuristic Techniques for Combinatorial Optimization*, (Blackwell, Oxford, 1993) pp. 152-196.

[70] C.C. Ribeiro and M.C. de Souza, Improved tabu search for the Steiner problem in graphs, Working paper, Catholic University of Rio de Janeiro (1998).

[71] J.M. Samaniego, Evolutionary methods for the Steiner problem in networks: an experimental evaluation, Working paper, University of the Philippines Los Banos (1997).

[72] C. Schiemangk, Thermodynamically motivated simulation for solving the Steiner tree problem and the optimization of interacting path systems, in: A. Iwainsky (ed.) *Optimization of Connection Structures in Graphs*, (Central Institute of Cybernetics and Information Processes, Berlin, 1985) pp. 91-120.

[73] L. Sondergeld and S. Voß, A multi-level star-shaped intensification and diversification approach in tabu search for the Steiner tree problem in graphs, Working Paper, TU Braunschweig (1996).

[74] C.C. de Souza and C.C. Ribeiro, Heuristics for the minimum rectilinear Steiner tree problem, Working Paper G-90-32, GERAD, Montreal (1990).

[75] G. Syswerda, Uniform Crossover in Genetic Algorithms, in: J.D. Schaffer (ed.), *Proceedings of the Third International Conference on Genetic Algorithms*, (Morgan Kaufmann, San Mateo, 1989) pp. 2-9.

[76] H. Takahashi and A. Matsuyama, An approximate solution for the Steiner problem in graphs, *Math. Japonica* Vol. 24 (1980) pp. 573-577.

[77] M.G.A. Verhoeven, *Parallel Local Search*, Ph.D. thesis, Eindhoven (1996).

[78] M.G.A. Verhoeven, M.E.M. Severens and E.H.L. Aarts, Local search for Steiner trees in graphs, in: V.J. Rayward-Smith, I.H. Osman, C.R. Reeves and G.D. Smith (eds.) *Modern Heuristic Search Methods*, (Wiley, Chichester, 1996) 117-129.

[79] S. Voß, *Steiner-Probleme in Graphen*, (Hain, Frankfurt/Main, 1990).

[80] S. Voß, Steiner's problem in graphs: heuristic methods, *Discrete Applied Mathematics* Vol. 40 (1992) pp. 45-72.

[81] S. Voß, On the worst case performance of Takahashi and Matsuyama's Steiner tree heuristic, Working Paper, TH Darmstadt (1992).

[82] S. Voß, Worst case performance of some heuristics for Steiner's problem in directed graphs, *Information Processing Letters* Vol. 48 (1993) pp. 99-105.

[83] S. Voß, Observing logical interdependencies in tabu search - methods and results, in: V.J. Rayward-Smith, I.H. Osman, C.R. Reeves und G.D. Smith (eds.), *Modern Heuristic Search Methods*, (Wiley, Chichester, 1996) pp. 41-59.

[84] S. Voß and C.W. Duin, Heuristic methods for the rectilinear Steiner arborescence problem, *Engineering Optimization* Vol. 21 (1993) pp. 121-145.

[85] S. Voß and K. Gutenschwager, A chunking based genetic algorithm for the Steiner tree problem in graphs, in: P.M. Pardalos and D. Du (eds.), *Network Design: Connectivity and Facilities Location*, DIMACS Series in Discrete Mathematics and Theoretical Computer Science 40, (AMS, Providence, 1998) pp. 335-355.

[86] A.S.C. Wade and V.J. Rayward-Smith, Effective local search techniques for the Steiner tree problem, *Studies in Locational Analysis* Vol. 11 (1997), pp. 219-241.

[87] B.M. Waxman and M. Imase, Worst-case performance of Rayward-Smith's Steiner tree heuristic, *Information Processing Letters* Vol. 29 (1988) pp. 283-287.

[88] P. Widmayer, Fast approximation algorithms for Steiner's problem in graphs, Habilitation thesis, Institut für Angewandte Informatik und formale Beschreibungsverfahren, University Karlsruhe (1986).

[89] P. Winter, Steiner problem in networks: a survey, *Networks* Vol. 17 (1987) pp. 129-167.

[90] P. Winter, Reductions for the rectilinear Steiner problem, Working Paper, Rutgers University (1995).

[91] P. Winter and J. MacGregor Smith, Path-distance heuristics for the Steiner problem in undirected networks, *Algorithmica* Vol. 7 (1992) pp. 309-327.

[92] D. L. Woodruff, Proposals for Chunking and Tabu Search, *European Journal of Operational Research* Vol. 106 (1998) pp. 585-598.

[93] R.T. Wong, A dual ascent approach for Steiner tree problems on a directed graph, *Mathematical Programming* Vol. 28 (1984) pp. 271-287.

[94] J. Xu, S.Y. Chiu and F. Glover, A probabilistic tabu search for the telecommunications network design, *Combinatorial Optimization: Theory and Practice* Vol. 1 (1996) pp. 69-94.

[95] J. Xu, S.Y. Chiu and F. Glover, Using tabu search to solve Steiner tree-star problem in telecommunications network design, *Telecommunication Systems* Vol. 6 (1996) pp. 117-125.

[96] A.Z. Zelikovsky, An $\frac{11}{6}$-approximation algorithm for the Steiner problem on graphs, in: J. Nesetril and M. Fiedler (eds.), *Fourth Czechoslowakian Symposium on Combinatorics, Graphs and Complexity*, (Elsevier, Amsterdam, 1992) pp. 351-354.

[97] A.Z. Zelikovsky, A faster approximation algorithm for the Steiner tree problem in graphs, *Information Processing Letters* Vol. 46 (1993) pp. 79-83.

[98] A. Zelikovsky, A polynomial-time subpolynom-approximation scheme for the acyclic directed Steiner tree problem, Working Paper, Institute of Mathematics, Moldova (1994).

[99] A. Zelikovsky, A series of approximation algorithms for the acyclic directed Steiner tree problem, *Algorithmica* Vol. 18 (1997) pp. 99-110.

[100] J. Zhongqi, S. Huaying and G. Hong ming, Using neural networks to solve Steiner tree problem, in: *IEEE TENCON '93*, Beijing (1993) pp. 807-810.